CHEMICALS FOR OIL FIELD OPERATIONS

CHEMICALS FOR OIL FIELD OPERATIONS

Recent Developments

Edited by J.I. DiStasio

NOYES DATA CORPORATION
Park Ridge, New Jersey, U.S.A.
1981

Library of Congress Catalog Card Number: 81-11052
ISBN: 0-8155-0861-1
ISSN: 0198-6880; 0270-9155
Printed in the United States

Published in the United States of America by
Noyes Data Corporation
Noyes Building, Park Ridge, New Jersey 07656

Library of Congress Cataloging in Publication Data

Main entry under title:

Chemicals for oil field operations.

(Chemical technology review ; 195) (Energy
technology review ; 69)
Includes indexes.
1. Oil fields--Production methods--Patents.
2. Drilling muds--Patents. I. DiStasio, J. I.
II. Series. III. Series: Energy technology
review ; 69.
TN870.C5125 622'.338 81-11052
ISBN 0-8155-0861-1 AACR2

FOREWORD

This comprehensive book provides information on chemicals useful in primary and enhanced oil recovery methods, workover and completion fluids, as well as additives beneficial to oil well maintenance. Based on U.S. patents issued since January 1979, this volume bridges the gap between technological advances and field operations. This title contains new developments since our previous title *Crude Oil Drilling Fluids* published in 1979.

This book is a data-based publication, providing information retrieved and made available from the U.S. patent literature. It thus serves a double purpose in that it supplies detailed technical information and can be used as a guide to the patent literature in this field. By indicating all the information that is significant, and eliminating legal jargon and juristic phraseology, this book presents an advanced commercially oriented review of recent developments in the field of chemicals for oil field operations.

The U.S. patent literature is the largest and most comprehensive collection of technical information in the world. There is more practical, commercial, timely process information assembled here than is available from any other source. The technical information obtained from a patent is extremely reliable and comprehensive; sufficient information must be included to avoid rejection for "insufficient disclosure." These patents include practically all of those issued on the subject in the United States during the period under review; there has been no bias in the selection of patents for inclusion.

The patent literature covers a substantial amount of information not available in the journal literature. The patent literature is a prime source of basic commercially useful information. This information is overlooked by those who rely primarily on the periodical journal literature. It is realized that there is a lag between a patent application on a new process development and the granting of a patent, but it is felt that this may roughly parallel or even anticipate the lag in putting that development into commercial practice.

Many of these patents are being utilized commercially. Whether used or not, they offer opportunities for technological transfer. Also, a major purpose of this book is to describe the number of technical possibilities available, which may open up profitable areas of research and development. The information contained in this book will allow you to establish a sound background before launching into research in this field.

Advanced composition and production methods developed by Noyes Data are employed to bring these durably bound books to you in a minimum of time. Special techniques are used to close the gap between "manuscript" and "completed book." Industrial technology is progressing so rapidly that time-honored, conventional typesetting, binding and shipping methods are no longer suitable. We have bypassed the delays in the conventional book publishing cycle and provide the user with an effective and convenient means of reviewing up-to-date information in depth.

The table of contents is organized in such a way as to serve as a subject index. Other indexes by company, inventor and patent number help in providing easy access to the information contained in this book.

16 Reasons Why the U.S. Patent Office Literature Is Important to You

1. The U.S. patent literature is the largest and most comprehensive collection of technical information in the world. There is more practical commercial process information assembled here than is available from any other source. Most important technological advances are described in the patent literature.
2. The technical information obtained from the patent literature is extremely comprehensive; sufficient information must be included to avoid rejection for "insufficient disclosure."
3. The patent literature is a prime source of basic commercially utilizable information. This information is overlooked by those who rely primarily on the periodical journal literature.
4. An important feature of the patent literature is that it can serve to avoid duplication of research and development.
5. Patents, unlike periodical literature, are bound by definition to contain new information, data and ideas.
6. It can serve as a source of new ideas in a different but related field, and may be outside the patent protection offered the original invention.
7. Since claims are narrowly defined, much valuable information is included that may be outside the legal protection afforded by the claims.
8. Patents discuss the difficulties associated with previous research, development or production techniques, and offer a specific method of overcoming problems. This gives clues to current process information that has not been published in periodicals or books.
9. Can aid in process design by providing a selection of alternate techniques. A powerful research and engineering tool.
10. Obtain licenses–many U.S. chemical patents have not been developed commercially.
11. Patents provide an excellent starting point for the next investigator.
12. Frequently, innovations derived from research are first disclosed in the patent literature, prior to coverage in the periodical literature.
13. Patents offer a most valuable method of keeping abreast of latest technologies, serving an individual's own "current awareness" program.
14. Identifying potential new competitors.
15. It is a creative source of ideas for those with imagination.
16. Scrutiny of the patent literature has important profit-making potential.

CONTENTS AND SUBJECT INDEX

INTRODUCTION

Skyrocketing energy costs in recent years have led to a surge of interest within the petroleum industry in new and more efficient methods of recovering oil and gas from subterranean formations. Many previously drilled wells that were considered "spent" or exhausted of their oil reserves are now being reexamined and reworked in light of new processes for recovering maximum amounts of oil in the most cost-conscious methods available.

Most modern wells are bored by a rotary process in which a drilling bit is made to turn at the lowest point of the hole while a drilling fluid or "mud" is pumped from the surface to lubricate the cutting action and flush away the rock fragments produced by the action of the bit. The mud also creates pressure inside the well, thus supporting the sides until a casing can be inserted. Drilling of an oil well is a complex and somewhat risky task, since some wells must be dug several miles deep before petroleum deposits can be reached. Many are now drilled offshore from platforms standing in the ocean bed, giving rise to pollution considerations when disposing of the oil-contaminated drill cuttings and also to the danger of leaks and spills during production.

Worldwide dwindling oil reserves, however, render these primary oil recovery processes less than satisfactory, since large amounts of oil are left trapped in the less accessible interstices of the petroleum-bearing formation. Through modern technology, it is now possible to recover increasingly greater amounts of these petroleum deposits using secondary and tertiary oil recovery techniques, also known as enhanced oil recovery, which result in much greater yield from each oil well. (Secondary and tertiary refer to the order in which these processes are employed. The same process may be considered secondary in one oil well, tertiary at another.)

The most widely used enhanced recovery technique is waterflooding. This type of process is aimed at flushing out the oil that is trapped in the pores of the rock strata. Great improvements have been made in this area by the inclusion of a wide array of polymers and surfactants, or combinations of the two, to improve the yield of recovered oil. There is also interest shown in steam injection techniques, and the use of expandable gases to displace the trapped oil.

The area of fracturing and acidizing of formations has also been the focus of attention. The hydraulic fracture fluid must exhibit advantageous viscosity and particulate solids transport properties. In the case of fracturing fluids which contain a natural resin, there is interest in breaker additives which effect a delayed action thinning out of the fluid medium. This facilitates the subsequent removal of the composition from the fractured formation.

This book contains summaries of over 200 processes which deal with the many chemical-related aspects of oil recovery. It is an invaluable source of information as to the state-of-the-art in this dynamic field.

PRIMARY OIL RECOVERY TECHNIQUES

DRILLING OPERATIONS

Drilling of Deep Boreholes

P.L. Guerber; U.S. Patent 4,185,703; January 29, 1980; assigned to Coyne & Bellier, Bureau d'ingenieurs Conseils, France describes a process of producing boreholes down to very great depths of several kilometers or tens of kilometers.

The process for drilling underground strata is preferably carried out in the following way: a plurality of superjacent drilling sections are filled respectively with a same plurality of different filling substances each chosen so as to remain liquid at all points of the respective sections of the borehole, taking into account the laws governing the increase in earth temperature and the increase in pressure with increasing depth in the borehole.

To accelerate the natural upward movement by flotation of the debris from the bottom of a very deep borehole, different means may be used.

First of all it is possible to maintain the borehole filling substance, at least in the vicinity of drilling tools, at a temperature slightly below its boiling point so that when its boiling commences this will result in the production of bubbles of gas some of which may be absorbed by solid particles of debris so as to accelerate the upward movement of such particles, and which all tend to create in the liquid filling substance an upwardly flowing stream carrying along solid debris towards the free surface of the filling substance.

According to a first variant, the liquid filling substance of the borehole is circulated during the drilling operation by pumping means which are set up at the ground surface and which comprise a suction duct immersed in the liquid filling the borehole and a delivery duct, and by a conduit arranged approximately vertically in the borehole and having an upper end connected to the delivery duct and a lower end opening near the bottom of the borehole. Since the upward movement of debris towards the surface of the liquid is brought about substantially by flotation, the pump may have a relatively low rate of through-flow and the conduit arranged in the borehole may be given a relatively small internal diameter. Since it is also possible to use a flexible conduit, its weight and expense are very much less than those of a line of tubes such as is used in the known rotary drilling process.

In a second variant, a fluid of a density below that of the borehole filling liquid is injected, under pressure if necessary, into the bottom of the borehole during the drilling operation. This fluid could be a liquid of lower density than that of the filling liquid; but it is preferable to use a gas and more particularly compressed air, which, with a minimum rate of flow, promotes the upward movement of solid debris towards the surface in the same way as the bubbles of filling liquid did in the case of the embodiment of this process mentioned above. Again in this second variant, the compressed air can be brought to the bottom of the borehole by a flexible conduit of small cross section and therefore of light weight and of an inexpensive type.

To constitute a drilling apparatus according to this process, such as that which is shown in Figure 1.1, a rotary tool **O**, having a cross section **A** is suspended by its attachment element **2** at the end of a supporting element **C** of great length, the cross section **L** thereof increasing from its lower end which is fixed to the tool **O**, to an upper end value **U**.

Figure 1.1: Apparatus for Drilling Deep Boreholes

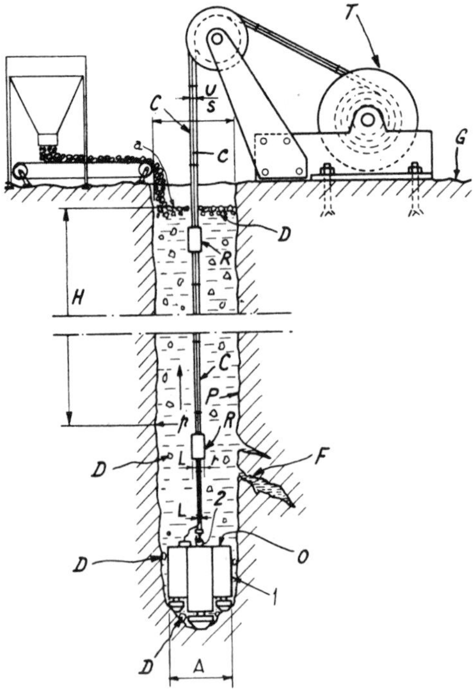

Source: U.S. Patent 4,185,703

This supporting element **C** may be constituted, for example, by a cable 15,000 m in length, whose cross section near the point of attachment **2** to the tool **O** amounts to, for example, 1 cm^2, and 8,000 m further on amounts to, for example, 12 cm^2, etc. Electrical resistances **R** of suitable values each having a cross-sectional area **r** are fixed at predetermined points of the cable **C**; the apparatus is also provided with electrical cables **c** for the supply of the abovementioned resistances **R** and the electric motor built into the drilling tool **O**. The dimensioning and the arrangement of the electrical resistances **R** will become clear from the following.

To excavate, for example, a borehole 15,000 m in depth by the process, using the apparatus which has just been described, the following procedure is adopted.

First a borehole of at the most 2,000 m in depth is drilled by conventional techniques, with a sufficient cross section **S** to allow the downward movement therein of the drilling tool **O** by means of a winch **T**. If, as is usually the case, the underground layers situated at a depth of more than 2,000 m have a mean density which itself is lower than 3, the bottom of the borehole, which has been contacted by the drilling tool **O** moved downwards at the end of the supporting element **C**, is then filled with solid pieces of antimony trichloride.

The temperature which prevails at the bottom of the borehole being near 70°C, the lower electrical resistance **R** of the drilling apparatus is supplied with a sufficient electrical power to bring the antimony trichloride to its melting point, which is only 72°C at atmospheric pressure. The electric motor of the drilling tool **O** is activated, and its rings or abrasive discs then attack the bottom of the borehole, filled with melted antimony trichloride **1**. To carry on with the drilling, it is advantageous then for at least the entire lower portion of the initial borehole to have also been filled with antimony trichloride in solid state, which is brought to melting point by a local supply of a suitable amount of heat from the electrical resistances **R** which are distributed along the supporting element **C** of the drilling tool **O**.

Thus, the work of the drilling tool **O** is facilitated by the fact that the tool is immersed in a liquid. On the other hand, earth or rock debris **D** and also any infiltrations which occur, generally water, move upwards naturally to the free surface **a** of the melted antimony trichloride because of their lower density (lower than 3). These debris and infiltrations **D** are then particularly easy to remove if the free surface **a** of the liquid antimony trichloride is near the surface **G** of the ground.

It is also possible, however, to skim the free surface **a** even where this is situated at a certain depth, by using simple well-known means. If some of the walls **P** of the lower portion of the borehole below 2,000 m, which is filled with liquid antimony trichloride, comprise cracks or collapsed portions **F**, the liquid antimony trichloride enters these immediately and, as soon as its temperature decreases, it solidifies therein and this has the effect of stopping any such cracks or fissures **F** and consolidating the walls **P** of the borehole, preventing or reducing any infiltrations.

Furthermore, as the drilling progresses, the liquid antimony trichloride exerts on the walls **P** of the borehole a pressure **p** which is proportional to the vertical distance **H** of the free surface and the density of the liquid antimony trichloride;

this hydrostatic pressure **p** opposes the pressure of the ground and therefore ensures the stability of the walls **P** of the borehole without any need to provide casing tubes for these walls. However, it should be noted that the increase in this hydrostatic pressure **p** with the depth **H** produces an increase in the melting temperature of the antimony trichloride, whose liquid state can be maintained only by providing an increasing amount of heat as the depth **H** increases.

The drilling in liquid antimony trichloride can be continued, if there are no leakages, up to a depth slightly less than that where there exists underground a temperature in the vicinity of its boiling point (230°C at atmospheric pressure) which again depends on the hydrostatic pressure **p** and consequently the depth **H**. The use of liquid antimony trichloride makes it possible to drill to a depth **H** of at least 4,500 m, where there is a temperature in the vicinity of 217°C.

In order to continue the borehole below this last depth, pieces of selenium which are solid under normal temperature and pressure conditions are dropped on to the free surface **a** of the liquid antimony trichloride. The density of selenium being in the vicinity of 4.6, i.e., greater than that of the antimony trichloride in liquid form, the selenium pieces fall slowly to the bottom of the borehole, which is then at a depth of about 4,500 m. The temperature of 217°C which the solid selenium encounters in this region makes the selenium melt, and makes it possible to continue the drilling work of the tool **O** in a liquid medium, possibly with local supply of small heat amounts by means of the electrical resistances **R** fixed to the supporting element or suspension cable **C** for the drilling tool **O**. The drilling in the liquid selenium can be continued up to a depth at which a temperature slightly below the boiling point of selenium (689°C at normal pressure) is reached; thus it is possible to reach a depth of about 9,500 m.

A final depth of about 15,000 m can then be reached by using no longer selenium but tellurium, which has a density of 6.2, that is higher still (melting point equal to 450°C and boiling temperature equal to 1390°C at atmospheric pressure).

In cases where the drilled strata have a particularly low average density and where the risks of pollution are negligible, it is possible in the first phase of the drilling process described hereinbefore to use not antimony trichloride which has a density of 3 but sulfur which has a density of 2; its melting temperature (120°C) and boiling temperature (444°C) at normal pressure are such that the liquid sulfur can be used for carrying out the drilling method at depths of between 3,000 and 7,200 m.

Drilling of Soft Formations Using Cold Drilling Liquid

It is known for holes to be drilled in the crust of the earth with use of cold drilling liquids so that the wall of the drill hole is frozen. Certain clay formations, which are unstable if noncooled drilling liquids are used and affect the drilling process detrimentally, retain their coherence and stay in place. Also, if these cold drilling liquids are used, cores can be extracted from unconsolidated rocks with a higher yield since these cores freeze and do not disintegrate easily during the drilling process.

It will be clear that the temperature of the circulating medium should be lower than the freezing point of the liquid contents of the formation to be frozen and equal to or higher than the freezing point of the circulating medium itself. For drilling liquids use may be made of various liquids, such as salt water, a salt-water-mud flush, oil, e.g., diesel oil, or a water/oil emulsion. Chilling of the drilling liquid may take place according to known methods; solid carbon dioxide is often used for this purpose. The drill pipes may have been provided with insulation material in order to effect transportation of as much cold as possible to the bottom of the drill hole.

A problem not connected with this is the drilling of holes having a small diameter (so-called slim holes) through formations comprising soft or plastic clays, or soft shales (slates). Under these conditions so-called insert drill bits, diamond drill bits or other drill bits suitable for drilling hard formations can make little progress as a result of the so-called balling-up effect. In order to keep the drill bit as long as possible at the bottom of the drill hole it is desirable, just because of the slight wear, to use one of these types of drill bit. Generally, the diamond drill bit will be preferred because of the absence of bearings.

W.H. van Eek and A.W.J. Grupping; U.S. Patent 4,191,266; March 4, 1980 describe a process to make the drilling of holes possible in formations comprising clays, soft shales or soft formations of this kind with use of drill bits more suitable for hard formations.

According to the process holes are drilled in formations comprising soft or plastic clays, soft shales (slates) or soft formations of this kind with a drill bit for harder rock formations if the cold drilling liquid always contains an amount of frigories sufficient to freeze at least part of the rock near the drill bit. The process especially relates to relatively small boreholes with diameters of 200 mm (8") or less (e.g., slim holes) but the method may also be applied for boreholes exceeding that diameter if necessary. By preference, a diamond drill bit is used for drilling.

It has appeared that, at least if the rock near the bottom of the drill hole has become frozen through withdrawal of an adequate amount of heat by the drilling liquid, the abovementioned balling-up effect does not occur. The disadvantage of using a diamond drill bit will then be changed into an advantage. As the diamond drill bit may be rotated faster (e.g., 400 rpm as against 100 to 200 rpm) and the more rapid the drill bit is rotated the smaller the advance per revolution needs to be in order yet to achieve the same drilling progress as is the case with drill bits of a different kind, there results, owing to the slight advance per revolution, a better chance of so freezing the clay and other little permeable formations of this kind in the vicinity of the drill bit head that the abovementioned balling-up effect does not occur. In this connection, it is noted that, as a rule, freezing of sands and similar permeable formations is not necessary to achieve the required drilling progress, seeing that, normally, these can be drilled with diamond drill bits without any difficulties.

The installation is characterized in that the drilling liquid system at the ground level comprises at least a cooler to prepare a cooled drilling liquid, an ice-machine to freeze part of the drilling liquid, and means to mix the cooled and frozen cooling liquid, force it through the system and feed it to the drilling line. Moreover, if so desired, a compressor or the like is present to add a gas to the drilling liquid. If desired, the cooler and the ice-machine may be combined into one apparatus. Finally, for control of the pressure in the drill hole, one or several packing glands and at least one adjustable throttling element, for instance in the connection before the separator, are present.

To illustrate the process, reference is made to Figure 1.2. A core drill bit **1**, in the process of drilling core **2**, is connected via flush motor **3** (dynamotor) and drill collars or stems **4**, to the hollow drilling string **5**, of which the pipes are provided with insulation material **6**. The motor **3** is driven by the drilling liquid. In this way a hole **7** is formed in the strata **8**.

Figure 1.2: Drilling of Borehole Under Freezing Conditions

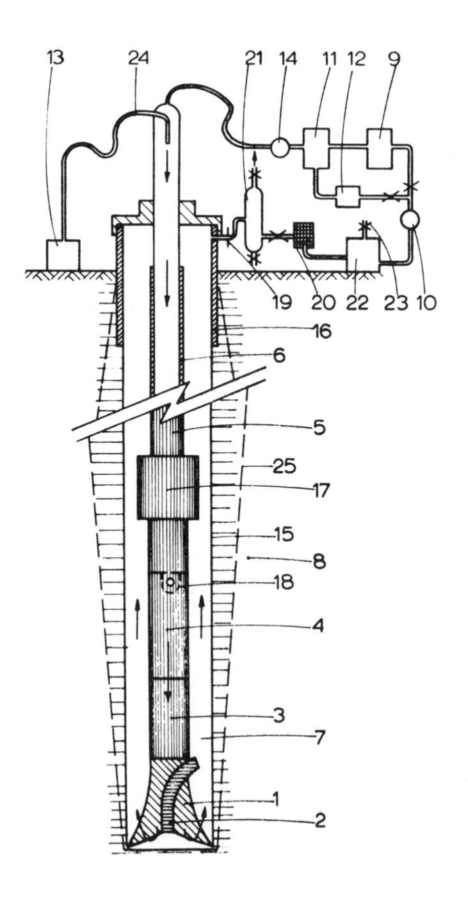

Source: U.S. Patent 4,191,266

The eutectic drilling liquid or mud is prepared by changing a portion of the liquid into ice in the ice-machine **9**, which ice is led to suction tank **11** to be there mixed with the aqueous salt solution which is cooled in the cooling device **12** and also transported to the tank **11**. The mixture is suctioned by another pump **14** and flows downwards through the string **5** and into bit **1**, through apertures and, subsequently, upwards via the annular circuit between string **5** and drill hole wall **15**, i.e., conductor **16**.

The drilling string is also provided with one or more packers **17** capable of shutting of the annular circuit against the frozen wall of the drill hole **15**. The hollow drill pipes can be shut off automatically with nonreturn valve **18**, if the liquid should try to flow back.

In the connection between the conductor **16** and shaking screen **20** a throttling element **19** is present, followed by a separator **21** in which gas is recovered from the liquid or mud. Drill cuttings or drill bit cores, and ice, if any, if these have not been collected in separator **21**, are removed from the drilling liquid on shaking screen **20**, whereupon the flush, via collecting tank **22**, can be pumped by means of pump **10** through the ice-machine **9** or, possibly, through the cooling device **12**. Extra flush and chemicals may be added via line **23** in the collecting tank **22**. Finally, air may be added via compressor **13** through line **24** in the drilling string **5**.

The frozen formation around the drill hole is indicated schematically by dotted line **25**. The drilling equipment is operated at the ground level by a conventional hoist, as is known for instance from the oil industry. Normally speaking, with use of this method this hoist may be of a smaller capacity at similar depths than is the case if the usual methods are applied.

Example: A drill hole having a depth of 1,500 m is drilled to a diameter of 127 mm (5") with the aid of a diamond drill bit lined with diamond board. The length of the drill bit, drill collars and dynadrill amounts to approximately 10 m (393"). The drilling line consists of lightweight pipes of about 3 kg/m (6.825 lb/ft), diameter 73 mm (2⅞"), provided with an insulation jacket measuring 12.7 mm (½") in thickness, so that the overall diameter is 98 mm (3⅞"). The insulation value of the jacket is equal to 0.3 W (mK) (50 x 10^{-6} Btu/sec/ft °F).

The drilling rate amounts on an average to 3 mm/sec (0.01 ft/sec). The rock temperature at the bottom, at a surface temperature of 10°C (50°F), is 55°C (131°F).

The drilling liquid has the following composition:

NaCl	296 g/ℓ of water
Attapulgite	40 g/ℓ of water
Floc gel	15 g/ℓ of water
Mass density (specific gravity)	1,200 kg/m^3
Plastic viscosity	7 mPa at +20°C
Plastic viscosity	21 mPa at -20°C
Apparent viscosity	9 mPa at +20°C
Apparent viscosity	24.5 mPa at -20°C

The drilling mud contains 10% by volume of ice of the same composition. The mud is circulated at a quantity of 8.2 ℓ/sec (130 gal/min), entering the drill hole having a temperature of -20°C (-4°F) and returning having a temperature of -10°C (14°F). The amount of heat to be discharged (= supply of frigories) should be about 530 kJ/sec (500 Btu/sec).

The liquid pressure at the base of the drill hole amounts to 17.6 MPa (2,550 psi) and to 14.8 MPa (2,150 psi) with addition of air or nitrogen. The pressure on the throttling element at the top of the annular space will amount to approximately 12 MPa (1,740 psi).

Method of Dynamically Killing a Well Blowout

A blowout may occur during well drilling operations when the well penetrates a high pressure gas-producing formation due to a number of circumstances. Thus, gas from a high pressure formation may enter the well and mix with the drilling mud so that its density is reduced by gas occlusion, thus reducing the hydrostatic head on the well to a value less than that of the formation pressure. A blowout may also occur during removal of the drill string from the well. Displacement of the drilling mud by the drill string may result in a decrease in the liquid level within the well with, again, a decrease in the hydrostatic head at the level of the high pressure formation.

When a blowout occurs, a number of remedial procedures are available to kill the blowout and bring the well under control. One technique involves the drilling of a relief well into a subterranean location near the blowout well. Communication between the relief well and blowout well is established and fluids then pumped down the relief well and into the blowout well in an attempt to impose a sufficient hydrostatic head to block the flow of gas from the formation into the well.

Communication between the wells may be established through the high pressure sand which caused the blowout or through a separate permeable zone penetrated by both the blowout and relief wells. The formation may be acidized in order to increase the fluid conductivity between the wells. Fracturing may also be employed although in most cases this is undesirable since most fractures tend to be naturally oriented in a generally vertical direction. This is particularly true in formations at depths of about 3,000 feet and more since at these depths the overburden pressure will usually exceed the horizontal stress characteristics of the formation.

E.M. Blount; U.S. Patent 4,224,989; September 30, 1980; assigned to Mobil Oil Corporation describes an improved technique for killing a blowout by the injection of fluid through a relief well. In carrying out the process, the fluid employed during the initial portion of the kill procedure is a low density fluid which produces a hydrostatic pressure component which is less than the static pressure of the formation.

The low density fluid is pumped down the relief well and into the blowout well at a rate to produce a frictional pressure component in the blowout well which, when added to the hydrostatic pressure component, is greater than the static formation pressure but less than the formation fracturing pressure. The injection of the low density fluid is continued at progressively increasing rates until a sufficient flow rate up the blowout well is achieved to block the flow of gas from the high pressure formation causing the blowout. Thereafter a high density fluid is introduced into the relief well which produces a hydrostatic pressure component which is greater than the static formation pressure. This high density fluid is pumped down the relief well and into the blowout well at a flow rate less than the maximum flow rate of the low density fluid. The sum of the frictional pressure component and the hydrostatic pressure component is less than the formation fracturing pressure.

Figure 1.3 illustrates a subterranean formation **2** which is penetrated by a well **4** which is blown out and a well **5** drilled as a relief well.

Figure 1.3: Blowout Well and Relief Well

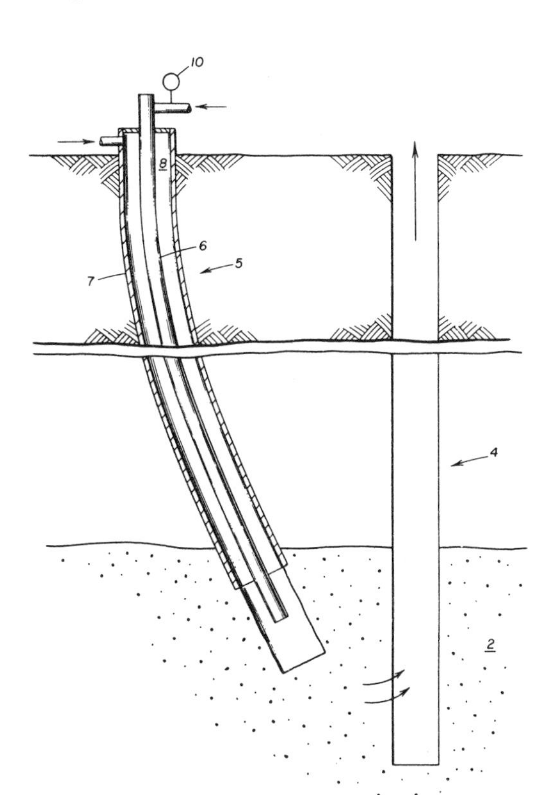

Source: U.S. Patent 4,224,989

The relief well is preferably equipped with a tubing string **6** and a well casing **7** which define an annulus **8** through which the kill fluids are injected. The tubing wellhead is provided with a pressure measuring means **10** which is employed to monitor the downhole pressure of the relief well as described hereinafter.

In killing the well with the dynamic kill fluid, the bottomhole pressure in the relief well must, of course, be greater than the bottomhole pressure of the blowout well in order to accommodate the frictional pressure loss of flow from the relief well to the blowout well. The bottomhole pressure of the relief well is equal to the sum of the wellhead pressure and the hydrostatic pressure minus the frictional pressure loss. The bottomhole pressure in the blowout well is equal to the sum of the hydrostatic pressure and the frictional pressure loss, it being assumed that the wellhead pressure of the blowout well is zero since the well is uncontrolled.

Once a single-phase flow condition in the blowout well is reached, the hydrostatic pressure components in the blowout and relief wells are substantially the same and thus the wellhead pressure on the relief well is equal to the sum of the frictional pressure components in the relief well, blowout well, and in the formation providing communication between the relief and blowout wells. These relationships may be expressed by the following equations:

$$BHP_b = BHP_r - FP_c \quad (1)$$

$$BHP_b = HP_b + FP_b \quad (2)$$

$$BHP_r = WHP_r HP_r - FP_r \quad (3)$$

$$WHP_r = FP_r + FP_b + FP_c \quad (4)$$

where WHP is the wellhead pressure, HP is the hydrostatic pressure, FP is the frictional pressure loss, BHP is the bottomhole pressure, and r, b and c denote the relief well, the blowout well, and the communication between these wells, respectively.

The frictional pressure components may be calculated by any suitable means as will be understood by those skilled in the art. In the case of annular flow, the frictional pressure loss, FP, in psi, may be defined by the following equation:

$$FP = (11.41fLpQ^2)/d_e^5 \quad (5)$$

where f is the fanning friction factor, L is the measured depth of the well in feet, p is the density of the fluid in lb/gal, Q is the flow rate in bbl/min, and d_e is the equivalent diameter in inches.

During the dynamic kill operation, the low density fluid is pumped down the relief well annulus at a sufficiently high rate to produce a bottomhole pressure in the well greater than the sum of the formation pressure and the frictional pressure loss in flow from the relief well to the blowout well. Thus, the low density fluid enters the blowout well at a pressure greater than the formation pressure. At the same time a fluid is pumped down the tubing string at a low rate to provide a substantially constant pressure differential from the wellhead to the bottom of the tubing string. That is, the fluid is pumped down the tubing string at a rate just sufficient to maintain fluid in the tubing string with negligible friction losses so that the pressure differential from the wellhead to the bottom is equal to the hydrostatic head. Thus the wellhead pressure at the tubing string may be measured and added to the calculated hydrostatic pressure in the tubing string to continuously monitor the bottomhole pressure in the relief well. The tubing string fluid may be the same as or different than the dynamic kill fluid, but in any event has density such that its hydrostatic head is less than the static formation pressure.

The pumping rate down the relief well annulus is progressively increased with fluid flowing from the relief well into and up the blowout well under turbulent flow conditions until the sum of the hydrostatic pressure components and the frictional pressure component in the blowout well exceeds the pressure at which gas enters the blowout well from the formation. Ultimately this blocks the flow of gas into the blowout well and the well begins the transition from two-phase flow to single-phase flow. During this procedure, the bottomhole pressure in the relief well is monitored to ensure that it does not reach the fracturing pressure of the formation. Once the steady-state flow condition is produced in the blowout well, the transition to a higher density fluid can begin. Preferably, the density of the fluid injected into the relief well annulus is progressively increased in at least two increments to the final fluid density desired for a static kill condition.

If the relief well capacity is not sufficient to produce a steady-state flow condition in the blowout well, one or more additional relief wells can be provided. Each relief well, of course, is operated in accordance with the abovementioned procedure in which the dynamic kill fluid is pumped down the relief well at a wellhead pressure which produces a bottomhole pressure greater than the formation pressure and less than that of the formation fracturing pressure.

A specific example of the process is provided by the following procedure employed to dynamically kill a blowout in a well cased with 8.535" i.d. casing and having 5" o.d. drill pipe in the hole. The measured total depth of the well was 10,210' and the well was blown out in a high pressure gas zone at a vertical depth of 9,650'. Reservoir engineering studies indicated that the gas zone had a static formation pressure, i.e., the pressure of the formation in the vicinity of the well before the blowout, of 7,100 psig. The formation fracturing pressure was estimated to be about 8,500 psig.

A directional relief well was drilled in the vicinity of the blowout well to a total measured depth of 10,900' (equivalent to a total vertical depth of 8,560'). The well was cased with 8.535" i.d. casing and equipped with a 3½" o.d. tubing. A directional survey indicated the relief well was about 27' from the blowout well at total depth.

In the dynamic kill procedure, fresh water was employed as the dynamic kill liquid. The water had a density of 8.33 lb/gal, equivalent to an incremental hydrostatic head of 0.433 psi/ft. Preliminarily to initiating the kill attempt, the drilling mud in the relief well was reversed out by pumping water down the annulus with mud returns through the tubing. Once the mud was completely displaced from the well, an acidizing procedure was started in order to increase the communication between the relief well and the blowout well. The acidizing procedure was carried out employing 15% hydrochloric acid which was pumped down the tubing at a flow rate of about 4 b/min. After injecting acid at this rate for about 40 minutes, the pump rate was reduced to about 3 b/min and shortly thereafter the wellhead pressure at the annulus decreased by 350 psi, indicating that communication from the relief well to the blowout well was established.

After pumping additional acid, the dynamic kill procedure was started by increasing the pumping rate down the annulus from an initial value of about 4.3 b/min at a wellhead pressure of 2,010 psig to a final value of 125 b/min at a wellhead pressure of 5,840 psig. Tubing injection was switched from acid to water and when the pumping rate down the annulus reached about 35 b/min, the rate down the tubing was reduced from 4 to 1 b/min and remained constant at that value throughout the kill procedure. This established a substantially constant hydrostatic head in the tubing and during the kill procedure the tubing wellhead pressure was measured in order to monitor the bottomhole pressure.

About 34 minutes after the start of the kill procedure, when the pumping rate down the annulus was at 85 b/min, the wellhead fire at the blowout was reported to be essentially out. Thereafter, the pumping rate was increased to 125 b/min and maintained at this value for about 15 minutes and then decreased to about 80 b/min at a wellhead pressure of 3,290 psig. During this interval, the blowout well reignited.

The total volume of water pumped down the annulus of the relief well during the dynamic kill procedure was 5,220 barrels. At the conclusion of this, the transition to an intermediate 14.5 lb/gal drilling mud was started with an initial pumping rate of 73 b/min at an annulus wellhead pressure of 3,460 psig. The pumping rate for the intermediate mud was stabilized at 83 b/min for a period of about 8 minutes during which the mud started to enter the blowout well.

Thereafter, the pumping rate down the annulus was progressively decreased to a value of about 49 b/min and, after the injection of 1,525 barrels of intermediate mud, the transition to a heavier 16.5 lb/gal mud was started. This heavier drilling mud was pumped down the annulus of the relief well at an initial rate of 49 b/min and thereafter reduced to about 15 b/min until sufficient mud was injected to fill the annuli of the relief and blowout wells. Thereafter, the pumping rate of the 16.5 lb/gal mud was reduced with variations to an ultimate rate of about 1½ b/min.

Cleanout of Slanted Wells

In the production of hydrocarbons from producing formations, it has become common to drill directional wells from the earth's surface to an underground producing formation. Directional drilling usually involves diverting a well from the vertical at some place below the earth's surface to direct it to a remote producing formation. The directional or slanted wells may deviate up to 90° from the vertical to reach a desired producing formation.

S.O. Hutchison and G.W. Anderson; U.S. Patent 4,187,911; February 12, 1980; assigned to Chevron Research Company describe a method for removing material such as sand from a directional well which deviates more than about 15° from the vertical over a substantial portion of its depth. A string of cleanout tubing is run into the directional well from the earth's surface to the position in the well where foam is desired. The tubing string is provided with centralizers to insure that the tubing string is in substantially coaxial alignment with the existing conduits in the well such as the casing string and the production liner or the production tubing if the cleanout is being done through production tubing. A gas and liquid foam having a gas-to-liquid ratio of at least 10 scfm/gal and not more than 30 scfm/gal is generated at the surface and injected down the well and circulated through the well at a circulation rate of at least 30 fpm to remove material from the well.

With reference to Figures 1.4a and 1.4b, a directional well is indicated at **10**. The well **10** extends from the earth's surface to penetrate a producing formation **12**.

A well liner **14** having slots **16** is positioned adjacent the producing formation **12**. The slots provide communication between the interior of the well **10** and the producing formation **12**. The oil produced through the slots is often accompanied by sand which may periodically need to be removed from the well. The upper portion of the well **10** is lined with a casing string **18** which may be cemented in place. A wellhead, indicated generally as **20**, closes off the upper portion of the casing and provides for connecting various flow lines to the well. Thus, an exit conduit **22** having a valve **24** is used to provide an opening from the well.

A cleanout tubing string **30** is run into the well. The tubing string **30** is centralized in the interior of the well by means of a plurality of centralizers **32**.

Figure 1.4: Method of Cleaning Slanted Wells

a.

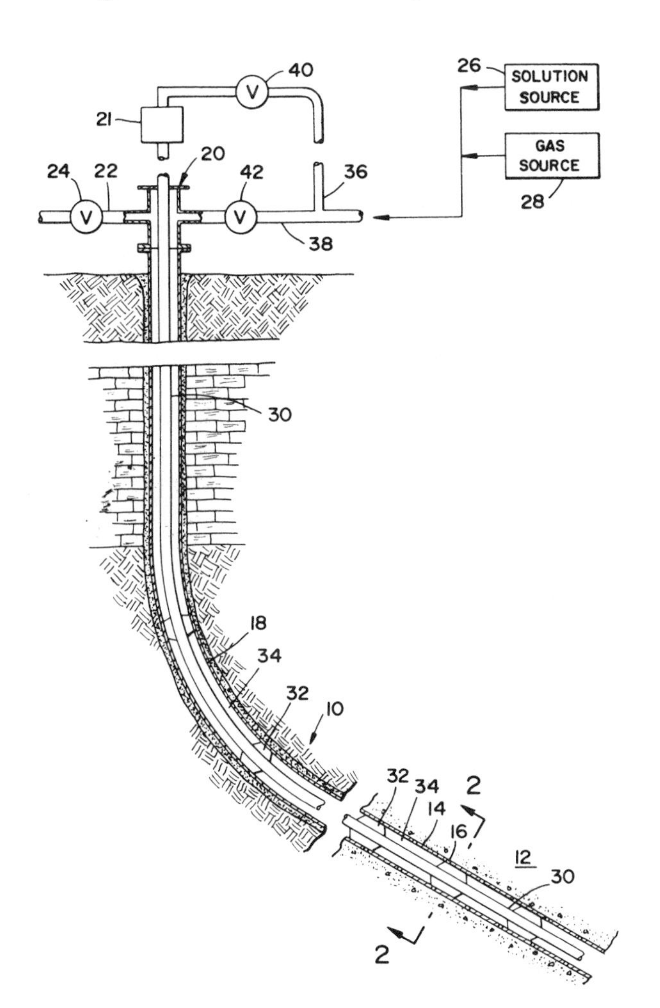

(continued)

Figure 1.4: (continued)

b.

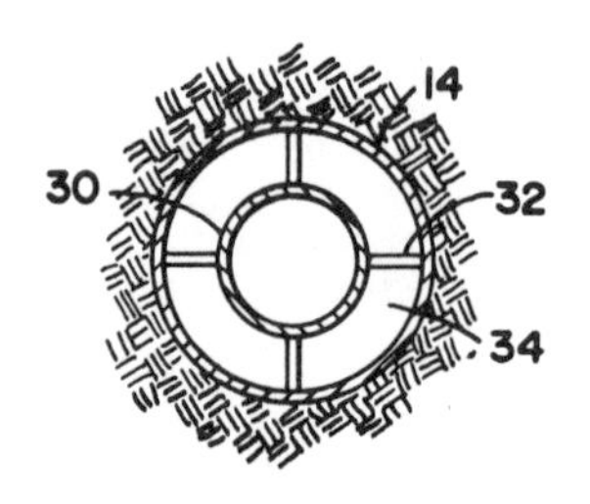

(a) Apparatus for use in process
(b) Enlarged sectional view of Line **2-2** of Figure 1.4a

Source: U.S. Patent 4,187,911

The tubing string **30** is centralized in coaxial alignment within the well conduits, in this case, liner **14** and casing string **18**. In this manner the tubing string **30** provides an inner passage from the surface to its lower end and an annular passage **34** between its outer surface and the well conduits **14** and **18** from the surface to its lower end.

A solution source **26** and a gas source **28** are connected by suitable conduits **36, 38** and valves **40, 42**, respectively, to the interior of the tubing string **30** and the tubing-casing annulus **34** through the wellhead **20**. Means are provided as shown at **21** for raising and lowering the tubing string. The solution source and gas source are sized to provide a foam for circulation in the well at a rate of at least 30 fpm with a ratio of gas-to-liquid of at least 10 scfm/gal of solution. Suitable foam compositions are described in U.S. Patent 3,463,231.

Previously the cleanout tubing was not centralized when cleaning directional wells. However, it has been found that when directional wells exceed 15° deviation from the vertical, efficient cleanout is not obtained when the cleanout string is not centralized. When the cleanout tubing string is not centralized, the sand in the well is moved only one layer at a time, producing movement similar to the movement of a sand dune. This sand dune effect is particularly noticeable in wells deviated more than 30° from the vertical. Centralizing the cleanout string greatly reduces the sand dune effect and permits efficient removal of sand when foam is circulated through the well with a velocity of at least 30 fpm and the ratio of gas to the liquid in the foam is maintained between 10 and 30 scfm/gal.

Method of Preventing Adherence of Oil to Surfaces

It frequently is desirable to make the surfaces of various tanks, vessels, or other equipment resistant to the adherence of various oily materials, particularly crude oil. One example is tanks and vessels that are used to contain or transport crude oil or refined products, particularly where contamination of the subsequent contents is a problem. Thus, the removal of substantially all of the oil upon draining of a tanker or other vessel or container poses several attractive advantages.

First, there is the advantage of avoiding wasted oil. Approximately 0.3% of the cargo of oil tankers is lost because it cannot be salvaged economically. A second advantage is that material is not left behind in the vessel which will contaminate subsequent cargoes. Such contamination can lead to great expenses in purifying subsequent cargoes. A third and principal advantage is the avoidance of pollution of the ocean. As is frequently the practice, residual oil washed out of the tanker with salt water is discharged into the open sea with ecologically disastrous results. Accordingly, if the surfaces of such tankers or vessels could be treated to minimize the adherence of oil to the walls thereof and to facilitate cleaning or washing thereof, such would be greatly beneficial.

A further example of equipment which is desirably resistant to the adherence of oil is oil spill cleanup equipment such as skimmers and booms. Such equipment is used periodically on oil-coated waters and, accordingly, must be cleaned as frequently. Further, the adherence of heavy oil, such as Bunker C, on various critical surfaces of this equipment can cause the equipment to malfunction. Thus, not only would treatment of the surfaces of this equipment reduce maintenance thereof but it would also ensure its efficient operation.

Yet another example of equipment which would desirably have surfaces resistant to the adherence of oil is that associated with oil wells. In some oil wells the crude is so viscous that production is limited by the speed of sucker rod descent during each stroke. Further, the oil tends to foul much of the production equipment that it contacts, resulting in added effort for recovery. Accordingly, if the various contacting surfaces could be rendered resistant to adherence by oil, production efficiency would be increased markedly.

P.E. Titus; U.S. Patent 4,160,063; July 3, 1979; assigned to Shell Oil Company describes the removal of residual oil from contacting surfaces by coating the surfaces with a film of a polyacrylamide crosslinked in the presence of water with a salt of a polyvalent metal.

Two very successful polymeric materials have been discovered for forming such oleophobic films: complexed polyacrylamide and complexed carboxymethylcellulose. These materials when dissolved in fresh or seawater are crosslinked with salts of polyvalent metals such as aluminum, chromium and iron. Typical salts include basic aluminum acetate, ferric chloride, ferrous chloride, chromic potassium sulfate, aluminum sulfate, aluminum nitrate, chromic acetate, aluminum potassium sulfate, aluminum ammonium sulfate. The speed of crosslinking varies markedly from less than a minute up to several hours depending on the type of metal available.

Polyacrylamide (Reten AO-1 from Hercules) has been found to work very suitably. Carboxymethylcellulose (CMC from Hercules, Incorporated) also performs well.

Polymer concentrations of about 0.25 to 2.0% by weight in both fresh water and seawater are found to be highly suitable. Crosslinking agent solutions of from about 1 to 5% by weight, basis water, have been used in a manner such that the resultant mixture, polymer plus crosslinking agent, is satisfactory for use. Depending upon the nature and size of the apparatus to be coated, the concentrations of these solutions can be further optimized as desired.

The coating may be applied in several different ways. In one case, where a large surface is to be covered, the polymer solution can be applied first, followed by an application of the complexing solution. In another case, the polymer solution and the complexing solution can be mixed simultaneously as they are being applied. In still another case, particularly where relatively small surfaces have to be covered, the polymer and complexing solutions can be mixed prior to application.

Application of the coating materials to surfaces can be made by various conventional means such as spraying or painting. Appropriately located nozzles may be employed which will coat the metal or other surfaces as the oil or other material is brought into contact therewith. The same nozzles can also be used to remove the oil or other material during the emptying or draining process. In addition, the polymer film may be put on the contacting surface by dipping, brushing or other means suitable for achieving full contact between the contacting surface and the film.

Where the vessel or barge has ribs inside its tank or other structural obstructions, the polymer may be applied as a foam so that it gets underneath the ribs or other obstructions by applying the foam to the top of the oil as the oil fills the container. The foam floats on the oil and touches the underside of ribs or other obstructions as it passes upwardly in the tank. Such polymer may be foamed in any conventional manner and may be foamed during or after addition of the complexing material.

If the polymer coating is allowed to dry, it will remain affixed to the coated surface. The film will regain its ability to resist oil if it is flushed with water prior to use. Where it is desired to retain the coating as a wet material, humectants such as glycerin or ethylene glycol may be added thereto to retard the drying out of the complexed polymer film when in contact with air for extended periods of time. Other low molecular weight polyalcohols function well as humectants. The introduction of oil to the polymer solution as a dispersion prior to complexing also serves to reduce drying of the coating when in contact with air.

Example: Three test solutions were made and labeled as follows:

Solution A = 1% by weight polyacrylamide (Reten A-01) dissolved in a 10% by weight sodium chloride solution;

Solution B = 1% by weight hydrated aluminum nitrate, $Al(NO_3)_3 \cdot 9H_2O$ in water; and

Solution C = 1% by weight chromic potassium sulfate, $Cr_2(SO_4)_3 \cdot K_2SO_4 \cdot 24H_2O$ in water.

Four steel test panels (4" x 8" x ⅛") were cleaned and coated as follows:

Panel 1 was coated by painting on a layer of Solution A followed by painting on a layer of Solution B;

Panel 2 was coated by painting on a layer of Solution A followed by painting on a layer of Solution C;

Panel 3 was coated by painting on a layer of material consisting of a mixture of 20 parts of Solution C and 100 parts of Solution A; and

Panel 4 was not covered by any coating and was used as a control.

After allowing the coated panels to cure for 20 minutes, they were immersed in No. 6 fuel oil. After 18 hours in the oil, the panels were removed and rinsed in a stream of tap water. The oil was readily removed from panels 1, 2 and 3, whereas panel 4 was covered by a thick layer of oil essentially unaffected by the water wash.

Chemically Activated Packer Element

It is frequently desired to restrict the flow of fluid through an annulus defined by the interior walls of a fluid conduit and the exterior of a tubular member within the fluid conduit. As used herein, fluid conduit includes elongated voids, such as defined by pipes, or by boreholes or mine shafts penetrating the earth, or the like structures having a substantially (i.e., disregarding small cracks, pores, and the like) closed cross-sectional perimeter; excluded from the term as used herein are fluid conduits which do not have a completely defined cross section, e.g., an open trough. Examples where such flow restriction is desired in wells include isolating a portion of an annulus between casing and the borehole or between concentric strings of casing or tubing, e.g., during the injection of treating fluids such as water or oil-based fluids, acids, cement slurries, sand consolidation slurries and the like.

G.P. Laflin and C.J. Durborow; U.S. Patent 4,137,970; February 6, 1979; assigned to The Dow Chemical Company describe an improvement in a tool for restricting the flow of fluid through an annulus defined by the interior walls of a fluid conduit and the exterior of a tubular member within the fluid conduit, which tool is of the type having an elongated mandrel, such as a tubular member, and radially surrounding at least a portion of the length of the mandrel, at least one unitary deformable packer element to prevent the flow of a fluid through the annulus. (By unitary is meant cast from a mold, machined, or the like, as distinguished from layers of strands or wrappings as in the interior of a golf ball.) The improvement lies in using as the deformable packer element, a chemically expandable packer element which expands upon contact with the fluid.

The choice of material for the packer element employed in this process depends upon the fluid the flow of which it is expected to restrict. The material must be one which will swell sufficiently upon contact with the fluid to form a fluid-restricting seal in the annulus. The degree of swelling required depends on several factors such as the pressure differential to be expected across the seal, the width of the annulus, the surface texture of the surfaces between which the seal is formed, and the like. Where the flow of an aqueous fluid, particularly water and other aqueous fluids such as brines naturally present in a subterranean formation is to be restricted, it is preferred to use the set compositions of U.S. Patent 3,502,149.

Example: A slurry was prepared using, per gallon of fluid, the ingredients in the proportionate amounts hereafter specified. 0.017 pound of sodium dichromate was dissolved in 0.086 gallon of water, which solution was then added to 0.604 gallon of a 1:4 volume blend of glycerin and diethylene glycol. To the liquid mixture was added 3.45 pounds of a high molecular weight (viscosity of

2% solution, 25 to 30 cp at 25°C) polyacrylamide nominally 1 to 4% hydrolyzed. The slurry was cast into a tubular plastic mold having an i.d. of 2" and an o.d. of 2.5". The slurry was allowed to cure at ambient temperatures into a rubber-like gel. When set, the mold was removed, and a 6" cylinder of the gel was cut from the cast piece. This cylinder was mounted on a segment of plastic pipe to form a packer. The packer was connected to a drain via another pipe and positioned in an overhead shaft penetrating an aquifer above a mine. As far as can be determined, the gel expanded upon contact with the formation water, and the aquifer was successfully drained with no detectable leakage of the seal.

TREATMENT OF DRILL CUTTINGS

Thermal Treatment of Drill Cuttings in Nonoxidative Atmosphere

On both offshore and inland drilling barges and rigs, drill cuttings are conveyed up the hole by a drilling fluid. Water base drilling fluids may be suitable for drilling in certain types of formations; however, for proper drilling in other formations, it is desirable to use an oil base drilling fluid. With an oil base drilling fluid, the cuttings, besides ordinarily containing moisture, are necessarily coated with an adherent film or layer of oily drilling fluid which may penetrate into the interior of each cutting. This is true despite the use of various vibrating screens, mechanical separation devices and various chemical and washing techniques. Because of pollution to the environment, whether on water or land, the cuttings cannot be permanently discarded until the pollutants have been removed.

T.E. Sample, Jr.; U.S. Patent 4,139,462; February 13, 1979; assigned to Dresser Industries, Inc. describes a safe means of treating oil or hydrocarbon coated or impregnated drill cuttings for disposal either on land or in a body of water, without incurring unacceptable environmental contamination or ecological upset.

The process involves collecting the cuttings in a heating chamber and vaporizing the volatile materials on the cuttings while under a nonoxidative atmosphere. This is a unitized continuous onsite process employing the principles of steam-stripping and nonoxidative thermal distillation to effect the simultaneous removal of hydrocarbons and water from the cuttings, leaving them in a condition sufficiently pollution-free as to be fit for direct disposal in waters adjacent to an offshore drilling platform. They can also be used as a landfill in the vicinity of a land-based operation. In addition, the hydrocarbons and water removed from the cuttings are collected in a state suitable for subsequent use as components of the drilling fluid system.

Generally, a conventional drilling derrick with its associated draw works, is mounted on a work platform for drilling a well into the earth formations lying beneath the ocean floor. A drill pipe having a drill bit at the lower end, is connected to a rotary table and draw works associated with the derrick. A mud pit is connected by way of a mud line and mud pump to a mud hose and swivel such that drilling mud is pumped into the top of the drill pipe down through the length thereof and into the bottom of the borehole through the drill bit. A portion of the borehole is cased with a cement sheath.

During the drilling operation, the mud is pumped down through the drill pipe and into the bottom of the borehole. Further pumping of the mud causes it to be pumped up, between the casing and drill pipe, and into a mud return pipe. As the drill bit cuts into the earth, the drill cuttings or portions of the rock and earth are carried back to the earth's surface via the mud.

Upon leaving the mud return line, the combined mud and drill cuttings are pumped to a storage or feed tank for processing. The oil-coated, damp, raw cuttings are separated from the hydrocarbon-containing drilling fluid through the use of one or more screens, i.e., a gravity screen or shale shaker, which is a vibrating screen, or combinations of both screens. For example, the material could pass through a gravity screen to make an initial 80 mesh separation. The solids and slurry from the gravity screen could then flow on to a vibrating screen of 120 mesh. From the vibrating screen, the cuttings could pass on to a washing screen of 160 mesh. The washing screen will have a continuous spray of a diesel oil solvent mixture to remove the oil mud that is adhered to the cuttings. In addition, centrifugal or other means of gross separation of the cuttings from the drilling fluid may also be employed.

After separation of the drill cuttings, they are transmitted through any suitable conduit, to a screw conveyor or the like. The screw conveyor moves the screened cuttings in a continuous manner to an enclosed, vented, heated vessel from which free oxygen or air is vigorously excluded. The cuttings pass through the heated vessel in such a manner that each incremental volume of cuttings passing through the vessel is exposed to a temperature of from about 500° to 700°F for a period of from about 5 to 15 minutes. The cleaned cuttings are then discharged from the bottom of the heated vessel. Means may optionally be provided for recovery and reuse of the heat removed from the cuttings.

The vapors from the drill cuttings are then collected in a suitable vessel and are condensed. Subsequently, the water is separated from the hydrocarbons, as well as any inorganic particulates from the cuttings, that may be entrained with the vapors as they are continuously vented from the vessel.

Heat is safely and conveniently provided to the cuttings cleaning vessel by use of external electrical resistance elements or, in the preferred form of this process, by means of a suitable thermal-transfer fluid in turn heated by an electrically energized auxiliary heat exchanger. The fluid is circulated in a closed path, continuous manner through conduits or enclosed channels within and/or surrounding the cuttings cleaning vessel and back through the heat exchanger.

It is critical to the success and safety of this process that during the period the cuttings are heated in the vessel, all oxygen and other oxidizing gases be rigorously excluded. If oxygen or air is allowed to contact the cuttings as they are heated, oxidation occurs and instead of a clean removal of oil and water from the cuttings being effected, and a dry friable solid devoid of free oil being produced, there results a sticky, dark, asphaltlike residue which is high in hydrocarbon content and totally unsuitable for clean disposal. There is also a danger of fire and explosion.

In the absence of oxygen and other oxidizing gases, the lighter hydrocarbon fractions in the oily cuttings are distilled directly away from the solid material, while the higher boiling hydrocarbon fractions are stripped away well below their normal distillation points, through the action of the water incorporated in the cuttings escaping as super-heated steam and carrying the heavier hydrocarbon vapor with it.

In order to ensure a nonoxidative atmosphere within the cuttings heating vessel, it is necessary, particularly while starting up and shutting down the heating process, that the interior of the vessel be purged with a nonoxidative gas, preferably carbon dioxide, nitrogen, or completely burned internal combustion engine exhaust gas.

Steam Treatment of Drill Cuttings

J. Kelly, Jr.; U.S. Patent 4,209,381; June 24, 1980; assigned to Mobil Oil Corporation describes a related process whereby a treating unit is located onsite and is adapted to receive contaminated drill cuttings after they have been separated from the mud. The treating unit is comprised of a conveyor means which receives the contaminated cuttings and carries them through a heating section of the treating unit. Positioned within the heating section are steam jets which spray the cuttings to heat the cuttings to a temperature of at least 212°F and preferably to a temperature of 300°F or more.

The steaming causes a steam distillation of the oil on the cuttings to take place. The steam and the oil, now entrapped in the steam, are withdrawn by means of a vacuum exhaust fan, or the like from the heating section and are passed to a cooling section of the treating unit where they are condensed. The condensed liquids are then removed from the cooling section and flow to a separation section in which the oil is separated from the water. The cuttings, now free of contaminants, pass from the heating section and are ready for disposal.

Washer/Separator System for Drilling Cuttings

J.D. Fisher; U.S. Patent 4,175,039; November 20, 1979 describes a washer/separator system for drilling cuttings which includes a receiving tank for the cuttings deposited from the wash trough. The cuttings fall through an agitation subchamber which insures thorough washing of the cuttings, while keeping solution agitation confined to the subchamber. The washed cuttings fall by the action of gravity through the bottom of the subchamber to the bottom of the tank. The bottom of the tank is sloped toward the center where a troughlike subextension of the tank houses a horizontally placed auger. As cuttings fill the tank, the solution level rises due to displacement. When the level has risen a predetermined amount, a level activated switch turns on a motor which drives the auger and at substantially the same time opens a drain valve at the outlet end of the auger. The auger pushes the cuttings through the valve where they may be deposited in the surrounding water. When the solution level then drops to a predetermined level, another level activated switch closes the valve and stops the auger, again at substantially the same time.

The intake for the return wash solution which washes the cuttings down the wash trough is placed substantially above the settled cuttings and away from the agitation subchamber yet below the lowest level of the wash solution. The return wash solution thus remains highly free of cutting debris. The advantages of the process are as follows:

(1) The device may be less than 8' tall, and in the preferred embodiment is, in fact, 7' tall, thus eliminating the need for drilling rig modification.

(2) Shut down due to pump clogging is virtually eliminated, since the return wash solution is highly free of cutting debris.

(3) Since the device does not use a shale shaker screen replacement and accompanying shut down is eliminated.

(4) Fresh water washing solution may be used because shale cutting break up does not affect the operation of the system.

(5) Since gravity, rather than screens, is used to cause separation, much finer filtration is effected and less abrasive sand remains in solution. Pump life is greatly extended.

In Figure 1.5 there is shown a front view of the preferred embodiment of the process. The main body of the device **10** is the receiving tank structure **20** filled with washing solution **21**. Cuttings to be washed drop through cutting intake port **40** into the agitation subchamber which forms an initial portion of the tank **20**. Agitation to enhance washing is provided by agitation blades **44** connected to a shaft **46** driven by an agitation motor **48**. A weir **50** separates the agitation subchamber **42** from the rest of the receiving tank **52**. Cuttings drop down through a gap **60** between the weir **50** and the bottom of the tank **62**, and collect in the auger trough **64**, which forms a cutting outlet feed section at the bottom of the tank.

The auger **66** is driven by auger drive motor **68**. The auger **66** is protected from large objects by a grill **70** placed over it.

Figure 1.5: Washer/Separator System for Drilling Cuttings

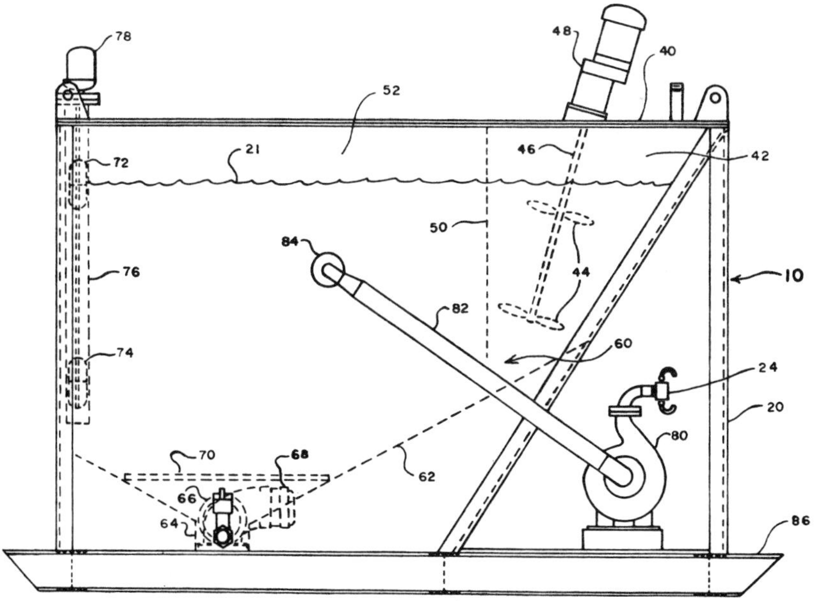

Source: U.S. Patent 4,175,039

The upper and lower level trip switches **72** and **74** are mounted in a liquid level control unit **76** on a side of the tank **52**. These switches are connected to auger discharge valve switching and sequencing unit **78**. This unit turns the auger drive motor **68** on and off and opens and closes the cutting discharge valve at the proper times and in the proper sequence as detected by the level trip switches **72** and **74**.

Return wash solution is drawn into the return wash pump **80** through return wash collecting pipe **82** which has its intake port **84** placed above the auger, as shown. The return wash is pumped out through return wash exhaust port **24**.

The tank is sized and the minimum fluid level for tripping the on switch for the auger **66** is selected so that the auger **66** is generally not activated until the auger chamber trough **64** is filled with washed cuttings. Thus in this process, when cuttings are discharged, they are discharged with a minimum amount of fluids. This desired result is further enhanced by the presence of the auger trough **64** in reasonably close conformity to the shape of the auger **66**. The entire receiving tank structure **20** is placed on a skid **86**.

CORROSION INHIBITORS

Substituted Pyridines

B.A. Oude Alink and N.E.S. Thompson; U.S. Patent 4,174,370; November 13, 1979; assigned to Petrolite Corporation describe substituted pyridines and a method for their preparation which comprises reacting an aldehyde, an amine, a carboxylic acid and oxygen. The overall reaction may be summarized as follows:

$$3RCH_2\overset{\overset{O}{\|}}{C}H + R'NH_2 + CH_3COOH + \tfrac{1}{2}O_2 \rightarrow \text{[pyridinium ring: R at 2-position, R at 4-position, } CH_2R \text{ at 3-position, } N^+\text{–}R'] \; CH_3COO^- + 4H_2O$$

Suitable aldehydes are those where R is alkyl, aryl, cycloalkyl, alkaryl, aralkyl, heterocyclic, etc. R is preferably alkyl, for example, having from 1 to 30 or more carbons, preferably from 1 to 12 carbons. Useful aldehydes include, for example, propionaldehyde, butyraldehyde, heptaldehyde, etc., as well as substituted aldehydes such as aldol, etc. R' is a substituted group, preferably a hydrocarbon group, for example, alkyl, cycloalkyl, aryl, alkenyl, heterocyclic, substituted derivatives of the above, etc.

These compounds can be used in a wide variety of applications and systems where iron, steel and ferrous alloys are affected by corrosion. They may be employed for inhibiting corrosion in processes which require a protective or passivating coating as by dissolution in the medium which comes in contact with the metal. They can be used in preventing atmospheric corrosion, underwater corrosion, corrosion in steam and hot water systems, corrosion in chemical industries, underground corrosion, etc.

These corrosion inhibitors find special utility in the prevention of corrosion of pipe or equipment which is in contact with a corrosive oil-containing medium, as, for example, in oil wells producing corrosive oil or oil-brine mixtures, in refineries, and the like.

The method of carrying out this process is relatively simple in principle. The corrosion preventive reagent is dissolved in the liquid corrosive medium in small amounts and is thus kept in contact with the metal surface to be protected. Alternatively, the corrosion inhibitor may be applied first to the metal surface, either as is, or as a solution in some carrier liquid or paste. Continuous application, as in the corrosive solution, is the preferred method, however.

This process finds particular utility in the protection of metal equipment of oil and gas wells, especially those containing or producing an acidic constituent such as H_2S, CO_2, air or oxygen, organic acids and the like. For the protection of such wells, the reagent, either undiluted or dissolved in a suitable solvent, is fed down the annulus of the well between the casing and producing tubing where it becomes commingled with the fluid in the well and is pumped or flowed from the well with these fluids, thus contacting the inner wall of the casing, the outer and inner wall of tubing, and the inner surface of all wellhead fittings, connections and flow lines handling the corrosive fluid.

Example: *1-Methyl-2-Propyl-3,5-Diethylpyridinium Acetate* – In a pressure reactor was placed 250.6 g of butyraldehyde and 69.7 g of acetic acid. To the mixture was added over a 5-minute period, 38.5 g of methylamine, while a reaction temperature of 67° to 75°C was maintained. Oxygen gas was added to 40 to 70 psi and the mixture stirred for 8 hours. The reaction mixture was cooled to ambient temperature and 150 g of water was added. The aqueous phase was extracted with ether, and evaporated under diminished pressure to yield 246 g of 1-methyl-2-propyl-3,5-diethylpyridinium acetate; NMR (solvent D_2O) δ in ppm, 8.68 m, 1H; 4.48, s, 3H; 3.00 m, 6H; 2.12, s, 3H; 1.78 m, 2H; 1.37 and 1.17 t's, 9H.

Quaternary Polyvinyl Heterocyclic Compounds

P.M. Quinlan; U.S. Patent 4,212,764; July 15, 1980; assigned to Petrolite Corporation describes quaternary polyvinyl heterocyclic compounds and the use thereof as corrosion inhibitors, particularly in acid systems.

More particularly, this process relates to quaternary polyvinyl heterocyclic compounds as illustrated by those derived from the following type of vinyl heterocyclics: vinylpyridine, vinylpyrazine, vinylpiperidine, vinylquinoline, alkylated vinylpyridine, alkylated pyrazine, alkylated vinylpiperidine, alkylate vinylquinoline, etc.

One type which may be used in the preparation of the herein described compound has been characterized for purposes of convenience as a nitrogen-containing vinyl-substituted heterocyclic. By nitrogen-containing vinyl-substituted heterocyclic is meant any chemical compound which has as a part of its structure a ring system containing nitrogen as a part of the cyclic system, and further has as a substitutent upon this cyclic unit a vinyl or a substituted vinyl group. This general specification includes a diverse group of materials. For instance, the heterocyclic ring may be an essentially aromatic ring such as pyridine or pyrazine, a fused ring system such as quinoline, or a nonaromatic ring such as piperidine. The essential structural element is the presence of one or more nitrogen atoms in the cyclic structure which are capable of entering into reaction with compounds such as halogen atoms capable of producing substituted nitrogen atoms or quaternary compounds. Further, there should be as a substitutent on the ring a vinyl or substituted vinyl group capable of inducing in the molecule a tendency toward polymerization by the usual vinyl polymerization mechanisms.

The following specific examples of compounds which may be employed for the purpose previously specified are as follows:

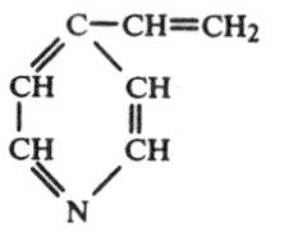

4-vinylpyridine

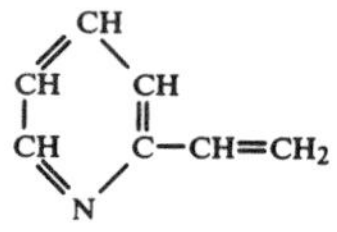

2-vinylpyridine

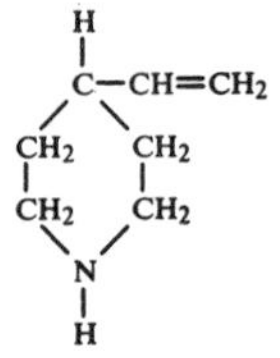

4-vinylpiperdine

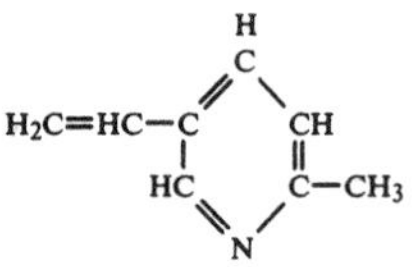

2-methyl-5-vinylpyridine

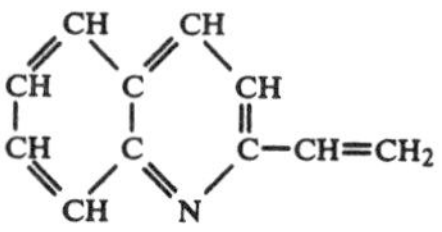

2-vinylquinoline

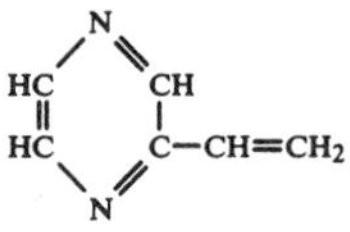

2-vinylpyrazine

As corrosion inhibitors these nitrogen groups are adsorptively active groups, i.e., are adsorbed on the metal. The preceding six formulae illustrate a number of suitable compounds which are particularly suited for use in this process. However, other well-known compounds can be substituted for these particular ones.

When 4-vinylpyridine is quaternized with alkyl halides spontaneous polymerization occurs to give the corresponding poly(4-vinylpyridinium) compound. For specific information on the preparation of polypyridiniums, reference is made to (1) V.A. Kobanov, et al., *J. Polym. Sci.,* Part C, 16, 1079 (1967); (2) V.A. Kobanov, et al., *J. Polym. Sci.,* Part C, 23, 357 (1968); and (3) I. Nielke and H. Ringsdorf, *Polymer Letters,* 9, 1 (1971).

Kobanov and Kargin have shown that at the temperatures above 5°C the reaction of 4-vinylpyridine with alkyl halides (Menshutkin's reaction) in organic media (benzene, nitrobenzene, acetonitrile, dimethylsulfoxide, propylene carbonate, methanol) leads not to the expected monomer salts, but to the polymers of the general formula:

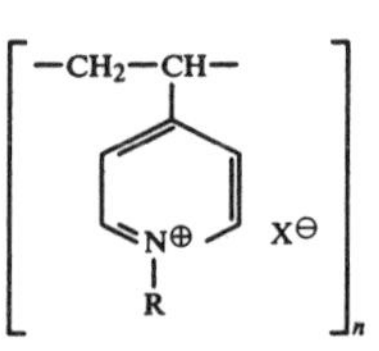

The amount of the composition employed in treating corrosive systems will vary with the particular compound employed, the particular system, the solids present in the system, the degree of corrosivity of the system, etc. A minor amount of the compound is generally employed sufficient to impart corrosion protection to the system. In practice, concentrations of about 1,000 to 2,000 ppm are employed.

Example: A mixture of 10.5 g of 4-vinylpyridine, 12.4 g of benzyl chloride, and 69 g of methanol was heated at reflux under a nitrogen blanket, for a period of 24 hours. The resulting polymer solution was viscous and water-soluble.

Quaternaries of Halogen Derivatives of Alkynoxymethylamines

P.M. Quinlan; U.S. Patent 4,187,277; February 5, 1980; assigned to Petrolite Corporation describes halogen derivatives of alkynoxymethylamines and uses for these compositions particularly as corrosion inhibitors. The alkynoxymethylamines are ideally illustrated by the following equation:

$$RNH_2 + 2CH_2O + 2R'OH \rightarrow RN(CH_2OR')_2 + 2H_2O$$

where R is a substituted group, preferably hydrocarbon such as alkyl, aryl, aralkyl, cycloalkyl, etc., and substituted derivatives thereof, etc., and R' is an alkynyl moiety.

The reaction is carried out by reacting the amine, aldehyde, and acetylenic alcohol under dehydrating conditions. The alkynoxymethylamine is formed. In practice the amine is gradually added to a mixture of formaldehyde and acetylenic alcohol in an azeotropic solvent at reflux until the theoretical amount of water is removed. Thereupon the product is separated from the reaction mixture, for example, by distillation under reduced pressure.

A wide variety of amines having at least one primary group can be employed. They include aliphatic, cycloaliphatic, aryl, heterocyclic, etc. amines. These amines may or may not contain other groups. The following are representative examples: n-butylamine; 2-ethylhexylamine; monoisopropanolamine; hexylamine; furfurylamine; monoethanolamine; 2-amino-4-methylpentane-4-amino-2-butanol; and 5-isopropylamino-1-pentanol.

The acetylenic alcohols employed may suitably include ethyloctynol, propargyl alcohol, hexynol and other acetylenic alcohols having the structural formula:

$$H{-}C{\equiv}C{-}R''{-}OH$$

where R'' is $-CR_3R_4-$ such as $H{-}C{\equiv}C{-}CR_3R_4{-}OH$ where R_3 is selected from the group consisting of CH_3 and H and R_4 is selected from the group consisting of hydrogen, alkyl groups having 1 to 18 carbon atoms, naphthalyl, phenyl, and alkyl substituted phenyls having 1 to 10 carbon atoms in the alkyl substituent. Examples of such alcohols include: methylbutynol, methylpentynol, hexynol, ethyloctynol, propargyl alcohol, benzylbutynol, naphthalylbutynol, and the like. Acetylenic alcohols which contain 3 to 10 carbons are preferred.

Although formaldehyde is preferred, other aldehydes or ketones may be employed in place of formaldehyde such as those of the formula, $R_1R_2C{=}O$, where R_1 and R_2 are hydrogen or a hydrocarbon group such as alkyl, i.e., methyl,

ethyl, propyl, butyl, etc.; aryl, i.e., phenyl, alkylphenyl, etc., benzyl; cycloalkyl, i.e., cyclohexyl, etc. Thus, the $-CH_2-$ group in the formula, $RN(CH_2OR')_2$, also may include substituted $-CH_2-$ groups, $RN(CR_1R_2OR')_2$, where the R_1 and R_2 are hydrogen or groups derived from the aldehyde or ketone.

The preparation of the halogen derivatives may be accomplished by the reaction of an alkali halide such as potassium iodide, sodium bromide and the like with sodium hypochlorite in the presence of the N,N-di(alkynoxymethyl)alkylamine. Thus the appropriate hypohalite is generated in situ in the presence of the acetylenic moiety to produce the desired halogen derivative. In brief a typical reaction would be:

$$-C{\equiv}CH + KI \xrightarrow{NaOCl} -C{\equiv}CI + NaOH + KCl$$

Typically the acetylenic moiety is mixed with an aqueous solution of the alkali halide. To this mixture is added, with cooling, the sodium hypochlorite. The mixture is then stirred at room temperature for prolonged periods of time. In most cases the acetylenic moiety is insoluble in the reaction medium. Thus the yields are decreased, and the length of the reaction is increased. The method may be improved considerably by adding an emulsifying agent, phenolethoxylates, sodium stearate, and the like, to the reaction mixture. This allows an intimate mixture of the acetylenic moiety and the alkali hypohalite to be obtained by agitation. The products are isolated by extraction and identified by analytical methods.

Monofunctional Quaternizing Agents: Any hydrocarbon halide, e.g., alkyl, alkenyl, cycloalkenyl, aralkyl, etc., halide which contains at least 1 carbon atom and up to about 30 carbon atoms or more per molecule can be employed to alkylate the products. It is especially preferred to use alkyl halides having between about 1 to 18 carbon atoms per molecule.

Difunctional Quaternizing Agents: X–Z–X may be a wide variety of compounds capable of joining amino groups, where Z may be alkylene, alkenylene, alkynylene, alkaralkylene, an alkylene-ether-containing group, an ester-containing group, etc., and X is a halide.

The following are representative examples of the quaternaries of this process. Compounds (1) are prepared by reacting 1 mol of monofunctional quaternizing agent with 1 mol of a halogen derivative of alkynoxymethylamines and Compounds (2) are prepared by reacting 1 mol of difunctional quaternizing agent with 2 mols of alkynoxymethylamines according to the equations:

$$(1) \quad RN(CH_2OR''X)_2 + R'A \rightarrow \underset{\displaystyle R'}{\underset{|}{RN^+}}(CH_2OR''X)_2 \; A^-$$

$$(2) \quad 2RN(CH_2OR''X)_2 + AZA \rightarrow \begin{matrix} R{-}N^+(CH_2OR''X)_2 \\ | \\ Z \\ | \\ R{-}N^+(CH_2OR''X)_2 \end{matrix} \quad 2A^-$$

Example 1: Into a 500 ml three-necked flask provided with a reflux condenser and stirrer were placed 18.4 g (0.04 mol) of $C_4H_9(CH_2OCH_2C{\equiv}CI)_2$, 5.7 g (0.04 mol) of methyl iodide, and 20 ml ethanol. The reaction mixture was heated to reflux and held there for 24 hours. The solvent was removed under reduced pressure on a rotary evaporator. The product had the following structure:

$$\left[\begin{array}{c} C_4H_9N^+(CH_2{-}O{-}CH_2C{\equiv}CI)_2 \\ | \\ CH_3 \end{array} \right] I^-$$

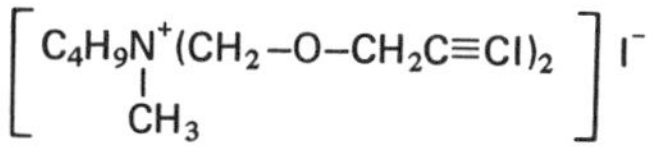

Example 2: Into a 500 ml three-necked flask equipped with a stirrer and reflux condenser were introduced 28.7 g (0.05 mol) of $C_{12}H_{25}N(CH_2OCH_2C{\equiv}CI)_2$, 5.4 g (0.025 mol) of 1,4-dibromobutane and 35.0 ml of n-propanol. The reaction mixture was heated at reflux for 72 hours. The isolated product had the following structure:

$$\left[\begin{array}{c} C_{12}H_{25}N^+(CH_2OCH_2C{\equiv}CI)_2 \\ | \\ (CH_2)_4 \\ | \\ C_{12}H_{25}N^+(CH_2OCH_2C{\equiv}CI)_2 \end{array} \right] 2Br^-$$

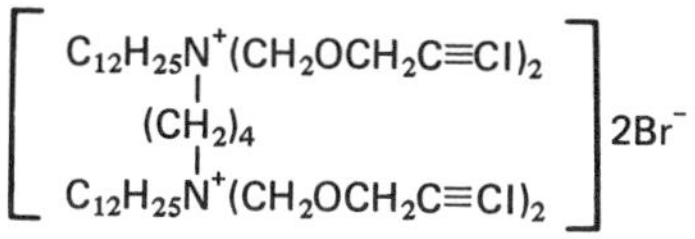

Thioether-Containing Quaternary Ammonium Derivatives of 1,4-Thiazines

P.M. Quinlan; U.S. Patent 4,188,359; February 12, 1980; assigned to Petrolite Corporation describes quaternary ammonium derivatives of 1,4-thiazines of the general formula

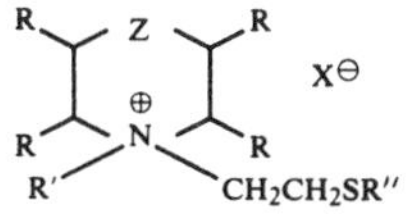

R, Z, R, X⊖, ⊕N, R, R, R′, CH_2CH_2SR''

where R is hydrogen, a hydrocarbon group such as alkyl, etc., or a substituted hydrocarbon group. R' is a hydrocarbon group such as alkyl, cycloalkyl, aralkyl, etc. or a substituted hydrocarbon group. R" is a hydrocarbon group such as alkyl, cycloalkyl, aralkyl, alkenyl, alkynyl, etc., or a substituted hydrocarbon group. Z is S, S→O, O←S→O, and X is an anion such as halide, sulfate, phosphate, nitrate, perchlorate or methyl sulfonate, ethyl sulfonate, vinyl sulfonate, etc. The preferred derivatives are those where R is hydrogen and it is further preferred that R' be an alkyl group having about 4 to 18 carbon atoms, a cyclohexyl group or a phenalkyl group and that R" be an alkyl group having 1 to 8 carbon atoms, a cycloalkyl group, an alkynyl group, or an alkenyl group.

These compounds are prepared by reacting a divinyl sulfur compound with a thioether substituted secondary amine. The reaction of secondary amines with divinyl sulfur compounds is described in U.S. Patent 3,770,732.

The reaction may be summarized by the following equation:

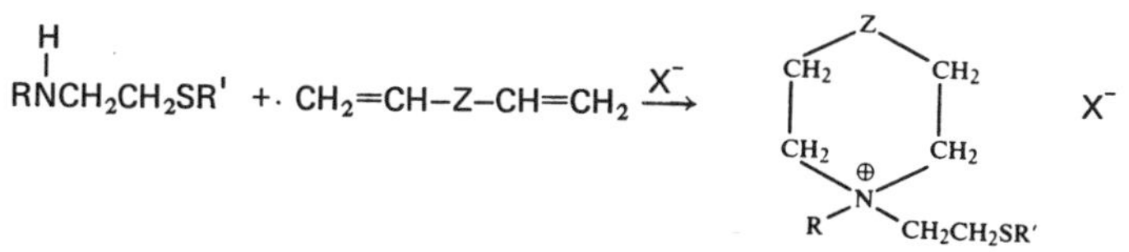

$$RN(H)CH_2CH_2SR' + CH_2{=}CH{-}Z{-}CH{=}CH_2 \xrightarrow{X^-}$$

Z, CH_2, CH_2, CH_2, CH_2, ⊕N, R, CH_2CH_2SR', X^-

where R is alkyl, cycloalkyl, aralkyl, etc. and R' is alkyl, alkenyl, alkynl, aralkyl, cycloalkyl, etc. Z is S, S→O, O←S→O.

Examples of the divinyl sulfur compounds are:

$$CH_2{=}CH{-}S{-}CH{=}CH_2$$

$$CH_2{=}CH{-}\overset{\overset{O}{\uparrow}}{S}{-}CH{=}CH_2$$

$$CH_2{=}CH{-}\underset{\underset{O}{\downarrow}}{\overset{\overset{O}{\uparrow}}{S}}{-}CH{=}CH_2$$

Examples of thioether substituted secondary amines capable of reacting with a divinyl sulfur compound to form the compounds of this process include, for example:

$$CH_3{-}CH_2{-}\overset{\overset{H}{|}}{N}{-}CH_2{-}CH_2{-}S{-}CH_2{-}CH_3$$

$$C_4H_9{-}\overset{\overset{H}{|}}{N}{-}CH_2{-}CH_2{-}S{-}C_4H_9$$

$$C_6H_5{-}CH_2{-}CH_2{-}\overset{\overset{H}{|}}{N}{-}CH_2{-}CH_2{-}S{-}C_4H_9 \quad (C_6H_5 = \text{phenyl})$$

$$C_4H_9{-}CH_2{-}CH_2{-}\overset{\overset{H}{|}}{N}{-}CH_2{-}CH_2{-}S{-}C{\equiv}CH$$

These compounds are obtained in good yield by utilizing a modification of the procedure described by O. Landini and F. Rolla, *Synthesis* 1974, p 565.

Suitable acids that may be employed to form the amine salts include hydrohalic acids such as hydrochloric, hydrobromic, hydroiodic, etc.; sulfuric, phosphoric, nitric, perchloric, methylsulfonic, ethylsulfonic, benzylsulfonic, and the like.

In carrying out the reaction it is preferred to form the amine salt in situ, i.e., in a solvent such as ethanol in which it is soluble. However, if desired, the salt may first be isolated and purified. To the solution of the amine salt in a suitable inert solvent is added the divinyl sulfur compound. The preferred temperature is about 20° to 50°C though higher or lower temperatures may be employed. A catalyst such as triethylamine may be used. In most instances the thiazine quaternary ammonium salt precipitates from the alcoholic medium and is purified by recrystallization. In some cases it is necessary to reduce the final volume in order to isolate the desired product.

While the compounds of this process are of themselves particularly good acid corrosion inhibitors, optionally they may be blended with acetylenic alcohols, dispersing and solubilizing agents such as ethoxylated phenols, alcohols, and fatty acids. They may also be blended with such known acid inhibitors as the quinoline or alkylpyridine quaternary compounds or synergists such as terpene alcohols, formamide, formic acid, alkyl amine, alkylene polyamines, heterocyclic amines, and the like.

These compounds may also be employed as microbiocides in both aqueous and hydrocarbon systems.

Example: 5.9 g (0.05 mol) divinyl sulfone was added dropwise, with stirring, to a solution of 9.6 g (0.05 mol)

$$C_4H_9-\overset{\displaystyle H}{\overset{|}{N}}-CH_2-CH_2-S-C_4H_9$$

in 25 ml of 4 N ethanolic hydrochloric acid. The mixture became warm and upon cooling crystals appeared. After 24 hours the crystalline product was filtered and washed several times with cold ethanol. The product was recrystallized from aqueous ethanol. Yield: 14.6 g (85%).

Analysis–Calculated for $C_{14}H_{30}ClNO_2S_2$: C, 48.91%; H, 11.46%; Cl, 10.32%; S, 18.61%; N, 4.07%. Found: C, 48.89%; H, 11.43%; Cl, 10.29%; S, 18.64%; N, 4.11%.

The product had the following structure which was confirmed by NMR spectrum:

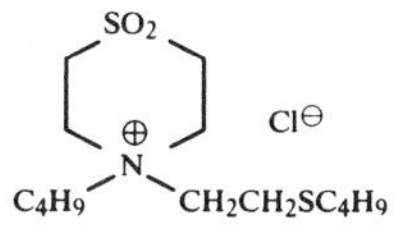

SO_2 ring, $\oplus N$, $Cl^{\ominus}$, C_4H_9, $CH_2CH_2SC_4H_9$

Sulfone-Containing Quaternary Ammonium Derivatives of 1,4-Thiazines

In a similar process *P.M. Quinlan; U.S. Patent 4,217,329; August 12, 1980; assigned to Petrolite Corporation* describes sulfone-containing quaternary ammonium derivatives of 1,4-thiazines of the general formula

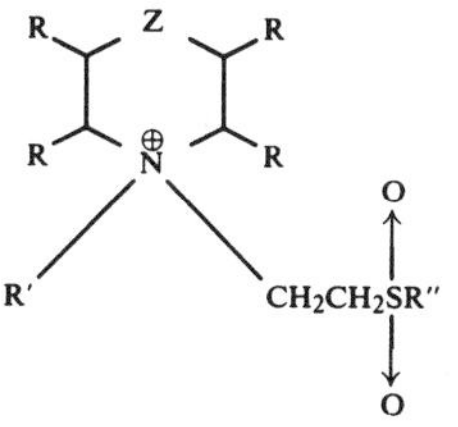

Z, R, R, R, R, $\overset{\oplus}{N}$, R', CH_2CH_2SR'' with O ← S → O

where R, R', R'', Z and X are as described in the previous process.

These compounds are prepared by oxidizing a thioether-substituted quaternary ammonium 1,4-thiazine in such a manner as to produce the corresponding sulfone.

The reaction may be illustrated by the following equation:

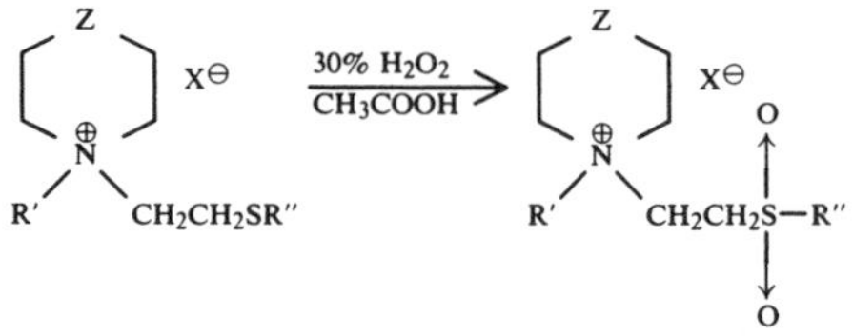

$$\text{[ring: Z, } \overset{\oplus}{N}\text{]}(R')(CH_2CH_2SR'')\ X^{\ominus} \xrightarrow[CH_3COOH]{30\%\ H_2O_2} \text{[ring: Z, } \overset{\oplus}{N}\text{]}(R')(CH_2CH_2S(\rightarrow O)_2-R'')\ X^{\ominus}$$

where R' is alkyl, cycloalkyl, aralkyl, etc. and R'' is alkyl, alkenyl, alkynl, aralkyl, cycloalkyl, etc. Z is S, S→O, O←S→O.

Examples of compounds suitable for conversion into sulfones are included in the following table.

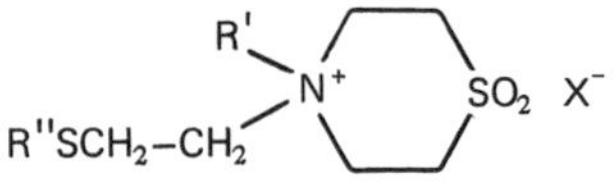

$$R''SCH_2-CH_2-N^+(R')\text{[ring]}SO_2\ X^-$$

R'	R''	X
C_4H_9	C_4H_9	Cl
$C_{12}H_{25}$	C_2H_5	Br
C_6H_{13}	C_6H_{11}	I
$C_6H_5CH_2CH_2$	$C_6H_5CH_2$	Cl
C_4H_9	C_6H_{11}	Br
$C_{10}H_{21}$	C_2H_5	Br
C_4H_9	C_8H_{17}	I
C_6H_{11}	C_6H_{11}	I
$C_{14}H_{29}$	CH_3	SO_3
$C_{18}H_{37}$	CH_3	I
C_4H_9	$CH{\equiv}C{-}CH_2$	Br
$C_{12}H_{25}$	CH_3	I
C_4H_9	$CH_2{=}CH{-}CH_2$	Br

Note: C_6H_5 is phenyl and C_6H_{11} is cyclohexyl.

As is well-known sulfones may be produced in high yields by treating the corresponding thioethers with a suitable oxidizing agent.

Perhaps the best general procedure is the addition of 30% hydrogen peroxide to a solution of the thioether in glacial acetic acid. At least 2 mols of H_2O_2 per mol of thioether are necessary for conversion into the sulfone. In most instances, in excess of 2 mols is utilized.

The peroxide solution is slowly added to the quaternary ammonium 1,4-thiazine dissolved in glacial acetic acid. It is preferred that the reaction temperature be kept below about 50°C. However, higher temperatures may be employed.

The sulfones are obtained in good yield and are characterized by NMR and IR spectra as well as other analytical means.

Example: Into a suitable reaction vessel equipped with a stirrer, addition funnel, and reflux condenser were placed 17.1 g (0.05 mol) of

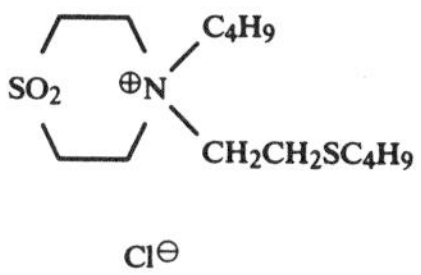

$$SO_2\langle\rangle\overset{\oplus}{N}(C_4H_9)(CH_2CH_2SC_4H_9)\quad Cl^{\ominus}$$

and 50 g of glacial acetic acid. The salt was stirred, with mild heating to 35°C, until it had completely dissolved. To this solution was added, over a period of 1 hour, 20 ml of 30% H_2O_2. After the addition was completed, the resulting mixture was stirred at room temperature for 2 hours.

The solvents were distilled off under reduced pressure. The resulting solid was washed with acetone, filtered, and the filtrate was washed several times with acetone. 16 g (85%) were recovered. The product was recrystallized from aqueous ethanol.

Analysis–Calculated: C, 44.7%; H, 8.0%; Cl, 10.6%; N, 3.7%; O, 17.0%; S, 17.0%.
Found: C, 44.4%; H, 7.9%; Cl, 10.4%; N, 3.69%; O, 16.8%; S, 17.0%.

The product was characterized by NMR and IR spectrum. It had the following structure:

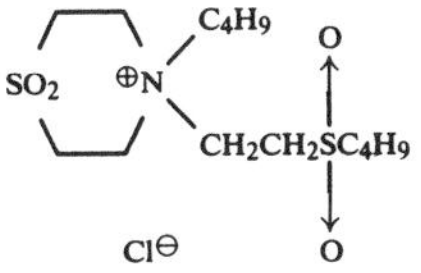

$$SO_2\langle\rangle\overset{\oplus}{N}(C_4H_9)(CH_2CH_2\overset{O\uparrow}{\underset{\downarrow O}{S}}C_4H_9)\quad Cl^{\ominus}$$

These products are useful as corrosion inhibitors and microbiocides.

Acylated Hydroxyalkylaminoalkylamides

In U.S. Patent 3,714,249 there is disclosed hydroxyalkylaminoalkylamide as illustrated by the formula

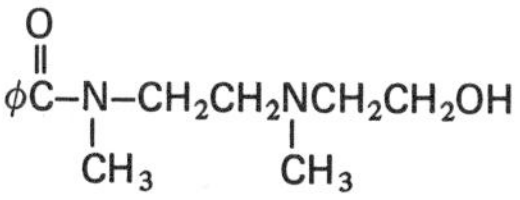

$$\phi\overset{O}{\overset{\|}{C}}-\underset{CH_3}{\underset{|}{N}}-CH_2CH_2\underset{CH_3}{\underset{|}{N}}CH_2CH_2OH$$

which is prepared by reacting under anhydrous conditions in the absence of a catalyst, (a) a nitrile of structure $R_1(CN)_n$ where R_1 is selected from the group of alkyl of 1 to 10 carbon atoms, alkylene of 2 to 10 carbon atoms, phenyl, naphthyl, phenylene, lower alkyl substituted phenyl of from 7 to 12 carbon atoms, phenylalkyl of 7 to 12 carbon atoms, and lower alkyl substituted phenylene of from 7 to 12 carbon atoms, and n is a small integer of from 1 to 3, with (b) an alkanolamine of structure as follows:

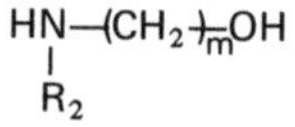

$$HN\text{-}(CH_2)_m\text{-}OH \quad (\text{N bears } R_2)$$

where R_2 is H or lower alkyl and m is an integer of 2 to 4, the reaction being conducted at a temperature of from about 100° to 200°C, at essentially atmospheric pressure, and at a mol ratio of alkanolamine per nitrile group exceeding 2:1, but less than about 20:1.

In U.S. Patent 4,060,553, the reaction of U.S. Patent 3,714,249 has been extended to include the reaction product of unsaturated nitriles with N-alkylalkanolamines to form similar compounds where in addition the N-alkylalkanolamine also reacts with the double bond of the unsaturated nitrile to form an N,N-di(alkylalkanol).

This is illustrated by the following equation:

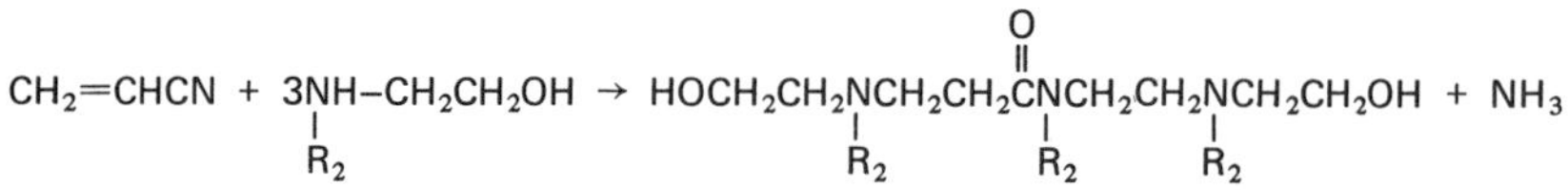

$$CH_2{=}CHCN + 3NH(R_2)\text{-}CH_2CH_2OH \rightarrow HOCH_2CH_2N(R_2)CH_2CH_2C(=O)N(R_2)CH_2CH_2N(R_2)CH_2CH_2OH + NH_3$$

This product can then be mono- or dialkylated.

D. Redmore, B.T. Outlaw and D.C. Scranton, Jr.; U.S. Patent 4,168,292; September 18, 1979; assigned to Petrolite Corporation have found that the compositions of U.S. Patent 3,714,249, having the formula

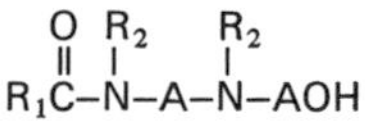

$$R_1C(=O)\text{-}N(R_2)\text{-}A\text{-}N(R_2)\text{-}AOH$$

where R_1 is the moiety of the original nitrile, e.g., alkyl, alkylene, aryl, aralkyl or a lower alkyl substituted aryl group, R_2 is hydrogen or a substituted group and A is alkylene, and the compositions of U.S. Patent 4,060,553 can be acylated with a wide variety of carboxylic acids, preferably fatty acids, or derivatives of carboxylic acids which act as carboxylic acid equivalents such as esters, etc., and that the resulting acylated compositions are particularly useful as corrosion inhibitors particularly where enhanced oil solubility is desired.

The chemistry of acylation may be presented as follows:

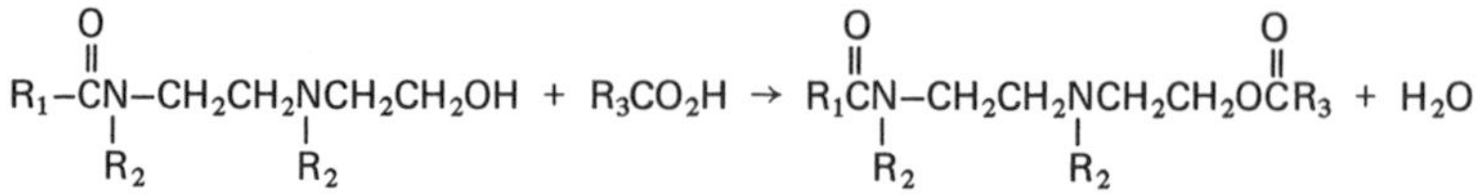

$$R_1\text{-}C(=O)N(R_2)\text{-}CH_2CH_2N(R_2)CH_2CH_2OH + R_3CO_2H \rightarrow R_1C(=O)N(R_2)\text{-}CH_2CH_2N(R_2)CH_2CH_2OC(=O)R_3 + H_2O$$

or in cases where R_2 = H amide formation also takes place

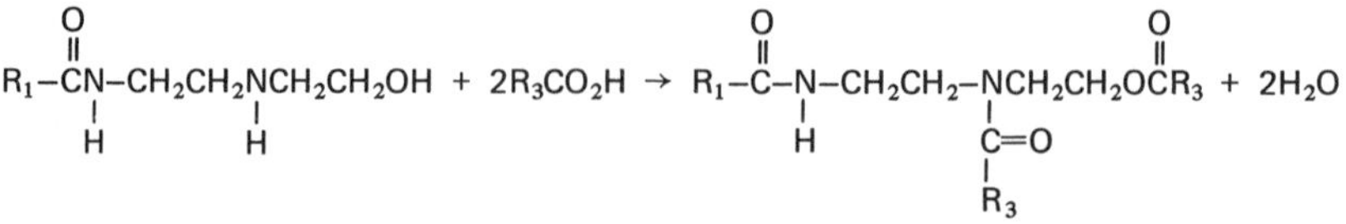

$$R_1\text{-}C(=O)NH\text{-}CH_2CH_2NHCH_2CH_2OH + 2R_3CO_2H \rightarrow R_1\text{-}C(=O)\text{-}NH\text{-}CH_2CH_2\text{-}N(C(=O)R_3)CH_2CH_2OC(=O)R_3 + 2H_2O$$

Although a wide variety of carboxylic acids produce excellent products, carboxylic acids having more than 6 and less than 40 carbon atoms but preferably 8 to 30 carbon atoms give most advantageous products. The most common examples include the detergent-forming acids, i.e., those acids which combine with alkalies to produce soap or soaplike bodies. The detergent-forming acids, in turn, include naturally-occurring fatty acids, resin acids, such as abietic acid, naturally-occurring petroleum acids, such as naphthenic acids, and carboxy acids, produced by the oxidation of petroleum. There are other acids which have somewhat similar characteristics and are derived from somewhat different sources and are different in structure, but can be included in the broad generic term previously indicated.

Suitable acids include straight chain and branched chain, saturated and unsaturated, aliphatic, alicyclic, fatty, aromatic, hydroaromatic, and aralkyl acids, etc.

In general, a minor but effective amount of the products are employed to inhibit corrosion such as at least about 5 ppm, but preferably from about 1 to 500 ppm.

Example 1: 58.7 g (0.57 mol) benzonitrile and 129 g (1.72 mol) 2-(methylamino)-ethanol were stirred at 150° to 180°C (reflux) for 30½ hours under a continuous sweep of nitrogen. Evolution of NH_3 was evident during the reaction. The resulting mixture was distilled under vacuum to remove the excess amine. The viscous reaction product (115.4 g) was found to have the structure

$$\phi\overset{\overset{\displaystyle O}{\|}}{C}-\underset{\underset{\displaystyle CH_3}{|}}{N}-CH_2CH_2\underset{\underset{\displaystyle CH_3}{|}}{N}CH_2CH_2OH$$

which on hydrolysis yields benzoic acid and the amine,

$$\underset{\underset{\displaystyle CH_3}{|}}{N}HCH_2CH_2\underset{\underset{\displaystyle CH_3}{|}}{N}CH_2CH_2OH$$

Example 2: To the condensate of Example 1 (23.6 g) was added Crofatol-P (a tall oil distilled, Crosby) (29 g) and the mixture heated at reflux in xylene using a Dean-Stark trap to collect water. After heating at reflux for 7 hours, 2 g of water had been collected and the esterification was complete.

Sorbitol, Benzotriazole and Phosphate

J.T. Jacob and V.R. Kuhn; U.S. Patent 4,202,796; May 13, 1980; assigned to Chemed Corporation describe a relatively nontoxic, nonchromate, nonzinc corrosion inhibiting composition which is capable of protecting ferrous metals from corrosion, the composition consisting essentially of (a) sorbitol, (b) benzotriazole or tolyltriazole, and (c) water-soluble phosphates, e.g., phosphoric acid, disodium phosphate, sodium tripolyphosphate, or tetrapotassium pyrophosphate. This mixture can be blended with any well-known scale inhibitors or dispersants.

In the composition, the preferred weight ratio of sorbitol:benzotriazole or tolyltriazole:water-soluble phosphate is 0.01 to 100:0.01 to 100:1. Even more preferably it is 0.1 to 10:0.1 to 10:1. These same ratios are applicable to levels of the compounds in water, where the phosphate is preferably maintained at about

0.01 to 5,000 ppm, and even more preferably about 0.1 to 50 ppm.

Preferred formulations are as follows:

Liquid Formulation

	Percent by Weight
Deionized water	12.8
Phosphoric acid (75%)	10.0
Ethane-1-hydroxy-1,1-diphosphonic acid (40%)	15.0
Sorbitol	10.0
Potassium hydroxide (45%)	46.2
Tolyltriazole	6.0

Powder Formulation

	Percent by Weight
Sodium phosphate (monobasic) monohydrate	15.84
Ethane-1-hydroxy-1,1-diphosphonic acid	8.04
Benzotriazole	9.00
Sorbitol	15.00
Sodium sulfate	36.47
Sodium carbonate	15.65

Esters of Carboxylic Acids and Aliphatic Alcohols

R.W. Jahnke; U.S. Patent 4,166,151; August 28, 1979; assigned to The Lubrizol Corporation describes protective compositions which may be easily and conveniently applied to metal surfaces.

The compositions are esters of aliphatic carboxylic acids containing about 10 to 25 carbon atoms and aliphatic alcohols containing about 15 to 40 carbon atoms. The acids are generally free from acetylenic unsaturation and include, for example, lauric, myristic, palmitic, stearic, arachidic, oleic, linoleic, linolenic and ricinoleic acids as well as mixtures thereof. Fatty acids containing olefinic unsaturation (e.g., oleic, linoleic, linolenic and ricinoleic) are preferred. Especially preferred are acids in the C_{14-20} range, notably oleic acid and commercially available fatty acid mixtures in which oleic acid is the principal constituent; examples of such mixtures are the tall oil fatty acids Acintol FA-2 and Sylfat 96.

Suitable alcohols are preferably free from acetylenic unsaturation and usually also from ethylenic unsaturation in more than about 10% by weight thereof. They include the hexadecanols, octadecanols, eicosanols and homologous higher alcohols having both straight and branched chains. Mixtures of such alcohols are useful and are often preferred from the standpoint of commercial availability.

Among the useful alcohol mixtures are those described in U.S. Patent 3,676,348, in which they are defined as comprising unsubstituted substantially saturated aliphatic monohydric normal alcohols containing about 20 to 34 carbon atoms per molecule and unsubstituted aliphatic monohydric or dihydric nonnormal alcohols containing carbon atoms in the same range and having less than about

5% unsaturated molecules, in combination with hydrocarbons also in about the C_{20-34} range. The alcohol content of these mixtures is at least 1% and preferably about 60 to 70%, and the weight ratio of normal to nonnormal alcohols is preferably between 2:1 and 1:2. The size of the hydrocarbon molecules is immaterial, since they do not form esters in appreciable amounts but merely serve as inert diluentlike substances.

A typical alcohol mixture suitable for preparation of the compositions is Epal 20+, characterized as comprising about 33% normal alcohols in the C_{18-34} range, about 34% branched (i.e., nonnormal) alcohols in the same carbon atom range, about 12% C_{24-40} paraffinic hydrocarbons, about 18% olefinic hydrocarbons in the same carbon atom range, and about 3% esters.

The esters may be prepared by known methods by reacting the alcohols with the acids or with functional derivatives thereof such as acid halides, anhydrides or lower alkyl esters. The reaction between the acid and the alcohol is usually effected in the presence of an acidic catalyst and a substantially inert diluent, many of which are known to those skilled in the art, at temperatures typically within the range of about 80° to 225°C. Approximately equivalent amounts of acid and alcohol are ordinarily used, but an excess of either (typically less than about a 100% excess) may be present without impairing utility. In particular, it is possible to treat an alcohol-hydrocarbon mixture as though it consisted entirely of alcohols, thereby using an excess of acid which may remain in the composition (as evidenced by an acidic neutralization number) without causing harm.

A further embodiment of this process is a solution or dispersion of the above-described ester composition in a substantially inert, normally liquid organic diluent. Suitable diluents include, for example, hydrocarbons, alcohols, ethers, esters and the like. The diluents are relatively volatile at ambient and normal storage temperatures, which means that they possess a vapor pressure at such temperatures high enough so that evaporation from a metal surface takes place within a relatively short time (e.g., 4 to 6 hours). Examples of suitable solvents are volatile naphtha fractions, Stoddard solvent, methanol, ethanol, various lower alkyl ethers, ethyl acetate and the like. Particularly preferred are hydrocarbon fractions such as the naphthas and Stoddard solvent, with the latter being especially desirable.

The solutions or dispersions normally contain about 5 to 75 pbw/100 parts of the solution or dispersion, of the ester composition. As is apparent from the definition thereof, either a solution or dispersion may be formed when the ester and diluent are blended. For example, Epal 20+ is soluble in Stoddard solvent only to about 10%, so that a composition comprising more than that percentage in combination with Stoddard solvent will be largely a dispersion. Such dispersions are suitable for use to protect metal workpieces.

Coating of the workpiece may be effected by any known method such as dipping, brushing, spraying, flow coating or the like and is normally effected at ambient temperatures, e.g., about 20° to 35°C. The workpiece may be ferrous metal, aluminum, or any other metal subject to corrosion, but the process is particularly useful for the treatment of ferrous metal (e.g., steel) workpieces.

This process is illustrated by an example in which an ester is prepared by reaction of 142 pbw of Acintol FA-2 with 280 parts of Epal 20+ in xylene solution at 188° to 190°C, in the presence of p-toluene-sulfonic acid. The xylene is removed by vacuum stripping and the resulting ester has an acid number of 9.5, using phenolphthalein as the indicator. A 60% (by weight) dispersion of this ester in Stoddard solvent is prepared by mixing at 65°C and is cooled to room temperature. Steel panels are dipped in the dispersion, removed and allowed to stand vertically for 4 hours at room temperature, whereupon the Stoddard solvent evaporates to leave a waxy film of the ester on the panels.

The panels are suspended over a 3.5 N hydrochloric acid solution and are observed periodically. After 4 days, only slight evidence of corrosion is noted, and after 5 days approximately 30% of the panel is coated with a very light stain. A control panel which has not been treated with the composition is much more seriously corroded.

Phosphorylated Oxyalkylated Polyols and Sulfites

Sulfite and bisulfite salts such as sodium or ammonium sulfites are effective oxygen scavengers in oxygen-containing systems thus reducing or inhibiting corrosion caused by the presence of oxygen in the system. However, such sulfites, when employed in hard or scale-producing water, are rendered less effective due to their precipitation as scale. Thus, hard water not only decreases the effectiveness of sulfites as oxygen scavengers but also creates problems due to scale formation.

T.J. Bellos and J.E. Davis; U.S. Patent 4,213,934; July 22, 1980; assigned to Petrolite Corporation describe a method of preventing the precipitation of sulfites as scale, which precipitation renders such sulfites less effective as oxygen scavengers, by a process which comprises the use of the phosphorylated oxyalkylated polyols in conjunction with the sulfites.

The oxyalkylated polyols which are phosphorylated according to this process are ideally represented by the formula $R[(OA)_nH]_x$, where R is an organic and preferably hydrocarbon moiety of the polyol, OA is the oxyalkylene moiety derived from the alkylene oxide, for example, ethylene oxide, propylene oxide, butylene oxides, etc., and mixtures or block units thereof, n is the number of oxyalkylated units and x represents the total number of units containing OH groups.

Preferred polyols include glycerol, polyglycerol, trimethanolethane, pentaerythritol, etc., mannitol, 1,2,3-hexanetriol, etc.

A number of processes are known in the art for preparing the phosphorylated polyols. A preferred process is to react polyphosphoric acid with a polyol. The polyphosphoric acid should have a P_2O_5 (i.e., phosphorus pentoxide) content of at least about 72%, preferably about 82 to 84%. A residue of orthophosphoric acid and polyphosphoric acid remains on completion of the reaction. This residue may be as high as about 25 to 50% of the total weight of the phosphorylated polyol. It may either be removed or left in admixture with the phosphorylated polyol.

Preferably the phosphorylated polyols produced by the process are prepared employing amounts of a polyphosphoric acid having about 0.5 to 1 M equivalents of P_2O_5 for each equivalent of the polyol used. Larger amounts of polyphosphoric acid can be used if desired. By equivalent of the polyol is meant the hydroxyl equivalents of the polyol. For example, 1 mol of glycerol is 3 equivalents thereof, and so forth. The phosphorylated polyols (acid esters) can be partially or completely converted to their corresponding alkali metal salts or ammonium salts by reacting with appropriate amounts of alkali metal hydroxides or ammonium hydroxide.

The compositions are polyfunctional acid phosphate esters of polyhydric alcohols, the esters having the formula $R\text{—}(OPO_3H_2)_x$ where R is the hydrocarbyl group of a polyhydric alcohol (i.e., R is any remaining organic residue of a polyhydric alcohol used as the starting material) and x is a number from 2 to 6, the esters often being referred to in the art as phosphorylated polyols.

The sulfite compounds useful in this process are selected from the group consisting of alkali metal sulfites, ammonium sulfite, alkali metal bisulfite, and ammonium bisulfite. The preferable sulfite compound is sodium sulfite, Na_2SO_3.

The concentration of sulfite compound in aqueous solution is not critical and may be any concentration up to saturation. Preferably the solution will contain from 5% up to saturation of the sulfite compound as lower concentrations require the handling of excessively large volumes of solution.

The concentration of the phosphorylated oxyalkylated polyols (POP) must be sufficient to prevent scale formation in the aqueous solution. Preferably the concentration of the POP is from 1 to 20% by weight based on the concentration of the sulfite compound.

It is preferred to prepare a sulfite composition which contains the sulfite compound in admixture with the POP prior to forming the aqueous solution. In this manner only one product needs to be shipped and handled to prepare an aqueous sulfite solution which is useful as an oxygen scavenger to reduce the harmful effects of oxygen in aqueous systems, particularly the formation of scale. Such aqueous systems may be at remote, relatively inaccessible locations where they are used as drilling fluids for oil and gas wells and waterflood treating solutions for increasing the recovery of oil from petroleum-containing formations.

Cobalt Catalyst for Sulfite/Oxygen Scavenging

B.L. Carlberg and R.A. Hart; U.S. Patent 4,231,869; November 4, 1980; assigned to Conoco, Inc. describe a method for the recovery of hydrocarbons from petroliferous formations wherein water is injected into the formation through a wellbore, the method comprising reducing oxygen levels by mixing the water with a source of sulfite ion, the mixture then being passed over a cation exchange resin having cobalt ions or other metallic catalyst adsorbed thereon.

Example 1: Dissolved oxygen in 1 liter of tap water was measured with a dissolved oxygen meter (membrane electrode type) and found to be about 4 mg/ℓ. Sodium sulfite in a ratio of 15 mg/ℓ to 1 mg/ℓ (ppm) of oxygen was added to the water, and time to reach minimum scavenged value (0.01 ppm oxygen) was measured. The experiment was repeated 12 times for an average uncatalyzed reaction time to 0.01 ppm oxygen of 333 seconds.

Example 2: A cation exchange resin (Dowex 50W) was placed into a solution of cobalt chloride (about 1 M) in distilled water. The resin was allowed to remain immersed in the water containing cobalt chloride at room temperature for a time sufficient to allow the cobaltous ions to become adsorbed upon the resin. The adsorption occurred fairly rapidly as the affinity for cobaltous ions seemed high using this resin.

The cobalt-containing beads formed were then added (approximately 100 beads) to 1 liter of water with sodium sulfite also being added in the same ratio as described in Example 1. The time to reach minimum scavenged value was measured by dissolved oxygen meter. As in Example 1 this test was repeated 12 times and the average time to reach 0.01 ppm oxygen was 182 seconds.

Example 3: Tap water containing 6 mg/ℓ oxygen was used in a continuously flowing system. Sodium sulfite in a ratio to O_2 of 8 to 1 (plus a 10 mg/ℓ excess) was added continuously. The treated tap water was continuously passed through a bed of cobalt-containing ion exchange resin prepared as described in Example 2.

A continuous recorded monitoring of dissolved oxygen levels was made downstream of the catalyst bed. Dissolved oxygen levels fell to 0.01 mg/ℓ in less than 1 minute. The system was maintained in constant operation for 1 month with no catalyst change whatsoever, with constant recording of oxygen levels. At the end of the experiment dissolved oxygen levels were 0.01 mg/ℓ. Occasional air bubbles in the system caused recording blips, but dissolved oxygen remained constant.

The treated water was tested for cobalt content. Less than 0.1 ppm cobalt (below detectable levels) was found.

Representative examples of strongly acidic cation exchange resins which are useful in the process are Dowex 50W, HCR-S, HCR-U, HGR, HGR-U (Dow Chemical Company), Amberlite 120, Amberlite 122, Amberlite 200, and Amberlite 252 (Rohm and Haas), and Duolite C-20, C-25 and ES-26 (Diamond Shamrock).

Catalysts are placed on the resin simply by making a water solution of the catalyst and immersing the cation exchange resin in the solution so formed. The catalyst can have counterions of any type which are water-soluble and which are not detrimental to the end use to which the water is to be put. Representative examples of such materials are cobaltous chloride, cobaltous bromide, cobaltous iodide and cobaltous nitrate.

Likewise, the sulfite may be added to the water by any means providing water-soluble counterions not detrimental to the end use. Representative examples of such forms are sodium sulfite, sodium bisulfite, potassium sulfite and lithium sulfite. Also, SO_2 can be sparged into the water to produce necessary sulfite levels. Sulfite to oxygen levels can range from stoichiometric to a three times stoichiometric excess of sulfite. Normally, a stoichiometric amount of sulfite plus about 10 ppm excess will be used and is preferred. Once the catalyst has been placed upon the cation exchange resin, water may be passed through the resin using any one of the systems well-known in the art, such as by placing the resin on a screen. The resin should be immersed in the solution containing the catalyst in ionic form for a sufficient period of time to allow the catalyst to adsorb onto the resin. Normally, this period of time will be at least 3 minutes. Preferred ranges are from 3 minutes to 1 hour and most preferred are from 5 to 30 minutes.

Injection of Nitrogen into Drilling Fluids

H.E. Mallory and J.W. Ward; U.S. Patent 4,136,747; January 30, 1979; assigned to Loffland Brothers Company describe a well bore drilling method and means whereby oxygen content of well drilling muds is replaced by nitrogen, exhaust gases, or gaseous mixtures of combustible products for reduction of corrosion.

Nitrogen, or any other suitable gas or mixture of gases, is injected into the drilling fluids to replace any oxygen in the fluids. The gas may be injected into the fluid in any suitable manner, such as injection into the fluids at a suitably vented station upstream of the pump suction, such as a vented tank, degasifier, or other vessel, and may be utilized not only for the removal of oxygen, but also to reduce or possibly eliminate the use of other chemicals in connection with the drilling fluids. Another method is to inject the gas directly into the mud pits.

Nitrogen is perhaps the preferable gas to inject into the drilling fluids in that it is plentiful and readily available. This method includes utilizing the nitrogen from the exhaust gases of the normal equipment, such as engines and the like, present at the well drilling site, and injecting the exhaust gases or the like into the drilling fluids through a degasser interposed between the exhaust system of the engines and the suction side of the drilling mud pumping equipment. The exhaust gases of engines using either natural gas or diesel fuel are substantially 87% nitrogen, and thus it will be apparent that substantially all types of exhaust gases are usable.

It is expected that perhaps 1,000 to 5,000 ft^3 of nitrogen will be used per hour during a typical well drilling operation, and as set forth above, since nearly all of the exhaust gases are usable as nitrogen, a plentiful supply of normally waste product is usually available at each well site.

The nitrogen is injected into the drilling fluids and replaces the oxygen in the drilling fluid; and not only is the oxygen content of the drilling fluid substantially eliminated or reduced to minute quantities, as for example, ½ ppm or less, for substantially eliminating corrosion of the drill pipe and other metallic elements used in the drilling operation, but also gases which are normally wasted are recovered for use, and the venting of engine exhaust gases into the atmosphere is greatly reduced for reducing environmental hazards. It is also considered that the nitrogen injected into the drilling fluids may reduce the catalytic effect the oxygen would have on hydrogen sulfide which may be present in the drilling fluids, thus further reducing any corrosive action.

Method of Coating a Borehole

Various means have been employed for coating the inside of a pipe in a well with anticorrosive chemicals or with other substances found desirable in any particular case, but perhaps the most commonly used is the method whereby a body of liquid anticorrosive chemical estimated to be sufficient to coat the inside of the pipe in question is injected into the upper end thereof and allowed to fall by gravity to the lower end.

This method involves a number of nondesirable features as follows:

(1) The interior surface coverage is likely to be incomplete. In practice the rather viscous liquid tends to flow down

one side of the pipe and adhere thereto as it flows instead of moving down as a complete plug or as a sheet of liquid completely covering the interior surface of the pipe.

(2) The liquid flowing over the surface of the pipe may not completely wet the entire surface of the pipe but may, after flowing past, withdraw from the surface of the pipe leaving it uncoated in spots.

(3) Particularly where the well is on an inclination to the vertical there is a strong tendency for the liquid to run down the lower side of the tubing or pipe and coat that side only, leaving the upper side of the interior of the pipe uncoated.

(4) Contrary to what might be expected, the movement of chemical downwardly within a tubular member such as a well tubing is very slow as a rule. The chemical seems to be inclined to form a piston within the tubing and move downwardly at a pace of perhaps not more than 1,000 fph, or a little over 16 fpm, requiring for deep wells from 12 to 24 hours for the chemical to move from top to bottom. This, of course, is highly undesirable in requiring an extremely long well shut down time.

D.W. Clayton; U.S. Patent 4,216,249; August 5, 1980; assigned to Champion Chemicals, Inc. describes a method for coating the inside of a hollow member such as a tubing or other pipe in a well in the earth, which method will result in a much more perfect job of complete coverage and intimate contact of the coating material with walls of the member and which may be much more rapidly carried out than the conventional method of providing such coating.

Another object of this process is to provide such a means which will be moved downwardly through a well at a much faster rate than that at which a liquid coating material can be moved through the well by gravity alone. It should be able to do this even in the presence of the rather viscous liquid material. The movement should be fast enough so that, for example, a 10,000' well, instead of requiring from 12 to 24 hours to treat, could be treated in about 2 hours.

In accordance with this process the body of coating material would be injected so as to inject into the upper end of the hollow member a sufficient body of coating material to coat the entire interior of the member. However, beginning immediately after the injection of the coating material into the hollow member, instead of waiting for it to fall by gravity and run down the walls of the member to coat them, a brush or brushes will be injected into the upper end of the hollow member and forced downwardly to force the coating material downwardly in the hollow member at a much more rapid rate than such material would move by itself. The device for doing this would consist of one or more bristle type brushes having bristles of such a character that they would bend and permit the brush to go through a somewhat strictured portion of the hollow member, but then resume their original shape so as to brush the interior walls of the full intended cross section of the hollow member. In case of the use of multiple such brushes, they would be employed in tandem and interconnected with each other by flexible joints which might be termed knuckle joints.

On the lower end of the brush, or on the lowermost of the brushes, if there be more than one, is preferably a device known as a "go-no-go" member which is essentially a gage sized to pass through the smallest opening which the brushes will pass through without likelihood of becoming stuck.

Carried preferably but not necessarily above the entire brush assembly would be a weight, preferably of a high density material so that as much weight as possible could be concentrated in as small a body as possible, and this weight would be of a cross-sectional area much smaller than the interior of the hollow member so as to permit it to fall readily through the body of liquid or other coating material through which the brush assembly is intended to be forced downwardly in the course of the coating operation. The downward forcing of the coating material in this fashion causes a small portion of the material to be squeezed between the brushes and the inner wall of the hollow member so as to brush such material well into the walls of the hollow member. Some of the coating material will squeeze past or extrude into the space above the brush assembly as the assembly moves downward. Then when the brush assembly reaches its lowermost position and starts up, the material on top of the brush assembly will be moved into the space between the brushes and the inner wall of the hollow member so as to again brush such material onto such walls.

In order to readily move the brush assembly upwardly in the well and complete the process the uppermost of the brushes is equipped with a means to receive a wireline or cable by which the brush assembly may be pulled upwardly.

The brushes are also preferably slidably mounted on mandrel sections, one mandrel section for each brush, so that if by chance one of the brushes did become wedged or stuck in the course of the operation, it would be possible to lower the mandrel assembly with the brush remaining in its tight position, and then to pull upwardly sharply on the mandrel assembly causing it to hammer against the brush and tend to dislocate it from its stuck position.

Insuring Well Treatment on Regular Basis

R.J. Harrison; U.S. Patent 4,132,268; January 2, 1979; assigned to Texaco Inc. describes a method for insuring oil well chemical treatment on a regular basis as scheduled and for eliminating personnel time which comprises the steps of:

(1) pumping a chemical into an injection line for delivery to the oil well;

(2) measuring out a precise slug of the chemical to the injection line with a first valve;

(3) controlling the first valve with a first timer for opening the first valve for a precise period of time for ejecting a precise slug of the chemical to the injection line for delivery to the well;

(4) flushing a high pressure flush liquid to the injection line for circulating the chemical slug throughout the well;

(5) measuring out a precise amount of flush liquid to the injection line for delivery to the well with a second valve; and

(6) controlling the second valve with a second timer for

> opening the second valve for a precise period of time for ejecting a precise amount of flush liquid to the injection line or slug directly behind and contiguous with the slug of chemical for circulating the chemical slug throughout the well.

Figure 1.6 is a vertical sectional view illustrating the chemical feeder system **10** for treating a producing oil well **11**. The oil well **11** is a conventional producing oil well having a casing **12** perforated at the bottom for oil to pass through it, the gravel pack **13**, and perforated production tubing **14** to the surface. Here the oil passes through a separator **15** from which oil is bled off in oil line **16** and water is bled off to a flush line **17**. Injection line **18** from the chemical feeder **10** injects chemical slugs as pushed forward by slugs of wash liquid down into an annulus **19** in the well.

Figure 1.6: Apparatus for Regularly Scheduled Well Treatment

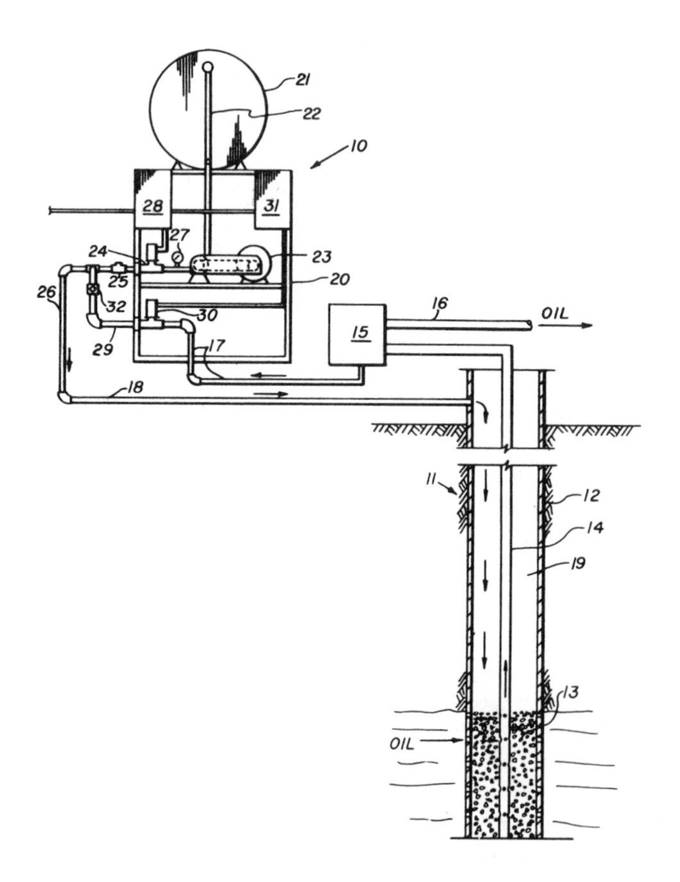

Source: U.S. Patent 4,132,268

The chemical feeder system **10** includes a steel rack **20** for supporting a 55 gallon (209 liter) chemical tank **21**, for example, with sight glass **22** for supplying the chemical to a ⅓ hp electric pump **23**. The chemical or chemicals from the pump pass through a conventional quick-acting or ½" motor snap valve, as a solenoid valve **24**, and then through check valve **25** to injection line **26** which injects the chemicals into the well **10**. A pressure gauge **27** is connected in the line between the pump **23** and the valve **24**. A timer or electric clock **28** is connected to valve **24** for metering out a predetermined precise amount or slug of chemicals, as a corrosion inhibitor, for the well every 7 days, for example.

A flush line **29** supplies a predetermined precise slug of flush liquid to the injection line **26** for pushing and carrying the chemical slug through the injection line to the well **11** for treating the well, as by coating all internal surfaces thereof with the corrosion inhibitor. This precision slug of flush liquid is metered from the flush line by a conventional quick action or solenoid or ½" motor snap valve **30** controlled by a second timer or 7-day electric clock **31** mounted on the steel rack **20**. Check valves **25** and **32** prevent any back flow of flush liquid or chemical in the injection line or the flush line, respectively.

While the flush line is illustrated as being supplied with water from the oil-water separator **15**, it may be supplied from any other suitable source.

Likewise, while the flush in the disclosed system comes from a separate line, it may emerge from the chemical feeder system already mixed with the chemical for being transported through the ejection line to the well. In the illustrated example, the flush may circulate the chemical or chemicals from 1 to 24 hours, as required, on any one or more days each week.

SCALE INHIBITORS AND REMOVERS

Disodium Maleate Scale Converter

In oil and gas well operations, water-insoluble scale is formed in tubing, casings, and associated equipment, as well as in the well bore and the formation itself, which carry, at least in part, water or brine waters. These waters can contain insoluble calcium, barium, magnesium, and iron salts. Such salts include calcium sulfate (gypsum), barium sulfate, calcium carbonate (limestone), complex calcium phosphate (hydroxyapatite), and magnesium salts.

This scale causes many problems in oil and gas well drilling and treating operations, particularly when it builds up in the piping. Generally, the scale is deposited or formed from a pressure or temperature change in the piping. Such scale deposits inhibit the flow of fluids, such as oil, water and/or other treating fluids through the piping, and if left unchecked will result in a complete blockage of the pipe. In addition, along with the scale, sand silicates, and other inert materials, and in some instances, heavier fractions of crude oil are deposited and entrapped therein.

The removal of such scale is conventionally accomplished by two basic methods. The first method includes treatment of the scale with a scale converter which converts the scale to an acid-soluble material followed by treatment with a mineral acid such as HCl. For example, insoluble sulfate scales are generally first reacted with a converter such as a carbonate to yield a water-insoluble-acid-soluble carbonate scale which is thereafter treated with the mineral acid. A second conventional method includes the use of chelating or sequestering agents, such as ethylenediaminetetraacetic acid or nitrilotriacetic acid.

Previously, scale converters have converted scale to either (1) an acid-soluble compound which is removed by an acid wash, or (2) a compound which can be removed by a sequestering agent.

R.L. Rybacki; U.S. Patent 4,147,647; April 3, 1979; assigned to Petrolite Corporation describes scale converters which convert scale directly into a water-soluble compound thus obviating the need for either an acid wash or a sequestering agent. These scale converters are water-soluble salts and preferably disodium maleate. Scale treated with the converters of this process is removed by an aqueous wash.

Sufficient converter should be added to dissolve the gypsum present. Theoretically at least a stoichiometric amount of converter is employed. The concentration of the converter employed can vary widely depending on many factors, such as the system from which it is to be removed, the agitation of the system, the thickness of the scale deposit, the temperature employed, the pH of the system, etc. Although concentrations of 1 to 90% by weight, such as from 25 to 75% can be employed, it was found that concentrations in excess of 35% are most effective with an optimum of about 45±5%.

Although the solution of this process is used as an aqueous solution, it does not preclude the use of dispersions or emulsions with an organic solvent such as an aromatic solvent such as xylene, a straight chain hydrocarbon such as kerosene or diesel to aid in removing paraffins and congealed oil which may be codeposited with the scale deposit.

Scale is removed by contacting the scale deposits with the solution for a time sufficient to convert the deposits to a soluble form so as to be removed by an aqueous wash. The details of the process will vary depending on the system from which the scale deposits are to be removed.

The process can be carried out at any operable pH. In general the pH of the salt is generally from about 7.5 to 12.0, but preferably from about 8.5 to 9.5.

Surface active materials chosen for their wetting and penetrating characteristics particularly in wetting scale are helpful when present in the solution, for example, in concentrations ranging from 1 to 10%, such as 1 to 5% or more, by weight but preferably 1 to 3%. Any suitable wetting agent can be employed providing it assists in wetting the scale.

When using the composition for dissolving calcium sulfate, temperatures substantially above ambient temperature are not required for purposes of efficiency. Temperatures as low as 50° to 60°F have been found satisfactory. However, an increase in temperature to a range such as normally is found under bottom hole conditions of producing oil wells which may range from about 90° to 160°F will cause a rapid increase in the reaction and, in turn, increase the rate of dissolving the calcium sulfate scale.

The composition is useful for removing calcium sulfate gyp scale or other scale encrustations, such as, sulfates or carbonates of calcium, barium, magnesium, which may be mixed with the gyp scale as accumulated deposits on oil, gas, or water well equipment, as well as industrial equipment. The composition will remove the gyp deposits from the rock formation face in open-hole type well completions. When pumped into the well, the composition by means of sufficient hydrostatic head pressure, or by applying additional positive pressure to overcome the existing reservoir pressure, will remove the deposited gyp scale in the producing channels of the reservoir rock, thus restoring permeability to the flow of oil and/or gas. The composition will also remove or solubilize gyp or anhydrite, which may be naturally present as part of the rock matrix, thus establishing new porosity and flow channels for increasing fluid flow into the well.

The following examples are presented for purposes of illustration. The compound tested was an aqueous solution of disodium maleate. Tests were carried out as follows. The compound or specified dilution thereof was added to a milk dilution bottle containing 4.0 g of calcium sulfate in the form of small chunks. After the material was allowed to soak for the indicated time period any free liquid was poured off into a weighted filter. The solid remaining was washed with one 100-ml and two 50-ml portions of tap water with any residue being transferred to the filter in the process. The filter was rinsed with about 20 ml of acetone, dried to constant weight and reweighed. The filter and residue was then further treated by addition of two 25-ml portions of 15% hydrochloric acid, rinsed with acetone, and again reweighed. Unless otherwise specified, the tests were run at 76°F. The results of this testing are given below.

Test*	Original $CaSO_4$ (g)	40% DSM Used (cc)	Further Dilution Media	Final Dilution (%)	Final Total (cc)	Treatment Duration (hr)	H_2O Solution (%)	Final HCl Total Converted (%)
A	4.00	10	None	–	10	24	95	97
B	4.00	25	None	–	25	24	99	99
C	2.00	10	None	–	10	24	99	99
D	4.00	10	**	50	20	24	73	75
E	4.00	10	***	50	20	24	79	84
F	4.00	10	***	50	20	24	86	86
G	4.00	10	***	50	20	15	78	79
H	4.00	10	***	50	20	8	74	75
I	4.00	10	***	50	20	4	45	48
J	4.00	10	***	50	20	24	86	88
X	4.00	10	**	–	10	24	36	80
Y	4.00	10	**	50	20	24	42	92
Z	4.00	10	***	50	20	24	41	91

Note: Tests A through E scale was W. Texas $CaSO_4$, benzene wash included in first wash; and tests F through J were gently agitated during the test.

*Tests A through J employ a 40% solution of disodium maleate (DSM) containing 2% wetting agent and tests X, Y and Z employ commercial compounds whch are employed to convert $CaSO_4$ to a compound which is then removed by an acid wash.
**Tap H_2O.
***5% brine.

Converter Type Scale Remover for Calcium Sulfate

L.W. Jones; U.S. Patent 4,155,857; May 22, 1979; assigned to Standard Oil Company (Indiana) describes a process and composition for removing calcium sulfate scale from a well (the well bore and the adjacent formation) using a fast-acting aqueous converter solution of sodium or potassium or ammonium gluconates, hydroxides, and carbonates to convert the calcium sulfate into compounds which can be readily removed by an acid. As this converter solution works at least about twice as fast as prior solutions (either converter or direct solvent), the solution generally need be maintained in the well for only about 4 to 12 hours and the lost production costs can be significantly reduced.

The aqueous converter solution has a weight ratio of hydroxide to carbonate of between about 3:2 and 5:1 and has a weight ratio of gluconate to combined hydroxide and carbonate of between about 2:1 and 5:1. The concentration of the combination of gluconate, hydroxide, and carbonate is between about 15 and 50% by weight of the aqueous solution introduced into the well bore.

Acid and Vanadium Catalyst to Dissolve Sulfur Scale

R.M. Ashby and F.S. Kaveggia; U.S. Patent 4,213,866; July 22, 1980 describe a method for beneficially treating passages through which petroleum values are to flow by reducing sulfur scaling of the passages.

This process uses a composition of a primary acid, a vanadium catalyst chosen from the class consisting of tetravalent or pentavalent compounds, and, for most applications, a secondary acid selected from one or more members of the class consisting of monocarboxylic acids, dicarboxylic acids, tricarboxylic acids and tetracarboxylic acids, and a wetting agent. Monocarboxylic acids include formic acid, acetic acid and derivatives of acetic. Dicarboxylic acid includes such acids as hydroxymalonic acid, dihydroxymalonic acid, malonic acid, and oxalic acid. Tricarboxylic acid includes citric acid. Tetracarboxylic acid includes ethylenediaminetetraacetic acid (EDTA).

By way of example, when formation material is to be dissolved and the formation is dolomite, the primary acid must react with the carbonates of the dolomite to form soluble magnesium and calcium salts. An example of such an acid and the preferred acid of this process is hydrochloric acid. Hydrochloric acid forms the soluble salts magnesium chloride and calcium chloride and magnesium bicarbonate [$Mg(HCO_3)_2$] and calcium bicarbonate [$Ca(HCO_3)_2$] upon reaction with the carbonates of the dolomite. Hydrochloric acid with vanadium ions present also dissolves shielding sulfur scale and to some beneficial extent other formation components which could act as shields or stabilizers preventing sufficient dissolution of the carbonates. This acid is preferred because it is more economical to use than other acids.

For the protection of drill string and well components suitable inhibitors may be employed. These inhibitors include copper and arsenic salts, and thiourea. Inhibitors are not as necessary with phosphoric acid, another suitable primary acid. While not critical, the pH of the acid system is less than 7 and preferably no more than about 3 to promote high reaction rate.

It is believed that the vanadium acts as a catalyst to promote a substantially greater reaction rate of sulfur scale and formation constituents with the primary acid than would be the case without its use. Additionally, it is preferred to use hydroxymalonic acid in the system for its promotion of more complete solutions. Hydroxymalonic acid and glycerin promote even more complete solution.

It is believed that these substances act as chelating or sequestering agents, even though this function is not well-known for them. Other agents instead of glycerin have proved only partially satisfactory.

The use of a wetting agent promotes penetration of the mineral and more rapid and perhaps more complete dissolution of the mineral. The wetting agent should be stable in acid and at the temperatures of reaction, i.e., about 175°C.

The preferred vanadium catalysts are pentavalent and tetravalent vanadium. Vanadium pentoxide (V_2O_5) is highly satisfactory and preferred because of cost. Vanadyl sulfate ($VOSO_4$) is also very effective, but not quite as economical as the pentoxide. Other vanadyls such as vanadyl chloride ($VOCl_2$) and vanadyl phosphate ($VOPO_4$) also work.

In drilling operations, the solution is merely added to the drilling hole, typically in the acid type drilling mud, to improve drilling rate. To improve recovery, the solution is merely pumped into a formation.

Example: A petroleum well in Louisiana in a dolomite formation was acidified with hydrochloric acid and no improvement in petroleum recovery was noted during about one month's time. The well was then treated with 8,000 gallons of a solution of hydrochloric acid, vanadium pentoxide, hydroxymalonic acid, and glycerin. The solution in weight percent was: vanadium pentoxide, 0.2%; hydroxymalonic acid, 0.1%; glycerin, 0.125%; and hydrochloric acid (5% by weight HCl in water, 30% technical grade HCl), balance.

After treatment, the well began to produce at commercially acceptable rates. The solution after treatment was heavy in sulfur and from this it was concluded that the well had been bound by sulfur scale.

Enhancing Effectiveness of Dialkyl Disulfides

F.T. Atkinson, S.P. Sharp and L.F. Sudduth; U.S. Patent 4,239,630; Dec. 16, 1980; assigned to Standard Oil Company (Indiana) describe a method for removing solid sulfur deposited in petroleum-bearing formations, oil wells, flowlines, etc., and more particularly a method for enhancing the rate at which dialkyl disulfides dissolve solid sulfur.

The sulfur solvent composition consists of a dialkyl disulfide to which has been added at least 1 pbw of an aliphatic saturated unsubstituted amine per 100 pbw dialkyl disulfide and from about 5 to 40 pbw of sulfur per 100 pbw dialkyl disulfide.

The dialkyl disulfides useful as starting materials in preparing the sulfur solvent of this process can be viewed as involving a pair of alkyl radicals (R and R') bonded to a disulfide unit as represented in the formula R–S–S–R'. Such compounds are also referred to in the chemical literature as alkyl disulfides, thus the terms should be considered equivalent for purposes of this process.

The dialkyl disulfides include such compounds as dimethyl disulfide, diethyl disulfide, dioctyl disulfide, ditertiary tetradecyl disulfide, and the like. One particularly useful starting material is a mixture of aliphatic disulfides in which the aliphatic group therein contains from about 2 to 11 carbon atoms; e.g., $(C_2H_5S)_2$,

$(C_{11}H_{23}S)_2$, etc., typically those disulfide mixtures produced as a product stream of the Merox process described in *The Oil and Gas Journal,* Vol 57, pp 73-78, October 26, 1959. Briefly, such mixtures of disulfides are produced by first contacting a refinery hydrocarbon stream containing aliphatic mercaptans with a caustic solution to produce corresponding sodium salt of the mercaptans. The latter are then converted to dialkyl disulfides by air oxidation, simultaneously regenerating the caustic.

The addition of the aliphatic amine in order to enhance the sulfur solvency properties of the dialkyl disulfide can be accomplished by any of the well-known methods found in the art including the method described in U.S. Patent 3,846,311 except the necessity of the aging step is viewed as being optional.

Preferably, the amount of amine activation will exceed 5 pbw amine per 100 pbw dialkyl disulfide. Examples of particularly useful amines include diethylamine, triethylamine, diisopropylamine, 2-ethylhexylamine, butylamine, hexylamine, octylamine, dodecylamine and the N-alkyl-1,3-propanediamines, e.g., Duomeen $(RNH_2CH_2CH_2CH_2NH_2)$.

To the amine activated dialkyl disulfide is added from about 5 to 40 pbw sulfur per 100 parts dialkyl disulfide, preferably from 5 to 20 pbw sulfur is used. The mixture is preferably agitated until the sulfur is dissolved which typically takes a period of 1 hour at 76°F. The rate of dissolution of the initially added sulfur may be increased by pregrinding the sulfur and the like or by heating the disulfide amine mixture, but due to the added expense of fuel, it is preferred to dissolve the sulfur at ambient temperatures. The resulting mixture is then immediately ready for use in removing sulfur from flow lines, wells, and rock formations penetrated by sour gas and distillate wells.

Bicyclic Macrocyclic Polyethers and Organic Acid Salts

F. De Jong, D.N. Reinhoudt and G.J. Torny-Schutte; U.S. Patent 4,215,000; July 29, 1980; assigned to Shell Development Company describe a process for dissolving a barium sulfate solid from a remote location into which fluid can be flowed, by flowing an aqueous solution into the remote location and into contact with the barium sulfate. The solution consists essentially of: water, a bicyclic macrocyclic polyether, a proportion of alkali metal salt of an organic acid which is less than that of the polyether but is sufficient to catalytically increase the rate of barium solid dissolving by the polyether, and enough dissolved alkaline inorganic alkali metal or ammonium compound to provide a solution pH of at least 8. The polyether is a bicyclic macrocylic polyether of the general formula (1):

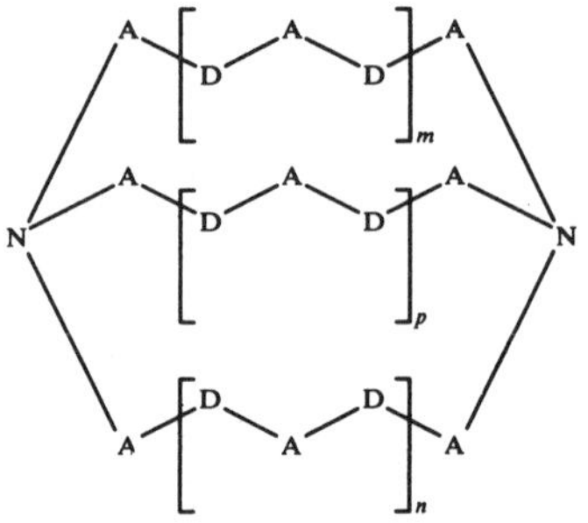

where each A represents a hydrocarbon radical having up to 12 carbon atoms and each D represents an oxygen or a sulfur atom or a hydrocarbon radical having up to 12 carbon atoms or $=NR^2$ group (in which group R^2 is a hydrogen atom or a hydrocarbon radical having up to 12 carbon atoms, a hydrocarbonsulfonyl radical having up to 12 carbon atoms, an alkoxycarbonylmethylene radical having fewer than 5 carbon atoms or a carboxymethylene radical), at least two of the D members being an oxygen or a sulfur atom or $=NR^2$ group, and m, n and p are integers from 0 to 5. The polyether contains an intramolecular cavity or crypt that is much more receptive to the cations of the alkaline earth metals than to the cations of the alkali metals.

The symbol A in formula (1) preferably represents a hydrocarbon radical having in the range of from 2 to 8 carbon atoms, such as ethylene, diethylene, triethylene and tetraethylene radicals. It preferably represents an ethylene radical. The A may also represent a 1,2-phenylene radical. The integers m, n and p are preferably in the range of from 1 to 3. The hydrocarbon radical R^2 may represent is preferably an alkyl group with 1 to 8, and particularly 1 to 4 carbon atoms. The hydrocarbon-sulfonyl radical R^2 may represent has preferably 1 to 8 and more preferably 1 to 4 carbon atoms. The alkoxycarbonyl radical R^2 may represent is preferably a methoxycarbonyl or an ethoxycarbonyl radical. A particularly preferred example is the 4,7,13,15,21,24-hexaoxa-1,10-diazabicyclo-[8.8.8]hexacosane (compound A). Other suitable compounds of formula (1) are those in which A represents an ethylene group and D, m, p and n are:

D	m	p	n
Oxygen	2	1	1
Oxygen	2	1	2
Oxygen	2	2	2

Further examples of compounds of formula (1) are: 16-methyl-4,7,13,21,24-pentaoxa-1,10,16-triazabicyclo[8.8.8]cosane; and 13,16-dimethyl-4,7,21,24-tetraoxa-1,10,13,16-tetraazabicyclo[8.8.8]hexacosane.

Suitably, the concentration of the bicyclic macrocyclic polyether is at least 0.01 mol/ℓ and is preferably higher than 0.05 mol/ℓ, for example, 0.3 mol/ℓ.

The molar ratio of the salt of an organic acid to the bicyclic macrocyclic polyether is not critical and may vary within wide limits. This molar ratio is preferably in the range of from 0.001 to 1, particularly from 0.01 to 0.5. If desired, this molar ratio may be higher than 1 but the proportion of the salt is preferably less than that of the polyether.

The aqueous compositions may contain well-treating or cleaning solution additives which are compatible with the compositions. Such additives include, for example, surfactant materials. These surfactant materials may be ionic or nonionic. Examples are alkali metal salts of alkylaryl sulfonates such as sodium dodecylbenzene sulfonate, alkali metal salts of sulfates of fatty alcohols such as sodium lauryl sulfate and such materials having a polyoxyethylene chain. The surfactant material may be present in the composition in a concentration in the range of from, for example, 1 to 3% by weight.

The contact time for dissolving the barium sulfate solid with this aqueous composition will vary not only with the conditions but also with the relative proportions of the constituents in the composition and will generally be in the range of from 1 minute to 3 hours; for example, from 15 minutes to 1.5 hours. In general, when using this aqueous composition for dissolving a barium sulfate solid, there is no need to employ temperatures substantially above ambient temperature. Temperatures as low as 10° to 20°C have been found satisfactory. However, the relatively high temperature (i.e., 30° to 70°C or higher) prevailing in an oil producing formation into which a production well penetrates, greatly enhances the rate of dissolution of the solid.

The cation of the salt of the organic acid present in the aqueous composition may be any cation, for example, a lithium, sodium, potassium or ammonium ion which is relatively incompatible with the bicyclic macrocyclic polyether of formula (1). The organic acid is preferably a polycarboxylic acid and particularly an aminopolycarboxylic acid, which may be a mono-, but is preferably a polyaminopolycarboxylic acid. Among the polyaminopolycarboxylic acids the alkylenepolyaminopolycarboxylic acids are preferred. Particularly preferred are alkylenepolyaminopolycarboxylic acids of the general formula (2):

$$(HO-\underset{\underset{O}{\|}}{C}-R^3-)_2N[(CH_2)_xN(-R^3-\underset{\underset{O}{\|}}{C}-OH)]_y-R^3-\underset{\underset{O}{\|}}{C}-OH$$

where each x and y is an integer of from 1 to 4 and R^3 represents an alkylene group with 1 to 3 carbon atoms and where up to x of the carboxyalkyl groups can be replaced by β-hydroxyethyl groups. Examples of suitable alkylenepolyaminopolycarboxylic acids are: ethylenediaminetetraacetic acid (R^3 is $-CH_2-$; $x = 2$; $y = 1$); ethylenetriaminepentaacetic acid (R^3 is $-CH_2-$; $x = 2$; $y = 2$). Ethylenediaminetetraacetic acid (EDTA) is particularly preferred.

Other examples of aminopolycarboxylic acids are: N-hydroxyethyliminodiacetic acid; monoethanolethylenediaminetriacetic acid; 2,2',2''-nitrilotriacetic acid; and tri(carboxymethyl)amine.

In another preferred embodiment of the process the polyaminopolycarboxylic acid is 1,10-di(carboxymethyl)-1,10-diaza-4,7,13,16-tetraoxacyclooctadecane. This compound is hereinafter referred to as compound B.

Compound B may be prepared by reacting 1,10-diaza-4,7,13,16-tetraoxacyclooctadecane with a salt of 2-chloroacetic acid in the presence of a base. Compound B has the formula:

```
           9       8          6       5          3       2
          CH2-----CH2        CH2-----CH2        CH2-----CH2
   O     /           \      /           \      /           \        O
   ||   /10           \ 7  /             \ 4  /             \       ||
HO-C-CH2-N              O                  O                 N1-CH2-C-OH
         \              O                  O                 /
          \           /  13\             /  16\             /
          CH2-----CH2        CH2-----CH2        CH2-----CH2
          11      12         14      15         17      18
```

The experiments described below were carried out in a cylindrical glass vessel having a height of 2.5 cm and an internal diameter of 0.8 cm. The vessel was charged with the starting materials and then fixed lengthwise to a horizontal shaft parallel to its central axis. The shaft was rotating at a speed of 180 rpm.

The temperature of the vessel was 20°C.

Examples 1 and 2 and Comparative Experiments A, B and C: The vessel was charged with 0.3 mmol barium sulfate, an aqueous liquid (1 ml) and 5 stainless steel balls having a diameter of 0.32 cm. 5 experiments were carried out, each with a different aqueous liquid. The table below states the compositions of the aqueous liquids.

	. . . . . .Starting Concentration (mol/ℓ)		
	Compound A	EDTA	KOH
Experiment A*	0.1	0	0
Experiment B*	0.1	0	0.1
Experiment C*	0	0.005	0.05
Example 1	0.1	0.005	0
Example 2	0.1	0.005	**

*Comparative.
**So much KOH was added that the pH reached a value of 11.

At various intervals the rotating shaft was brought to a standstill, the suspension was allowed to separate by settling, a sample was drawn from the aqueous layer and the amount of barium sulfate dissolved therein was determined. The following table presents these amounts for the 5 experiments.

	. . . . .$BaSO_4$ Dissolved in the Aqueous Liquid (g/ℓ)				
Time	. . Comparative Experiment . .			. . Examples . . .	
(hr)	A	B	C	1	2
0.5	8.7	8.5	0.9	16.0	20.4
1	10.9	11	0.9	18.9	21.1
3	18.2	18.5	0.9	19.7	21.8
5	20.4	20.5	0.9	19.7	21.1
7	21.1	21	0.9	19.7	21.8
16	21.1	21	0.9	19.7	21.1

From the amounts of 0.9 and 21.1 g $BaSO_4$/ℓ, dissolved in Comparative Experiment C and Example 2, respectively, it can be calculated that the loadings of EDTA in Comparative Experiment C and of Compound A in Example 2 are 77% and 90%, respectively.

Example 3: *Preparation of Compound B* – A solution of sodium hydroxide (80 mmol) in water (12 ml) and 1,10-diaza-4,7,13,16-tetraoxacyclooctadecane (20 mmol) were consecutively added to a solution of 2-chloroacetic acid (40 mmol) in water (8 ml). The mixture formed was stirred for 4 hours at 60°C, cooled to 25°C and acidified with a concentrated solution of hydrochloric acid (specific gravity 9.19) until a pH of 1 was reached.

The water was evaporated from the acidified mixture at a pressure of 2 kPa, leaving a residue to which methanol (40 ml) was added. The suspended sodium chloride was filtered off, the methanol was evaporated from the filtrate and the residue formed was taken up in a mixture of ethanol (6 ml) and chloroform (18 ml) at reflux temperature. The solution formed was cooled to 20°C and the precipitated 2HCl salt of compound B (4.9 g, yield 55%) was filtered off. This salt had a melting point of 106° to 107°C. Its PMR spectrum showed the following absorptions (60 MHz, D_2O):

	δ, ppm
4 H singlet	0.58
8 H triplet, broad	0.88
16 H multiplet	1.10

The 2HCl salt of Compound B (201.7 mg) was dissolved in water (7 ml), 1 drop of phenolphthalein was added and the solution was titrated with a 0.09 N aqueous solution of sodium hydroxide (10.95 ml) until the color changed; in this way a solution of Compound B was obtained.

In the cleaning method according to the process, the aqueous cleaning composition adapted for dissolving barium sulfate is contacted with the surfaces to be cleaned for a period of time sufficiently long to remove at least a portion of the barium sulfate scale on the surfaces. Hereby, the composition may be circulated over the surfaces to be cleaned. Thus, when cleaning equipment in a well, the composition may be circulated through the tubular goods in the well, such as by being pumped down through the production tube and being returned to the surface through the annular space between the production tubes and the casing (or vice versa).

The composition may also be pumped down through the production tubing and into the formation, thereby cleaning the well and the formation pore space by dissolving barium sulfate present therein while flowing along the surfaces that need cleaning. The spent composition of such once-through-dynamic wash procedure is later on returned to the surface by the fluids that are produced through the well after the cleaning operation.

In an alternative manner, the cleaning composition may be applied batchwise. The composition is then pumped down in the well and optionally into the pore space of the formation parts to be cleaned and kept in contact in nonflowing condition with the surfaces that are covered with barium sulfate scale, over a period of time sufficiently long to dissolve at least a considerable part of the scale. If desired, portions of the cleaning composition in which barium sulfate has been dissolved can be acidified to protonate the bicyclic macrocyclic polyether so that it can be recovered and reused.

Salts of Carboxymethyl Monocyclic Macrocyclic Polyamines

R. De Jong, G.J. Torny-Schutte and D.N. Reinhoudt; U.S. Patent 4,190,462; February 26, 1980; assigned to Shell Oil Company describe the use of monocyclic macrocyclic polyamines in aqueous solutions which are free of any bicylic macrocyclic ether, as described in Example 3 of the previous process using Compound B as an illustration.

Example 1: The vessel was charged with barium sulfate (0.5 mmol), water (1 ml), Compound B (0.1 mmol) and enough lithium hydroxide to provide an aqueous solution pH of 11. After 24 hours of rotation the suspension was allowed to separate by settling. A sample of the aqueous liquid layer contained dissolved barium and dissolved sulfate, both in amounts of 15.2 g/ℓ, calculated as barium sulfate. Hence, the lithium salt of Compound B had been used with an efficiency of $15.2/233 \times 0.1 \times 100 = 65\%$.

Example 2: This experiment differed from Example 1 in that Compound B was replaced by 1,10-diaza-4,7,13,16-tetraoxacyclooctadecane, in which the carboxymethyl groups were replaced by hydrogen atoms. The sample drawn from the aqueous layer contained dissolved barium and dissolved sulfate, both in amounts of 0.09 g/ℓ, calculated as barium sulfate. Hence, the 1,10-diaza-4,7,13,16-tetraoxacyclooctadecane had been used with an efficiency of 0.09/233 x 0.1 x 100 or only 0.4%.

Addition of Aldehyde to Acid Cleaners

Hydrogen sulfide gas produced during a cleaning operation leads to several problems. First, hydrogen sulfide is an extremely toxic gas and previous techniques have required the entire system to be vented to an appropriate flare system (in which the gas is burned) or to a sodium hydroxide scrubbing system. Neither of these alternatives is very attractive because the sulfur dioxide and sulfur trioxide formed during the burning of hydrogen sulfide are substantial pollutants in and of themselves. The sodium sulfide produced during the scrubbing system is a solid that presents disposal problems. It can be landfilled or put into disposal ponds but only under conditions such that the sodium sulfide does not contact acid. Sodium sulfide reacts rapidly with acids to regenerate hydrogen sulfide.

Second, aside from the toxic nature of hydrogen sulfide, the material causes operational problems as well because it is a gas. The volume of gas produced can be substantial. The gas takes up space within the unit being cleaned and can prevent the liquid cleaning solution from coming in contact with all of the metal surfaces. This can occur, for example, in cleaning a horizontal pipeline where the gas can form a pad over the top of the flowing liquid and prevent the liquid from filling the pipeline and cleaning the entire surface. The gas produced can also cause the pumps used in the system to cavitate, lose prime, and/or cease to function efficiently. Of course, if enough gas is generated in a confined vessel, the vessel can rupture.

W.W. Frenier, M.D. Coffey, J.D. Huffines and D.C. Smith; U.S. Patent 4,220,550; September 2, 1980; assigned to The Dow Chemical Company describe a method of chemically cleaning sulfide-containing scale from metal surfaces. The process utilizes aqueous acid cleaning solutions containing an aldehyde in amounts sufficient to prevent or substantially prevent the evolution of hydrogen sulfide gas.

The aqueous acid cleaning solutions are well-known. Normally, these acid cleaning solutions are aqueous solutions of nonoxidizing inorganic and/or organic acids and more typically are aqueous solutions of hydrochloric acid or sulfuric acid. Examples of suitable acids include, for example, hydrochloric, sulfuric, phosphoric, formic, glycolic, citric, and the like. Aqueous solutions of hydrochloric acid or sulfuric acid are preferred. Aqueous solutions of sulfuric acid are most preferred. The acid strength can be varied as desired, but normally acid strengths of from about 5 to 40% are used.

The aldehydes are likewise known as a class of compounds having many members. Any member of this known class can be used herein so long as it is soluble or dispersible in the aqueous acid cleaning solution and is sufficiently reactive with hydrogen sulfide produced during the cleaning process that it prevents or substantially prevents the evolution of hydrogen sulfide gas under conditions of use. Examples of suitable aldehydes include formaldehyde, paraformaldehyde, acetaldehyde, glyoxal, β-hydroxybutyraldehyde, benzaldehyde, methyl-3-cyclohexene carboxaldehyde, and the like. Of these, formaldehyde and acetaldehyde are preferred based on economics and performance, and formaldehyde is most preferred. Commercial solutions of formalin or alcoholic solutions of formaldehyde are readily available and may be used.

The aldehydes are included in the system in an amount to prevent or substantially prevent the evolution of hydrogen sulfide gas during the cleaning process. The amount of acid soluble sulfide in the scale can be normally determined experimentally before the cleaning job is done and a stoichiometric amount of aldehyde can be determined (i.e., equimolar amounts of aldehyde and hydrogen sulfide). It is preferred, however, to use excess formaldehyde. By excess is meant amounts beyond stoichiometric required and up to 1 equivalent weight of aldehyde or more per equivalent weight of acid.

The aqueous cleaning solution may also comprise additional additives, if desired. For example, acid corrosion inhibitors (e.g., acetylenic alcohols, filming amines, etc.), surfactants, mutual solvents (such as alcohols and ethyoxylated alcohols or phenols, etc.) can be included as desired. Corrosion inhibitors usually will be required to limit acid attack upon the base metal. Amine-based corrosion inhibitors, such as those described in U.S. Patent 3,077,454, are preferred.

The aqueous acid cleaning solution is normally a liquid system but can be used as a foam. Liquid cleaning solutions are preferred in most instances.

The cleaning compositions used in the process can be formulated external to the item or vessel to be cleaned. Alternatively, the item or vessel to be cleaned can be charged with water or an aqueous solution or dispersion of the aldehyde to be used and the acid added subsequently. This technique has the advantage of permitting the operator to ascertain the circulation of liquid within the system prior to loading the active cleaning ingredient. This will therefore represent a preferred embodiment for cleaning many systems.

The temperature utilized during the cleaning process can be varied but is normally selected in the range of from ambient up to about 180°F for the mineral acids and even up to about 225°F for the organic acids. The upper temperature is limited only by the stability of the aldehyde and/or the ability to control acid and/or ferric ion corrosion with appropriate inhibitors. Preferred temperatures are normally in the range of from about 140° to 160°F.

Example: A finely ground iron sulfide (FeS; 9.7 g) was placed in a 250 ml flask fitted with a magnetic stirring bar, thermometer, and gas outlet. Water (84 ml) was added and the mixture heated to 150°F. At this point, a mixture of 47 ml 37.5% hydrochloric acid and 19.11 ml of 37% formalin (a twofold molar excess) was added. The gas outlet port was immediately connected to a water displacement apparatus to measure the volume of any gas which was given off during the reaction. Normally, there was a temperature rise of approximately 10°F attributable to the heat generated by the heat of diluting hydrochloric acid. This increase in temperature also accounted for a collected gas volume of approximately 1.6 ml due to expansion of gas in the system.

During the 4 hour reaction time, 83.5% of the calculated iron available was dissolved with the final solution containing approximately 3.17% by weight iron. There was a steady but very slight evolution of gas which in part contained hydrogen sulfide (as detected by lead acetate paper). The volume of gas generated and collected accounted for approximately 1% or less of the total hydrogen sulfide produced by this reaction. The remainder of the hydrogen sulfide generated was present essentially as trithiane, a white crystalline solid remaining in the liquid. The trithiane was identified by infrared analysis.

Substantially equivalent results were achieved using 37% formalin or paraformaldehyde in hydrochloric acid or sulfuric acid (at acid concentrations of 5, 10 and 15%). Likewise, substantially equivalent results were achieved using acetaldehyde or phenylacetaldehyde in 15% hydrochloric acid. The concentration of aldehyde in this system was the same as set forth above (2 molar excess) or a 4 molar excess. Little advantage was realized by going from 2 to 4 molar excess of aldehyde.

Likewise, substantially similar results were achieved using 37% formalin and 5% formic acid, 5% phosphoric acid, or a 5% acid mixture having 2 parts of glycolic acid for each part of formic acid. A 2 molar excess of aldehyde was used.

In other experiments, it was observed that β-hydroxybutyraldehyde, glyoxal, benzaldehyde, salicylaldehyde, acrolein, and 2-furfuraldehyde in hydrochloric acid (5 or 15%) gave good results in preventing or substantially preventing the elimination of hydrogen sulfide gas under the above experimental conditions.

Similar results were achieved when the iron sulfide in the experiment was replaced with zinc sulfide or sodium sulfide as the source of hydrogen sulfide.

Zinc Tetraammonium Carbonate as H_2S Scavenger

V.R. Tisdale; U.S. Patent 4,147,212; April 3, 1979; assigned to The Sherwin-Williams Co. has found that a water-soluble zinc ammonium carbonate complex provides nearly quantitative removal of hydrogen sulfide by intimately contacting the carrier thereof with substantially stoichiometric quantitites of the complex in aqueous solution.

While the quantity of zinc complex used will vary widely depending upon individual conditions, it is a rule of thumb that about 3.7 lb of zinc tetraammonium carbonate is equivalent to 1 lb of H_2S and, consequently, major amounts are not generally required to remove anticipated quantities of H_2S. A probable range may have less than 1 lb/bbl of well fluid (as an example of utility) to, in rare instances, 20 to 25 lb/bbl where the packer fluids are left in the well for a long period of time, and gradual accumulation of H_2S down the hole may occur through gas seepage from porous formations or bacterial action, etc.

Phosphate Esters of Oxyalkylated Urea

D.K. Durham; U.S. Patent 4,155,869; May 22, 1979; assigned to Standard Oil Company (Indiana) describes a method of using compositions involving phosphate esters derived from an oxyalkylated urea which are effective in inhibiting scale deposition. These phosphate esters of oxyalkylated urea are particularly suitable for oil well applications and the like.

The scale inhibitors are synthesized by first oxyalkylating urea in a manner such that at least 2 mols alkylene oxide are reacted per mol of urea. After oxyalkylation, the terminal hydroxyl groups are phosphorylated resulting in the phosphate esters of an oxyalkylated urea. This product can then be partially neutralized by the addition of a base or the like producing the water-soluble scale inhibitor compositions.

The oxyalkylation reaction can be carried out by any of the methods well-known in the art. The oxyalkylation as described in U.S. Patent 2,842,523 is particularly useful provided about 2 to 20 mols of alkylene oxide are used per mol of urea. The infrared spectra of the oxyalkylation products indicate a mixture is produced including a significant contribution from the amino alkyl carbamate structure. Significant scale inhibition is observed with as little as 2 mols of alkylene oxide being added per mol of urea. The desired scale inhibition can be observed with as many as 20 mols of alkylene oxide per mol of urea present. The 4 mol alkylene oxide adduct is particularly suitable.

The phosphorylation of the oxyalkylated urea can be performed by any of the methods well-known in the art. The preferred method of synthesis is to react the oxyalkylated urea in the presence of a stoichiometric excess of polyphosphoric acid or its equivalent. The degree of phosphorylation ranges from about 1 to 6 or more mols of phosphate per mol of urea. In the cases where 4 or more mols of alkylene oxide are used per mol of urea, about 6 mols of phosphate per mol of urea represents a particularly advantageous stoichiometric excess in that it tends to drive the esterification reaction to completion while the remaining excess polyphosphoric acid need not be removed but can be carried along for additional corrosion inhibition. The reaction occurs essentially spontaneously after the reactants are heated. Since the reaction is exothermic, controlled addition of the reactants along with appropriate cooling to prevent overheating can be employed during the reaction.

Having synthesized the desired phosphate ester of an alkoxylated urea, it may be used directly as is or it may be partially neutralized by the addition of base and then added to the aqueous solution requiring treatment. A pH range of about 3 to 7 is effective for scale inhibition with a pH of about 4.5 being preferred. The addition of sodium, potassium or ammonium hydroxide, anhydrous ammonia, a water-soluble amine or their mixtures to achieve the desired pH is useful. Based on ppm of the nonneutralized form, virtually total prevention of $CaCO_3$, $CaSO_4$ and $BaSO_4$ precipitation is achieved at as low as about 10 ppm.

Example: To a stirred 1,500 ml oxyalkylation reactor fitted with an external cooling system was added 364.17 g of urea and 2.25 g of potassium hydroxide. The contents were then heated to 130°C to melt the urea and then purged with nitrogen slowly for 1 hour until the water content was 1,000 ppm or less and sealed under 25 psi. Ethylene oxide, 533.28 g, was then reacted with the urea over a 4-hour period at a maximum pressure of 60 psi and a temperature range of 120° to 150°C. The ethoxylated urea was then cooled to 30°C and removed from the reaction chamber. The yield was 99.8%. The final product was a viscous water white liquid.

To phosphorylate the above product, 287.16 g of polyphosphoric acid was added to a 400 ml beaker equipped with a stirrer and heated to 60°C using a hot plate. Then 250 g of ethoxylated urea described above was dripped into the polyphosphoric acid over a 60-minute period with stirring. Due to the exotherm the reaction was cooled to maintain a temperature between 85° and 100°C. The viscosity increased with the increase in the amount of ethoxylated urea added. After the

250 g of ethoxylated urea was added, the reaction temperature was held at 95°C for 30 minutes. The clear water white, viscous liquid was cooled to 60°C after which 273.06 g of H_2O was added. The product was then neutralized with 185.9 g of 50% sodium hydroxide to a pH of 4.5 with cooling to control the exotherm from the neutralization. The resulting product was 50% active and the yield of reaction was 99.0%.

OTHER CLEANING OPERATIONS

Removal of Asphaltenic and Paraffinic Deposits

In some petroleum producing areas, deposits containing both asphaltenic and paraffinic compounds build up on the faces of producing formations as well as in tubular goods, production equipment and related apparatus whereby production is decreased requiring frequent and expensive remedial procedures.

W.G.F. Ford and T.R. Gardner; U.S. Patent 4,207,193; June 10, 1980; assigned to Halliburton Company describe a process for removing asphaltenic- and paraffinic-containing deposits comprised of an aqueous carrier liquid having dispersed therein a hydrocarbon solvent, a base selected from ammonium hydroxide, organic bases and acidic salts thereof and mixtures thereof wherein organic bases are selected from the group consisting of pyridine, morpholine and primary, secondary and tertiary amines defined by the general formula:

$$\begin{array}{l} R'' \\ | \\ N{-}R \\ | \\ R' \end{array}$$

where R, R' and R'' represent members selected from the group consisting of hydrogen, alkyl radicals having 1 to 4 carbon atoms, alkyl amine radicals having from 1 to 4 carbon atoms, cycloalkyl radicals having 3 to 6 carbon atoms and mixtures thereof, and a surfactant, preferably a nonionic alkylated aryl polyether alcohol.

Base materials of the type defined above which are preferred for use herein are organic bases having no more than 3 nitrogen atoms per molecule. Compounds which are within the scope of the above formula having more than 1 nitrogen atom per molecule are those which do not include direct nitrogen-to-nitrogen bonding. Examples of organic bases which are particularly useful are pyridine, morpholine, and low molecular weight primary, secondary and tertiary amines such as n-butylamine, ethylenediamine, diethylenetriamine, dimethylaminopropylamine, diethylaminopropylamine and cycloalkylamines such as cyclohexylamine.

Base concentrations in the range of from about 0.04 to 2.5% by weight of the aqueous composition are effective in enhancing the removal of asphaltenic compounds by the composition. Preferably, the amine is selected from the group consisting of n-butylamine, ethylenediamine, diethylenetriamine and mixtures thereof and is present in the composition in an amount in the range of from about 0.1 to 0.5% by weight. Most preferably, the amine is ethylenediamine present in the composition in an amount of about 0.25% by weight.

Examples of hydrocarbon solvents which are particularly useful are benzene, xylene, toluene, naphtha, kerosene and mixtures of such hydrocarbons. The hydrocarbon solvents or mixtures thereof can be included in the aqueous composition in an amount in the range of from about 3 to 9% by weight, and preferably, in an amount of about 7 to 8.5% by weight. Most preferably, the hydrocarbon solvent is xylene, or a mixture of xylene and toluene present in the composition in an amount of about 8% by weight.

The surfactants which are useful in this process are those which function as a dispersant and have solubility in both hydrocarbons and water; i.e., the surfactant must be capable of dissolving to some extent in either water or hydrocarbons. Preferred such surfactants are those defined by the general formula:

$$CH_3(CH_2)_x\text{—}C_6H_4\text{—}O(CH_2CH_2O)_yH$$

where x has a value ranging from about 2 to 11 and y has a value ranging from about 10 to 40.

The alkyl aryl portion of the surfactant, i.e., the hydrocarbon soluble portion, can thus contain from about 9 to 18 carbon atoms. The oxyethylene portion of the compound, i.e., the water-soluble portion, can contain from about 10 to 40 mols of ethylene oxide.

Of the various surfactants within the scope of the above formula, ethoxylated nonylphenol having an ethylene oxide content of from about 10 to 40 mols and ethoxylated octylphenol having an ethylene oxide content of from about 10 to 40 mols are preferred. The surfactant is included in the aqueous composition in an amount in the range of from about 0.04 to 4% by weight, and more preferably, in an amount in the range of from about 0.2 to 1.5% by weight. The most preferred surfactant is ethoxylated nonylphenol having an ethylene oxide content of about 30 mols present in the composition in an amount of about 0.6% by weight. Surfactants of the type described are generally commercially available dissolved in an aromatic hydrocarbon solvent such as toluene in an amount of about 12% by weight of the solution.

A variety of aqueous carrier liquids present in the range of from about 85 to 97% by weight of the aqueous composition can be utilized herein. Examples of aqueous carrier liquids useful herein include acids, fresh water, brine and aqueous solutions containing chemicals useful for conducting other treatments in addition to the removal of deposits. In this regard, in the performance of production stimulation treatments in subterranean oil and gas producing formations, e.g., fracturing and/or acidizing the formations to increase the permeability thereof, the aqueous composition can be utilized so that deposits contained on the formation surfaces and in tubular goods are removed while the stimulation treatment is carried out.

A specific preferred aqueous composition for removing asphaltenic- and paraffinic-containing deposits from surfaces is comprised of about 91.15% by weight of water, a hydrocarbon solvent selected from the group consisting of toluene, xylene and both toluene and xylene present in the composition in an amount of about 8% by weight, ethylenediamine present in the composition in an amount of about 0.25% by weight and ethoxylated nonylphenol having an ethylene oxide content of about 30 mols present in the composition in an amount of about 0.6% by weight.

Another preferred composition is comprised of about 90.95% by weight water, toluene present in the composition in an amount of about 4% by weight, xylene present in the composition in an amount of about 4.2% by weight, ethylenediamine present in the composition in an amount of about 0.25% by weight and ethoxylated nonylphenol having an ethylene oxide content of about 30 mols present in the composition in an amount of about 0.6% by weight.

In preparing the compositions, the surfactant, preferably dissolved in toluene or xylene, is first added to the aqueous carrier fluid while the carrier fluid is being agitated. Any additional hydrocarbon solvent or hydrocarbon solvents utilized are next slowly added to the mixture while it is being agitated followed by the addition of the amine utilized. The aqueous composition is agitated to homogeneously disperse the hydrocarbon solvent, amine and surfactant in the aqueous carrier fluid. If acid is to be included in the composition, a corrosion inhibitor, if used, is added to the composition after the addition of the amine followed by addition of the acid while continuously agitating the composition.

In carrying out the method, the composition is circulated over or otherwise brought into contact with the surface or surfaces from which deposits are to be removed. The composition can be heated if desired, but it is normally utilized at ambient temperatures. As the composition contacts the deposits, the asphaltenic and paraffinic constituents thereof are stripped from the surfaces in finely divided particles and suspended in the aqueous carrier fluid. If the deposits include inorganic scale constituents and one or more acids are included in the composition, the inorganic scale is simultaneously removed with the asphaltenic or paraffinic compounds.

TRANSPORTATION OF HEAVY CRUDE

Anionic Surfactant and Alkalinity Agent

The transportation of heavy crudes by pipeline is difficult because of their low mobility and high viscosity. The usual methods to facilitate the flow of heavy crudes have included cutting them with ligher fractions of hydrocarbons. However, the procedures involve the use of relatively large amounts of expensive hydrocarbon solvents to transport a relatively cheap product. The practice also necessarily requires the availability of the cutting hydrocarbon solvents which, in some instances, is inconvenient.

Another method to assist the flow of hydrocarbon in pipeline is the installation of heating equipment at frequent intervals along the pipeline, whereby the crude is heated to reduce its viscosity and thereby facilitate its transport.

Heaters employed for this purpose can be operated by withdrawing some of the crude being transported for use as fuel. However, this procedure may result in the loss of as much as 15 to 20% of the crude being transported.

Other methods to facilitate transport of heavy crudes have employed thermal viscosity breaking, which, however, produces substantial amounts of gas.

It is known that substantial amounts of water may be introduced into a pipeline containing a stream of viscous crude flowing therethrough to reduce the drag on the stream and thus facilitate the flow through the pipeline. This has been done by the addition of water together with crude into the pipeline such that a water-in-oil emulsion is formed.

K.H. Flournoy, R.L. Cardenas and J.T. Carlin; U.S. Patent 4,152,290; May 1, 1979; assigned to Texaco Inc. describe a method for transporting viscous hydrocarbons such as crude oil in which the hydrocarbon together with an aqueous solution of an anionic surfactant or a mixture of anionic surfactants and an alkalinity agent and, optionally, a guanidine salt or an oxyalkylated, nitrogen-containing aromatic compound, is introduced into the pipeline with mixing. During the mixing operation an oil-in-water emulsion is formed which is stable in hard water and salt tolerant.

In this process the aqueous solution added to the viscous hydrocarbon will generally range from a minimum of about 10% by volume based on the volume of the hydrocarbon introduced into the pipeline up to a maximum of about 40% or more by volume with the preferred amount being about 20 to 30% by volume on the same basis. In the aqueous solution the concentration of the anionic surfactant will range from about 0.01 to 2.0% by weight and the alkalinity agent from about 0.01 to 1.0% by weight. Optionally, the aqueous solution can contain from about 0.01 to 0.50% by weight of a material selected from the group consisting of (a) guanidine salts, (b) oxyalkylated, nitrogen-containing aromatic compounds, and (c) mixtures of (a) and (b) above.

A particularly useful class of anionic surfactants comprises compounds selected from the group consisting of water-soluble salts of alkyl sulfates having from 6 to 20 carbon atoms and water-soluble salts of unsaturated aliphatic carboxylic monobasic acids having from 6 to 20 carbon atoms.

Anionic surfactants suitable for use in the aqueous solution include, for example, compounds of the formula:

$$CH_3{-}(CH_2)_d{-}CH{=}CH{-}(CH_2)_e{-}COOM$$

where d and e are integers and the sum of d + e is from 2 to 16 and M is selected from the group consisting of monovalent ions as exemplified by Na^+, K^+, L^+, NH_4^+, etc. An example of materials of this type is sodium oleate.

Another class of compounds which may be employed as the anionic surfactant has the general formula:

$$CH_3{-}(CH_2)_f{-}SO_3M$$

where f is an integer of from 5 to 19 and M is a monovalent cation such as Na^+, K^+, Li^+, NH_4^+, etc. Examples of compounds of this type include sodium dodecyl sulfate, potassium cetyl sulfate, sodium decyl sulfate, sodium tetradecyl sulfate, etc.

Another group of anionic surfactants suitable for use in this process include sulfate compounds of the formula:

$$RO(CH_2CH_2O)_gSO_3M$$

where R is selected from the group consisting of alkyl of from 8 to 25 carbon atoms and

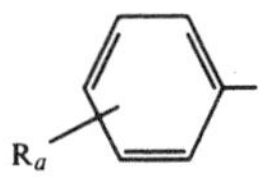

R_a

where R_a is alkyl of from 8 to 20 carbon atoms, g is an integer of from 1 to about 20 and M is a metallic cation such as sodium, potassium, lithium or the ammonium ion. For example, sodium tridecylpolyoxyethylene sulfate and potassium nonylphenol polyoxyethylene sulfate are preferred sulfate type surfactants of this class.

Another group of anionic oxyalkylated surfactants which are especially useful in the process include sulfonate compounds of the formula:

$$R_dO(CH_2CH_2O)_hCH_2CH_2SO_3M$$

where R_d is selected from the group consisting of alkyl of from 8 to 20 carbon atoms;

R_e

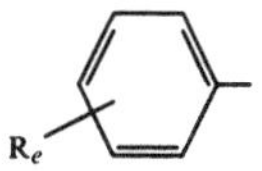

where R_e is alkyl of from 8 to 20 carbon atoms and

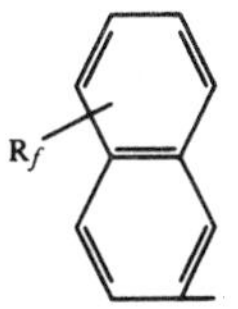

R_f

where R_f is alkyl of from 1 to 10 carbon atoms; h is an integer of from 1 to 24 and M is a metallic cation such as sodium, potassium, lithium or the ammonium ion.

Another group of anionic surfactants which are suitable for use in the process include block-type sulfonate compounds of the formula:

$$R_kO(C_2H_4O)_r-(C_3H_6O)_s-(C_2H_4O)_tCH_2CH_2SO_3M$$

where R_k is selected from the group consisting of alkyl of from 8 to 20 carbon atoms

R_m

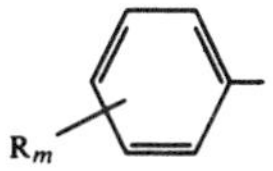

where R_m is alkyl of from 8 to 20 carbon atoms and

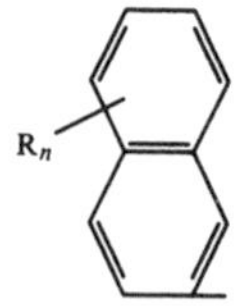

R_n

where R_n is alkyl of from 1 to 10 carbon atoms; r is an integer of from 1 to about 18, s is an integer of from 1 to about 12, t is an integer of from 1 to about 20 and the sum of r + s + t is not more than 30, wherein at least 60% of the oxyalkylene units are oxyethylene units and M is a metallic cation selected from the group consisting of sodium, potassium, lithium or an ammonium ion.

Block-type sulfonate surfactants as described above may be prepared by first condensing ethylene oxide with a suitable initiator in the presence of, for example, about 0.12% by weight of sodium hydroxide in a stirred autoclave maintained at 95° to 100°C. After devolatizing the resulting product to remove low boiling products, if desired, a second condensation reaction is conducted with propylene oxide under the same conditions and finally a third condensation is conducted with ethylene oxide. After the block-type oxyalkylated precursor has been prepared, it is reacted with sulfurous oxychloride (i.e., $SOCl_2$) to replace the terminal hydroxyl group with chlorine and this intermediate may then be reacted with sodium sulfite to form the desired sulfonate.

The alkalinity agent can be selected from the group consisting of sodium hydroxide, potassium hydroxide and lithium hydroxide.

Guanidine salts useful in preparing the aqueous solution employed include guanidine hydrochloride, guanidine acetate, guanidine sulfate, guanidine carbonate, guanidine thiocyanate, guanidine nitrate, etc., and mixtures thereof.

An especially useful group of the water-soluble, oxyalkylated, nitrogen-containing aromatic compounds are of the formula:

$$R_v(OR_w)_xOH$$

where R_v is selected from the group consisting of:

O_2N-

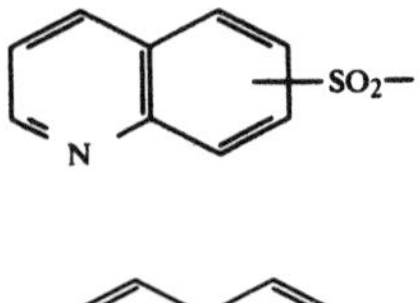

$-SO_2-$

N

N

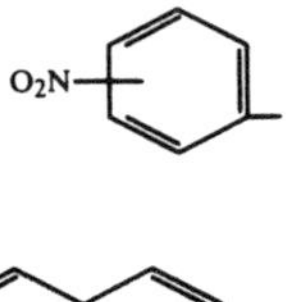

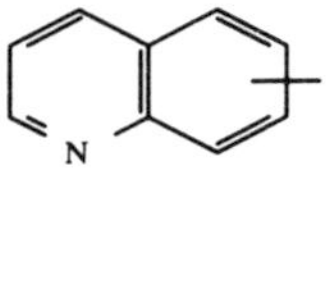

where R_W is alkylene of from 2 to 4 inclusive carbon atoms and x is an integer of from about 5 to 50 and, preferably, from about 5 to 20.

The water-soluble oxyalkylated products of this process can be conveniently prepared by a number of processes well-known in the art. For example, the alkylene oxide can be reacted with the initiator dissolved in a suitable solvent throughout which an alkaline catalyst, such as potassium hydroxide or sodium hydroxide, is uniformly dispersed. The quantity of the catalyst utilized generally will be from about 0.05 to 0.1% by weight of the reactants. Preferably, the reaction temperature will range from about 80° to 180°C, while the reaction time will be from about 1 to 20 hours or more depending on the particular reaction conditions. This process is described in U.S. Patent 2,425,845. The preparation of such oxyalkylated, nitrogen-containing aromatic compounds such as exemplified by ethoxylated 8-hydroxyquinoline, propoxylated nitrophenol, ethoxylated 8-quinoline sulfonic acid, etc., is more completely described in U.S. Patent 3,731,741.

Useful oxyalkylated, nitrogen-containing aromatic compounds include, for example:

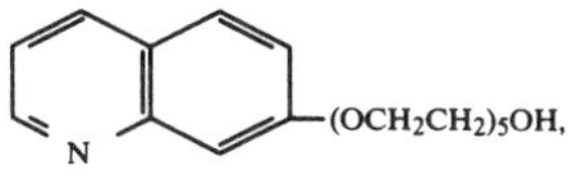

N $-(OCH_2CH_2)_5OH$,

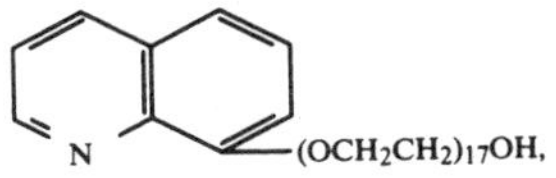

N $-(OCH_2CH_2)_{17}OH$,

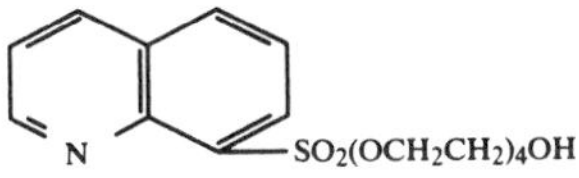

N $-SO_2(OCH_2CH_2)_4OH$

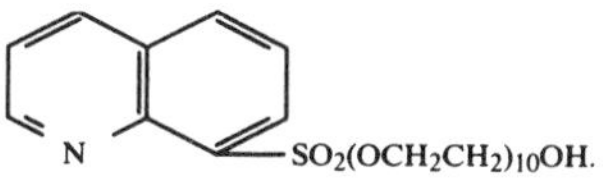

N $-SO_2(OCH_2CH_2)_{10}OH$.

Example: An aqueous solution containing 0.5% by weight of sodium dodecyl sulfate, 0.4% by weight of a sulfonate surfactant having the formula:

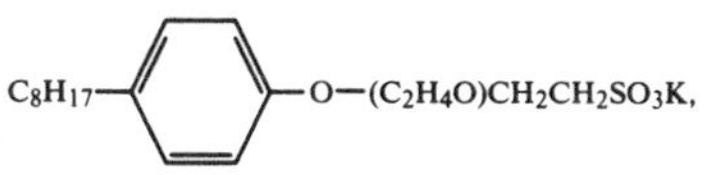

$C_8H_{17}-C_6H_4-O-(C_2H_4O)CH_2CH_2SO_3K$,

0.10% by weight of guanidine hydrochloride, and about 0.10% by weight of sodium hydroxide was prepared by adding with mixing the abovementioned ingredients to water having a salinity of about 2.03% by weight at a temperature of about 24°C after which the thus-prepared solution is introduced with mixing into a large diameter pipeline together with sufficient heavy California crude oil, to give an oil-in-water emulsion in which the amount of aqueous solution is about 20% by volume based on the volume of crude. The horsepower requirement for transporting the formed oil-in-water emulsion through the pipeline at the rate of 1,400 bpd is found to be substantially less than the horsepower requirement for transporting the same volume of the heavy California crude under the same conditions.

Sulfonate Surfactant and Rosin or Naphthenic Acid Soap

K.H. Flournoy, R.B. Alston and W.B. Braden, Jr.; U.S. Patents 4,192,755; March 11, 1980; and 4,134,415; January 16, 1979; both assigned to Texaco Inc. describe the introduction into a pipeline of a viscous hydrocarbon or mixture of hydrocarbons together with an aqueous solution of an organic sulfonate, a rosin soap or a naphthenic acid soap and, optionally, a coupling agent such as ethylene glycol monobutyl ether, whereby a low viscosity, salt-tolerant, oil-in-water emulsion is formed which facilitates movement of the heavy oils through the pipeline. If desired, the solution may contain an added alkalinity agent.

In this process the aqueous solution added to the viscous hydrocarbon will generally range from a minimum of about 15% by volume based on the volume of the hydrocarbon introduced into the pipeline up to a maximum of about 35% or more by volume with the preferred amount being about 20 to 30% by volume on the same basis. In the aqueous solution the concentration of the sulfonate surfactant will range from about 0.01 to 2.0% by weight; the rosin soap or the naphthenic acid soap from about 0.01 to 1.8% by weight and the optionally present coupling agent from 0.02 to about 1.0% by weight. Mixtures of the abovementioned ingredients may also be used.

Sulfonate type surfactants which may be utilized in the process include alkyl sulfonates and alkoxylated alkyl or alkaryl sulfonates.

One type of useful sulfonates includes compounds of the following general formula:

(1) $$R{-}SO_3{-}Y$$

where R is an alkyl radical, linear or branched, having 5 to 25 and, preferably, from 8 to 14 carbon atoms and Y is a monovalent cation such as sodium, potassium or the ammonium ion. For example, if R is linear dodecyl and Y is the ammonium radical, then the compound is ammonium dodecyl sulfonate.

Another group of sulfonate surfactants which are especially suitable includes compounds of the formula:

(2) $$R_aO(CH_2CH_2O)_xCH_2CH_2SO_3M$$

where R_a is selected from the group consisting of alkyl of from 8 to 26 carbon atoms and

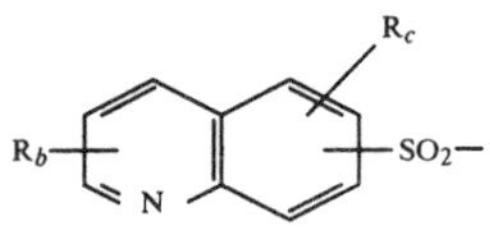

where R_b is selected from the group consisting of hydrogen and alkyl of from 1 to 5 carbon atoms, R_c is selected from the group consisting of hydrogen and alkyl of from 1 to 5 carbon atoms and x is an integer of from 1 to 20 and M is a metallic cation such as sodium or potassium or ammonium ion.

One method of preparing the type (2) sulfonates described above is as follows. A polyethoxylated alkanol is first reacted with sulfurous oxychloride ($SOCl_2$) in order to replace the terminal hydroxyl group with a chlorine, which may then be reacted with sodium sulfite, Na_2SO_3, to form the desired polyethoxylated alcohol sulfonate.

A wide variety of rosin soaps may be utilized in the method as exemplified by sodium abietate, potassium abietate, ammonium abietate, etc. Mixtures of these same rosin soaps may be employed if desired. Suitable naphthenic acid soaps include sodium naphthenate, potassium naphthenate, ammonium naphthenate, etc. and mixtures thereof.

Glycol monoalkyl ethers useful in the process include compounds of the formula:

$$R_cO(R_dO)_mH$$

where R_c is alkyl of from 1 to 6 carbon atoms, R_d is alkylene of from 2 to 3 inclusive carbon atoms and m is an integer of from 1 to 3 inclusive. Examples of glycol monoalkyl ethers suitable for use in this process include ethylene glycol monomethyl ether, diethylene glycol monoethyl ether, triethylene glycol monopropyl ether, propylene glycol monomethyl ether, dipropylene glycol monoethyl ether, triethylene glycol monobutyl ether, tetrapropylene glycol monoethyl ether, etc.

The alkaline agent, if employed, is selected from the group consisting of the alkali metal hydroxides as exemplified by sodium hydroxide, potassium hydroxide, mixtures thereof, etc. which are generally added to the aqueous solution in an amount of from about 0.05 to 1.0% by weight.

Example: An aqueous solution containing 0.8% by weight of sodium dodecyl sulfonate, 0.10% by weight of potassium abietate, and about 0.35% by weight of diethylene glycol monobutyl ether is prepared by adding with mixing the abovementioned ingredients to water having salinity of about 1.73% by weight at a temperature of about 25°C, after which the thus-prepared solution is introduced with mixing into a large diameter pipeline together with sufficient heavy California crude to give an oil-in-water emulsion in which the amount of aqueous solution is about 22% by volume based on the volume of the crude. The horsepower requirement for transporting the formed oil-in-water emulsion through the pipeline at the rate of 2,000 bpd is found to be substantially less than the horsepower requirement for transporting the same volume of this California heavy crude under the same conditions.

SHALE OIL RECOVERY

Purification of Off-Gases During Shale Oil Recovery

The presence of large deposits of oil shale in the Rocky Mountain region of the United States has given rise to extensive efforts to develop methods of recovering shale oil from kerogen in the oil shale deposits. It should be noted that the term oil shale as used in the industry is in fact a misnomer; it is neither shale nor does it contain oil. It is a sedimentary formation comprising marlstone deposit interspersed with layers containing an organic polymer called kerogen, which upon heating decomposes to produce liquid and gaseous products. It is the formation containing kerogen that is called oil shale herein, and the liquid product is called shale oil.

A number of methods have been developed for processing the oil shale which involve either first mining the kerogen-bearing shale and processing the shale on the surface, or processing the shale in situ. The latter approach is preferable from the standpoint of environmental impact since the spent shale remains in place, reducing the chance of surface contamination and the requirement for disposal of solid wastes.

The recovery of liquid and gaseous products from oil shale deposits has been described in several patents, one of which is U.S. Patent 3,661,423. This patent describes in situ recovery of liquid and gaseous carbonaceous materials from a subterranean formation containing oil shale by fragmenting such formation to form a stationary, fragmented, permeable body or mass of formation particles containing oil shale within the formation, referred to herein as an in situ oil shale retort. Hot retorting gases are passed through the in situ oil shale retort to convert kerogen contained in the oil shale to liquid and gaseous products, thereby producing retorted oil shale.

The off-gas, which contains nitrogen, hydrogen, carbon monoxide, carbon dioxide, methane, ethane, ethylene, propane, propylene and other hydrocarbons, water vapor, and hydrogen sulfide must be disposed of in an ecologically sound manner. This is primarily because its low fuel value, i.e., less than about 70 Btu/ft^3, can make it uneconomical to use as a fuel. Added to this is the difficulty encountered in initiating and maintaining combustion of fuels with such a low Btu content. Since environmental considerations prohibit discharge of such gas directly to the atmosphere, there is a need for an economical method of purifying the off-gas from an in situ oil shale retort.

C.Y. Cha; U.S. Patent 4,156,461; May 29, 1979; assigned to Occidental Oil Shale, Inc. describes a process to accomplish such purification while taking advantage of the previously wasted sensible heat remaining in an in situ oil shale retort at the conclusion of the retorting operation.

According to the process, the hydrocarbon, hydrogen and carbon monoxide concentrations of a gas are reduced by reacting these constituents of the gas with an oxygen-bearing material in the presence of a fragmented permeable mass of particles containing oil shale. The oil shale promotes the oxidation of the hydrocarbons, hydrogen and carbon monoxide to water and carbon dioxide. Preferably the oil shale has been treated to remove organic materials and retains a portion of the sensible heat generated during such treatment. Such gas with relatively lower hydrocarbon, hydrogen and carbon monoxide concentrations is then withdrawn from the fragmented permeable mass of oil shale.

An additional feature of the process is that at least a portion of the carbon dioxide generated by the reaction of the hydrocarbon and carbon monoxide constituents of the gas can be reduced by reacting it with alkaline earth metal oxides contained in the treated oil shale.

Referring to Figure 1.7, an already retorted in situ oil shale retort **8** is shown in the form of a cavity **10** formed in an unfragmented subterranean formation **11** containing oil shale. The cavity contains an explosively expanded and fragmented permeable mass **12** of formation particles. The cavity **10** can be created simultaneously with fragmentation of the mass of formation particles **12** by blasting using any of a variety of techniques. Methods of forming an in situ oil retort are described in U.S. Patents 3,661,423, 4,043,595, 4,043,596, 4,043,597 and 4,043,598. A variety of other techniques can also be used.

Figure 1.7: Oil Shale Retort

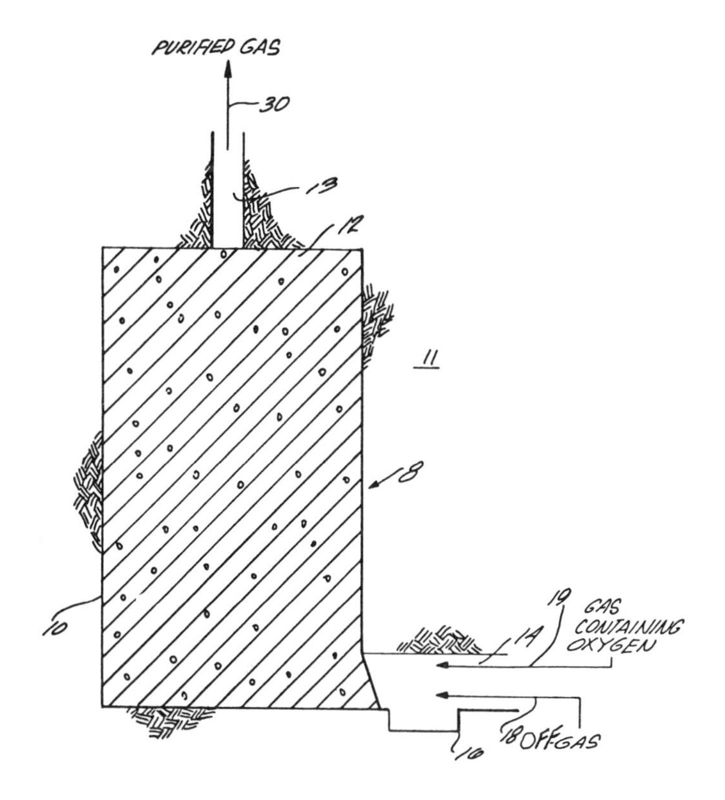

Source: U.S. Patent 4,156,461

A conduit **13** communicates with the top of the fragmented mass of formation particles. During the retorting operation of the retort **8**, a combustion zone is established in the retort and advanced by introducing a gaseous feed containing an oxygen-supplying gas, such as air or air mixed with other gases, into the in situ oil shale retort through the conduit **13**. As the gaseous feed is introduced to the combustion zone, oxygen oxidizes carbonaceous material in the oil shale to produce combusted oil shale and combustion gas. Heat from the exothermic oxidation reactions, carried by flowing gases, advances the combustion zone downwardly through the fragmented mass of particles.

Combustion gas produced in the combustion zone, any unreacted portion of the oxygen-supplying gaseous feed and gas from carbonate decomposition are passed through the fragmented mass of particles on the advancing side of the combustion zone to establish a retorting zone on the advancing side of the combustion zone. Kerogen in the oil shale is retorted in the retorting zone to yield retorted oil shale and liquid and gaseous products, including hydrocarbons.

There is a drift **14** in communication with the bottom of the retort. The drift contains a sump **16** in which liquid products are collected to be withdrawn for further processing. The off-gas containing gaseous products, combustion gas, gas from carbonate decomposition, and any unreacted portion of the gaseous combustion zone feed are also withdrawn from the in situ oil shale retort **8** by way of the drift **14**. The off-gas can contain large amounts of nitrogen with lesser amounts of hydrogen, carbon monoxide, hydrocarbons such as methane, ethane, ethylene, propane, propylene and higher hydrocarbons, water vapor and sulfur compounds such as hydrogen sulfide. The off-gas also can contain particulates and hydrocarbon-containing aerosols.

A process gas stream **18** containing off-gas from an active oil shale retort, and a gas stream **19** containing oxygen, such as air, are introduced through the drift **14** to the already treated retort **8**. By active retort is meant a retort in which combustion and retorting operations are being conducted. It will be understood that although the oxygen-containing gas is ordinarily ambient air, other composition variations are included within the term. Thus, for example, if desired, pure oxygen or air augmented with additional oxygen can be used so that the partial pressure of oxygen is increased. Similarly, air can be diluted with an oxygen-free gas such as nitrogen.

The off-gas and oxygen-containing gas can be introduced separately into the retort, or can be substantially homogeneously mixed prior to introduction into the retort. Mixing can be accomplished by any of a number of methods, such as with jet mixers, injectors, fans and the like.

Preferably the off-gas and the oxygen-containing gas are introduced to the hottest portion of the fragmented permeable mass in the retort to minimize pressure drop through the retort and the cost of passing gas through the retort. By introducing the gases to the hottest portion of the retort, heat is transferred by the flowing gases to the cooler portions of the retort, with the result that the fragmented permeable mass eventually has a substantially uniform temperature gradient, and no exceptionally hot region, with the temperature decreasing in the direction of movement of the gases. This results in reduced pressure drop across the retort because the volumetric flow rate of the gases through the retort decreases as the temperature of the fragmented permeable mass increases due to thermal contraction of the formation particles as the mass of particles cools. Thus, the cross-sectional area available for flow of gases through the retort increases.

For economy, the conduit used for introducing oxygen-supplying gaseous feed to the retort **8** during the retorting operation is utilized to withdraw an effluent gas **30** of reduced hydrocarbon, hydrogen, carbon monoxide and hydrogen sulfide concentration from the retort. Similarly, the drift **14** used for withdrawing off-gas from the retort **8** during the retorting operation is utilized for introducing the gas streams **18** and **19** to the retort. The effluent gas **30** has a relatively lower hydrocarbon, hydrogen, carbon monoxide, hydrogen sulfide and total sulfur concentration than the gas **18** introduced into the retort **8**.

As the off-gas stream **18** and the oxygen-containing gas stream **19** pass through the hot spent retort, hydrocarbons are oxidized to carbon dioxide and water, hydrogen is oxidized to water, carbon monoxide is oxidized to carbon dioxide, and hydrogen sulfide is oxidized to sulfur and oxygen-bearing compounds, including sulfur dioxide. Although not essential, it is preferred that there be a stoichiometric excess of oxygen. It was found that these oxidations readily occur, notwithstanding the relatively low fuel value and the prior difficulties encountered with combusting off-gas.

It has been found, however, that at temperatures below about 600°F, the rate of conversion of the hydrocarbons, hydrogen and carbon monoxide to carbon dioxide and water can be too slow and/or the flow rate of off-gas can be too great to achieve adequate removal in a single retort. The off-gas **18** can then be passed with an oxygen-containing gas through additional retorts, in series and/or parallel, containing oil shale treated to remove organic materials, or recirculated several times in a single retort to achieve maximum removal.

Oil shale contains large quantities of alkaline earth metal carbonates, principally calcium and magnesium carbonates, which during retorting and combustion are at least partly calcined to produce alkaline earth metal oxides. Thus combusted oil shale particles in the retort **8** can contain approximately 20 to 30% calcium oxide and 5 to 10% magnesium oxide, with smaller quantities of less reactive oxides present.

The carbon dioxide and sulfur dioxide produced from the reactions can combine with these constituents of the oil shale to yield solid materials such as carbonates and sulfites.

Therefore, when an oil shale retort containing treated oil shale is used, not only can the hydrocarbon, hydrogen, carbon monoxide and hydrogen sulfide content of a gas stream be reduced, but also the total concentration of sulfur compounds in the gas stream can be reduced.

DRILLING, WORKOVER AND COMPLETION FLUIDS

AQUEOUS, CLAY-BASED DRILLING FLUIDS

Titanium or Zirconium Lignosulfonates

A well fluid for use in rotary drilling must have sufficient viscosity that it easily carries rock chips and material loosened by the drill bit out to the surface of the ground by flow of the fluid and it should be thixotropic so that when drilling is stopped at any time, the fluid will gel and prevent chips from settling around the drill bit.

The apparent viscosity or resistance to flow of drilling muds is the result of two properties, plastic viscosity and yield point. Each of these two properties represents a different source of resistance to flow. Plastic viscosity is a property related to the concentration of solids in the fluid, whereas yield point is a property related to the interparticle forces. Gel strength, on the other hand, is a property that denotes the thixotropy of mud at rest. The yield point, gel strength, and, in turn, the apparent viscosity of the mud, commonly are controlled by chemical treatment with materials such as complex phosphates, alkalies, mined lignites, plant tannins, and modified lignosulfonates.

P.H. Javora and B.Q. Green; U.S. Patent 4,220,585; September 2, 1980; assigned to Dresser Industries, Inc. describe viscosity-controlling additives for drilling fluids composed of complex lignosulfonates containing titanium and/or zirconium. Commercially available zirconium compounds typically contain a small percentage of hafnium. The additives are in many cases more effective viscosity-controlling agents than the chromium or chromium-iron lignosulfonates disclosed in the prior art and widely used in the drilling industry. They have the additional advantage of avoiding the toxic nature ascribed to chromium lignosulfonate. In alternate forms of the process, mixed metal forms of lignosulfonate are employed, specifically titanium or zirconium in combination with iron.

Generally, lignosulfonates are prepared by reacting lignin liquors obtained from the pulping of wood with salts of the desired metal or metals and when necessary, removing any precipitated material. When the material is to be oxidized, the oxidation can be any one of the process steps. If necessary, the product can be sulfonated to produce additional sulfonate groups in the product.

Viscosity-controlling agents prepared by reacting lignosulfonate liquor with a zirconium salt or complex are found to be effective when the zirconium content of the product is from at least about 1 to 9% by wt; the preferred zirconium content appears to be in the range of about 4 to 6% by wt. Incorporation of greater amounts of zirconium does not appear to improve the viscosity-controlling properties of the product. In the case of zirconium-iron materials, compositions containing from 1 to 3% iron, the total metal content (iron plus zirconium) of from about 4 to about 6% appears to be particularly effective. If the lignin component of the lignosulfonate is partly in oxidized form, it is preferred to employ about 1 to 5% by wt of oxidant with the preferred amount being from about 3 to about 5%.

For the titanium lignosulfonate products, the titanium content should be from about 0.5 to 5% with the preferred amount being from about 1.5 to about 3.5%. Preferably titanium-iron lignosulfonate materials contain from about 1 to about 3% iron with a total metal content (iron plus titanium) from about 2 to about 6%. Oxidized products in this case appear to be most effective when prepared by using from about 1 to about 6% by wt oxidant with the preferred amount being in the range of from 2 to 5%. An especially effective composition is a titanium-iron lignosulfonate containing 2.3% by wt of titanium and 1.5% by wt of iron. The lignin component of this particular composition was partially oxidized by the addition of 4.5% by wt of oxidant to the reaction mixture.

In the following example all measurements of the parameters apparent viscosity, plastic viscosity, yield point and gel strength were made in accordance with *API Recommended Practice 13B Standard Procedure for Testing Drilling Fluids,* 6th Edition published by the American Petroleum Institute, April 1976.

Example: A solution of lignosulfonic acid was prepared from 834 g of spent soft wood sulfite liquor (obtained from Consolidated Papers, Inc. and containing 55% by wt solids and 2% by weight calcium) diluted with 300 g of water and 20 ml of concentrated sulfuric acid. This mixture was warmed to 140°F for 3.5 hr and filtered to isolate the filtrate from the precipitated gypsum. To 351 g of the resultant lignosulfonic acid solution, diluted with 100 g of water, was added 47 g of zirconium acetate-acetic acid solution containing the equivalent of 22% zirconium dioxide. After a 30 minute reaction time, 11.7 g of hydrated ferric sulfate containing 78.5% $Fe_2(SO_4)_3$ was added, and dissolved in several hours. The pH was adjusted to 4.0 with 20% sodium hydroxide and the solution was spray-dried to yield the brown solid powdered material composed of zirconium-iron lignosulfonate containing 4.9% by wt zirconium and 1.5% iron.

This material was then tested for viscosity-controlling and dispersant and/or deflocculent properties by addition to a standard base drilling fluid prepared from clayey material containing 25% by wt sodium bentonite (montmorillonite), 50% by wt X-ACT clay (a calcium montmorillonite) and 25% by wt grundite. In each of these tests the zirconium-iron lignosulfonate was compared to a base mud made up by the use of 68 lb/bbl of the clayey solids in water. Comparisons were made with mud to which no dispersant had been added, for example, to which 6 lb/bbl of chromium lignosulfonate had been added and a sample to which

6 lb/bbl of zirconium-iron lignosulfonate had been added. Each of the slurries tested was aged 18 hours at a temperature of 200°F. The results are given in the table below.

Base	Apparent Viscosity (cp)	Plastic Viscosity (cp)	Yield Point	. . . .Gel Strength. . . . 1 Minute	10 Minutes
			. . . . ($lb/100\ ft^2$)		
–	48	12	72	61	85
Chromium lignosulfonate	13	10	6	6	10
Zirconium-iron lignosulfonate	13	12	2	3	8

From the foregoing data it can be seen that the thinning and dispersant properties of zirconium-iron lignosulfonate compare favorably with those chromium lignosulfonates wihout the use of any chromium in the added material.

Magnesium Oxide, Bentonite and Ferrochrome Lignosulfonate

J.W. Forster and L.E. Roper; U.S. Patent 4,209,409; June 24, 1980; assigned to Phillips Petroleum Company have found that adding magnesium oxide to a slurry of prehydrated bentonite and ferrochrome lignosulfonate, the slurry having a pH of about 9.5 to 12, increases the life of this slurry when later diluted with salty water. By salty water is meant an aqueous drilling fluid containing at least 8 and generally at least 10 wt % sodium chloride. This process thus provides a slurry composition which comprises fresh water, bentonite, ferrochrome lignosulfonate, and magnesium oxide, and which has a pH of about 9.5 to 12.

The additive mixture contains the three main ingredients in the following ranges in parts by weight: bentonite, 15 to 40; ferrochrome lignosulfonate, 2.5 to 3.5; and magnesium oxide, 0.5 to 3. If NaOH is present in the mixture, the mixture will contain 0.5 to 2 pbw NaOH to achieve the proper pH of the slurry made from that mixture. This slurry, when added to saltwater drilling fluid, increases the yield point thereof considerably, which means the drilling fluid containing the slurry carries a larger quantity of drill cuttings out of a drill hole.

Preferred is a slurry composition containing about 4 lb/bbl of ferrochrome lignosulfonate and about 0.75 lb/bbl of magnesium oxide. In case sodium hydroxide is used as the basic material in the slurry, the slurry will broadly contain 0.5 to 2 lb of sodium hydroxide per barrel of fresh water, more preferably 0.75 to 1 lb.

The slurry composition and the drilling fluid are prepared as follows: First the ingredients bentonite, ferrochrome lignosulfonate, magnesium oxide, and the basic material, if necessary, are mixed together in the proper quantities in fresh water to form a slurry. This slurry is yielded. Yielding means to allow the bentonite particles to hydrate, i.e., to separate, soak up water and gain size; the yielding or hydrating preferably is done while stirring the slurry. This yielding, when carried out in the field, will take place in a period of time of about six hours. After this yielding, a slurry composition is ready to be mixed with the salty drilling fluid.

Example: This calculated example shows the use of a slurry for cleaning a drill hole.

To 100 bbl of fresh water in a prehydrating pit, 3,000 lb bentonite, 300 lb ferrochrome lignosulfonate, 75 lb sodium hydroxide, and 100 lb magnesium oxide are added via a centrifugal pump to the pit. The mixture is sheared with e.g., a Lightnin mixer or a mud gun. After the shearing the mixture is allowed to stand for yielding of the bentonite for about one hour. Additional yielding will occur if the mixture is allowed to stand for a longer period of time.

The resulting slurry is added to a salty drilling fluid as necessary to provide the viscosity for drill hole cleaning. For this purpose intermittent slugs of about 20 bbl of the slurry prepared as described in this example are added to the drilling mud system to clean the drill hole of excess cuttings. This is advantageously done before a round trip is carried out, e.g., to exchange the bit. Several slugs of the slurry composition may be necessary to clean the hole.

Humate Thinning Additives

W.C. Firth, Jr.; U.S. Patent 4,235,727; November 25, 1980; assigned to Union Camp Corporation describes the use of humates to thin water-based drilling fluids.

The humate compositions employed in the process are naturally occurring compositions of matter found in association with rutile sands. Rutile sand deposits are found in several places throughout the world. In the United States rutile sand deposits are located in Florida, Georgia and South Carolina. The rutile sands are in a formation commonly referred to as "hard pan." The hard pan comprises rutile sands bound together by a coating of humate. It is this humate which is employed in the process.

The desired humate may be separated from the rutile sand deposits by first breaking up the deposit formation of hard pan into a ground ore of a convenient size for handling. Much of the humate in the ground ore can be washed off with water to effect the desired separation. Additional humate can be obtained by washing the concentrated ore with aqueous sodium hydroxide. The aqueous mixture containing the free humates may then be treated with, for example, a strong mineral acid such as sulfuric acid or alum to regain the natural pH of the humate and facilitate settling out of the suspended humate. The separated humate may then be dried in the sunlight or by artificial means.

In general the humate compositions employed in the process have compositional make-ups which provide a carbon to hydrogen ratio (weight to weight) of from 9.5 to 17.5:1.0; a carbon to oxygen ratio of 1.0 to 2.0:1.0; an aluminum content of 2.8 to 8.4% by wt, a titanium content of 0.5 to 1.5% by wt and a calcium content of less than 0.5% by wt.

The humate material may also be chemically (and physically) modified for improved performance as a drilling mud additive. For example, it can be treated with sulfite, bisulfite or sulfur dioxide in an alkaline medium to increase its solubility in water.

The method is carried out by providing the abovedescribed humate, separated from its association with rutile sand deposits and dispersing a viscosity-reducing proportion of the humate in a water-based drilling mud. Dispersion may be carried out employing conventional mixing and agitating equipment, employed conventionally for dispersing like additives in drilling mud compositions. The proportion of humate dispersed in the drilling mud may be varied over a wide range, i.e., from about 0.05 to about 6.5% by wt of the drilling mud (dry weight of humate/wet weight of drilling mud).

Extender for Sepiolite Clay

During the initial make-up of drilling fluids viscosified with sepiolite clay minerals, the slurry is very thin because the fibers of the clay particles have not separated sufficiently. In order to provide a drilling fluid with sufficient viscosity to support drill solids and weighting materials, large amounts of shearing energy by mechanical mixers must be used in the mud pits to separate these bundles of fibers. In lieu of such mechanical mixers, the shearing energy can be achieved by circulation of the drilling fluid through the drill pipe, bit and annulus. This is not always practical because of drilling fluid weight considerations. Alternatively, small amounts of bentonite, up to 5 lb/bbl can be used. This is sometimes not desirable because of the subsequent inferior performance of bentonite as the mud reaches 400+°F and the necessity of fresh water prehydration of the bentonite.

L.D. Barker and J.K. Bannerman; U.S. Patent 4,201,679; May 6, 1980; assigned to Dresser Industries, Inc. describe a clay extender which will impart to a slurry of sepiolite clay during initial make-up, a viscosity of sufficient nature to support a weighting agent without the need of excessive shearing energy. The viscosifier consists essentially of sepiolite and a partially hydrolyzed polyacrylamide polymer extender. The polymer has the general formula $(CH_2CHCONH_2)_x$, where x ranges between about 5,600 and 14,900.

The extender, which has a molecular weight between about 400,000 and 900,000 is present in an amount between about 0.01 and 0.25 lb/15 lb of sepiolite.

The extender may contain, in addition, an inert substrate, such as calcined kaolinite. The substrate is preferably present in an amount between about 15 and 35%.

The extender is usually hydrolyzed in an amount between 20 and 80%.

Sepiolite is a mineral having the formula

$$Mg_4(Si_2O_5)_3(OH)_2 \cdot 6H_2O.$$

The mineral is a soft, lightweight absorbent, white to light gray or light yellow clay mineral.

Chemically, it contains approximately 55% SiO_2, 0.7% Al_2O_3 + Fe_2O_3, 25% MgO, 0.5% CaO, 0.4% alkalies and the balance H_2O. Its density is approximately 2.1.

The extender and inert substrate, if desired, can be added to the sepiolite in any reasonable manner. From a practical standpoint, it will ordinarily be preferred to add the extender to the clay after the clay has been ground and dried. The resulting product can then be sacked in a conventional manner. However, if desired, the clay and extender and inert substrate can be separately added to water to form a slurry, the extender and clay being added in any desired order. This type of wet mixing could be employed, for example, at the well site, but ordinarily it would not be preferred because the mixing in the dry form prior to sacking permits closer quality control and a more uniform product. More practical, the contents of the package would be added to water to make a slurry at the well site.

Reaction Product of Fatty Vegetable Oil and 4,4'-Thiodiphenol

One of the more important functions of the drilling fluid is to reduce the considerable torque on the rotating drill pipe caused by friction between the outside of the drill pipe and the wall of the well. When adverse drilling conditions, such as drilling through offsets, highly deviated holes, and dog legs result in increased frictional forces, the lubricating properties of aqueous drilling fluids are generally insufficient to prevent the drill pipe from tolerating excessive torque, thus leading to costly delays and interruptions in the drilling process. Thus, there is needed an additive for aqueous drilling fluids which will impart enhanced lubricating properties to the fluid.

S.H. Elrod and W.B. Nance; U.S. Patent 4,181,617; January 1, 1980; assigned to Milchem Incorporated describe a lubricant for use in aqueous drilling fluids, the lubricant being the reaction product of a fatty vegetable oil with 4,4'-thiodiphenol. The composition is added to aqueous drilling fluids and circulated down the bore hole of a well being drilled, for the reduction of torque and drag experienced by the drill pipe.

The lubricant is prepared by reacting together from about 0.5 to 2% by wt of 4,4'-thiodiphenol with from about 99.5 to 98% by wt of a fatty vegetable oil. Preferably, the fatty vegetable oil will be a member selected from the class consisting of castor oil, coconut oil, corn oil, palm oil and cottonseed oil. Even more preferably, the fatty vegetable oil will be cottonseed oil, inasmuch as it is readily apparent, is comparatively inexpensive, and is easily reactable with 4,4'-thiodiphenol.

It has been found preferable to carry out the reaction in the presence of heat from about 110° to about 180°F. Even more preferably, the reaction is conducted by application of heat and low shear to the reactants.

Drilling Fluid Containing Non-Newtonian Colloidal System

J. Bretz and L.S. Cech; U.S. Patent 4,230,586; October 28, 1980; assigned to The Lubrizol Corporation describe aqueous drilling fluids and muds which exhibit enhanced lubricating properties and thereby reduce the torque requirements for drilling operations.

The aqueous well-drilling fluids comprise:

(A) at least one non-Newtonian colloidal disperse system comprising:

- (1) solid metal-containing colloidal particles at least a portion of which are predispersed in
- (2) at least one liquid dispersing medium; and
- (3) as an essential component, at least one organic compound which is soluble in the dispersing medium, the molecules of the organic compound being characterized by a hydrophobic portion and at least one polar substituent; and

(B) at least one emulsifier.

These aqueous drilling fluids also comprise clay/water slurries. They are not water-in-oil emulsions which, in general, are more costly and less convenient than the aqueous drilling fluids of this process.

The non-Newtonian colloidal disperse systems used in the additives of this process are well known to the art and are described, for example, in U.S. Patents 3,492,231, 3,242,079, 3,027,325, 3,488,284, and 3,372,114.

The solid metal-containing particles are metal salts of inorganic acids and low molecular weight organic acids (such as formic, acetic and propionic acids), hydrates thereof, or mixtures of two or more of these. These salts are usually alkali and alkaline earth formates, acetates, carbonates, hydrogen carbonates, hydrogen sulfides, sulfides, sulfates, hydrogen sulfates and halides. Magnesium, calcium and barium salts are typical examples. Mixtures of two or more of any of these can also be present. Typically the metal particles are solid metal-containing colloidal particles consisting essentially of alkaline earth metal salts, these salts being further characterized by having been formed in situ and predispersed.

Colloidal disperse systems used in the fluids of this process also comprise at least one liquid dispersing medium. The identity of the medium is not critical as the medium serves primarily as a liquid vehicle in which the solid particles are dispersed. Normally it consists of one or more substantially inert, nonpolar organic liquids, that is, liquids which are substantially chemically inactive in the particular environment in question. The liquid dispersing medium may be substantially volatile or nonvolatile at standard temperatures and pressures. Often the non-Newtonian disperse system is prepared in such a manner that a mixture of such volatile and nonvolatile organic liquids is used as the dispersing medium thus permitting easy removal of all or a portion of the volatile component by heating. This is an optional and often desirable means for controlling the viscosity or fluidity of the disperse system.

Typical dispersing media include materials such as mineral oils and synthetic oils. Other organic liquids such as ethers, alkanols, alkylene glycols, ketones, and the like are useful as dispersing mediums. In addition, ester plasticizers can also be used.

Typical ester plasticizers are chosen from the group consisting of phthalates, phosphates, adipates, azelates, oleates, and sebacates. Specific examples are the dialkyl phthalates such as di(2-ethylhexyl) phthalate, dibutyl phthalate, diethyl phthalate, dioctyl phthalate, butyl octyl phthalate, dicyclohexyl phthalate, butyl benzyl phthalate; diaryl phosphates such as tricresyl phosphate, triphenyl phosphate, cresyl diphenyl phosphate; trialkyl phosphates, such as trioctyl phosphate and tributyl phosphate; alkoxyalkyl phosphates such as tributoxyethyl phosphate; alkylaryl phosphates such as octylphenyl phosphate; alkyl adipate such as di(2-ethylhexyl) adipate, diisooctyl adipate, octyldecyl adipate; dialkyl sebacates such as dibutyl sebacate, dioctyl sebacate, diisooctyl sebacate; alkyl azelates such as di(2-ethylhexyl) azelate and di(2-ethylbutyl) azelate and the like. Analogous esters can be made from citric and salicylic acids. Esters of monocarboxylic acids such as benzoic acid with, for example, diethylene glycol, dipropylene glycol, triethylene glycol, and the like are also useful as are toluene sulfonamides.

Mixtures of two or more of the abovedescribed dispersing media are also useful and often cheaper and more efficient.

From the standpoint of availability, cost and performance, liquid hydrocarbons and particularly liquid petroleum fractions represent particularly useful dispersing media. Included within these classes are benzene and alkylated benzenes, naphthalene-based petroleum fractions, paraffin-based petroleum fractions, petroleum ethers, petroleum naphthas, mineral oil, Stoddard solvent, and mixtures thereof. Typically the dispersing medium is mineral oil or at least about 25% of the total medium is mineral oil. Often at least about 50% of the dispersing medium is mineral oil. As noted, mineral oil can serve as the exclusive dispersing medium or it can be combined with some nonmineral oil organic liquid such as, for example, esters, ketones, etc.

In addition to the solid metal-containing particles and the dispersing medium, the aforedescribed non-Newtonian colloidal disperse systems include at least one organic compound which is soluble in the dispersing medium and whose molecules are characterized by the presence of a hydrophobic portion and at least one polar substituent. While the types of suitable organic compounds are extremely diverse and include generally oil-soluble organic acids such as phosphorus acids, thiophosphorus acids, sulfur acids, carboxylic acids, thiocarboxylic acids and the like, as well as their corresponding alkali and alkaline earth salts, the alkaline earth and alkali metal salts of oil-soluble petrosulfonic acids, mono-, di- and trialiphatic hydrocarbon sulfonic acids and oil-soluble fatty acids, are, for reasons of economy, availability and performance particularly suitable.

Broadly speaking, the non-Newtonian colloidal disperse systems used in the drilling fluids are prepared by treating a single phase homogeneous Newtonian system of an overbased organic compound corresponding to one or more of the organic compounds described hereinabove with a conversion agent which is usually an active hydrogen-containing compound.

Typical active hydrogen-containing conversion agents include lower aliphatic carboxylic acids, water, aliphatic alcohols, alicyclic alcohols, phenols, ketones, aldehydes, amines, boron acids, and phosphorus acids. Oxygen, air and carbon dioxide can also be used as conversion agents. Often a mixture of water and alcohols (e.g., a lower alkanol) is used. Such mixtures usually have weight ratios of alcohol to water of from about 0.05:1 to about 24:1. Water and carbon dioxide mixtures are also very useful conversion agents.

The treating operation is simply a thorough mixing together of the two components, i.e., homogenization. This homogenization is generally achieved by vigorous agitation of the components at or near the reflux temperature of the mixture. Usually this temperature ranges from about 25° to about 200°C, typically it is no more than about 150°C. This treatment converts these single phase systems into non-Newtonian colloidal disperse systems.

The second component of the additives of this process is (B) at least one emulsifier. These emulsifiers function primarily to emulsify the noncolloidal disperse system in drilling fluid or mud so that it does not substantially separate during periods of storage or low agitation. They thus also serve to stabilize the emulsions formed.

Many such emulsifiers are known to the art, for example, the discussion in Kirk-Othmer, *Encyclopedia of Chemical Technology,* Volume 8, pages 127-134.

The emulsifiers used can be ionic or nonionic; typically they are nonionic or cationic emulsifiers.

Among the useful emulsifiers are alkylene oxide condensates (i.e., alkoxylates) with active hydrogen compounds such as alcohols, phenols, amides and amines. The amides are often fatty acid amides such as oleyl amides. A particularly useful class of emulsifiers are the ethoxylated amines wherein the amine has at least 12 carbon atoms. Such cationic emulsifiers can be represented by the general formulas:

$$RN\begin{cases}(CH_2CH_2O)_xH \\ (CH_2CH_2O)_yH\end{cases}$$

and

$$\begin{array}{l}(CH_2CH_2O)_zH \\ | \\ RNCH_2CH_2CH_2N\end{array}\begin{cases}(CH_2CH_2O)_xH \\ (CH_2CH_2O)_yH\end{cases}$$

wherein R is an aliphatic hydrocarbyl group with at least about 12 carbon atoms, x, y and z are integers of 0 to 40 and the sum x+y is between 2 and 50. Usually the aliphatic group R has a maximum of about 22 carbons. Often such R groups are fatty alkyl or alkenyl groups such as coco (C_{12}), stearyl (C_{18}), tallow (C_{18}), oleyl (C_{18}), and the like. Typically R is a tallow residue and the sum x+y is about 5. Homologous alkoxylated amines wherein the ethoxyl residue ($-CH_2CH_2O-$) is replaced, at least in part, by a propoxyl residue

$$\begin{array}{c}(-CH_2CH-O) \\ | \\ CH_3\end{array}$$

are also useful. Mixtures of one or more of the aforedescribed emulsifiers can be used.

Useful emulsifiers of the alkoxylated type include Ethofats, Ethomeens, Ethoduomeens, Ethomids and Ethoquads (Armak Company).

The additives used in the drilling fluids generally comprise about 10 to about 95% of (A) at least one non-Newtonian colloidal disperse system and (B) about 5 to about 90% of at least one emulsifier. Typically, they comprise about 70 to about 90% non-Newtonian colloidal disperse system (A) and about 30 to about 10% emulsifier (B).

These additives are present in drilling fluids and drilling muds in concentrations ranging from about 0.5 to about 10% of the drilling mud. They may be added directly to the drilling mud or they may be first diluted with water or substantially inert solvent/diluent such as the dispersing media described above. Mixtures of water and dispersing media or of dispersing media can be used.

Often it is found that the addition of the emulsifier (B) to the colloidal disperse system (A) makes the whole additive system more easy to handle by reducing its viscosity.

The drilling fluids and muds of the process can contain other materials which are known to be used in such applications, such as clay thickeners, density-increasing agents such as barites, rust-inhibiting and corrosion-inhibiting agents, extreme pressure agents, supplementary surfactants and acid or basic reagents to adjust the pH of the system. A typical drilling fluid or mud is made from a 5% bentonite clay slurry using well-known techniques.

Example 1: A non-Newtonian colloidal disperse system is made according to the procedure described in U.S. Patent 3,492,231 by gelling in the presence of a water/alcohol mixture a basic, carbonated calcium petrosulfonate (approximate molecular weight of the free sulfonic acid is 430) having a metal ratio of 1,200 and a 50% mineral oil content. The basic calcium petroleum sulfonate is made according to the procedure described in U.S. Patent 3,350,308.

Example 2: To 800 parts of the non-Newtonian colloidal disperse system described in Example 1 at about 90°C is slowly added with stirring 200 parts of an ethoxylated tallow amine containing approximately 5 ethylene oxide-derived units. This ethoxylated tallow amine is known as Ethomeen T/15 (Armak Corporation). The cooled additive mixture has a density of 9.15 lb/gal.

Example 3: 3 parts of the additive described in Example 2 is slowly combined with a drilling mud made from a 5% bentonite slurry using moderate stirring. The viscosity of the mud increases significantly as the additive is dispersed in it. A rust inhibitor is added. The resultant slurry exhibits a Timken value of 30 lb. A 5% bentonite slurry lacking the additive fails at a Timken value of less than 5 lb.

Polyethoxylated, Sulfurized Fatty Alcohols

T.O. Walker and K.W. Warren; U.S. Patent 4,141,840; February 27, 1979; assigned to Texaco Inc. provide an improved aqueous drilling fluid incorporating a minor amount of a water-soluble or water-dispersible polyethoxylated sulfurized fatty alcohol. The expression "water-soluble polyethoxylated sulfurized fatty alcohol" refers to an unsaturated fatty alcohol containing from about 14 to 18 C atoms, e.g., the various tetradecenols, hexadecenols, and octadecenols, including myristoleyl, palmitoleyl, oleyl, elaidyl, isooleyl, linoleyl, linolenyl, elaestearyl, ricinoleyl, etc; from about 15 to 30 mols of ethylene oxide per mol of alcohol; and from about 3 to 6% of sulfur, based on the weight of the unsulfurized polyethoxylated unsaturated fatty alcohol moiety. The preferred polyethoxylated, sulfurized unsaturated fatty alcohols contain from 16 to 18 C atoms, 20 to 25 ethoxy groups, and 4 to 6% by weight sulfur.

Polyethoxylated, unsaturated fatty alcohol precursors of the water-soluble polyethoxylated, sulfurized fatty alcohols of the process are available as Lipal 20-OA and 25-OA (Drew Chemical Corporation), respectively. Sulfurization of these water-soluble polyethoxylated fatty alcohol precursors can be achieved by heating the polyethoxylated fatty alcohol to a temperature in the range of from about 150° to 200°C and adding thereto slowly the requisite amount of elemental sulfur while maintaining the resulting heated admixture at the prescribed temperature for approximately 2 hours, followed by a cooling period to let the sulfurized product reach room temperature.

The amount of the water-soluble polyethoxylated sulfurized fatty alcohol additives added to the aqueous drilling fluid is a minor but sufficient amount to increase the lubricity of the drilling fluid as measured, for example, by torque reduction while at the same time not impairing the rheology of the drilling fluid. The amount to be added to the well drilling fluid is in the range of from 0.5 to 5 lb thereof, per barrel of drilling fluid, preferably from about 1 to 3 lb/bbl.

It has been found that the drilling fluids containing the additives of the process exhibit a high degree of lubricity with little or no abnormal distortion of mud properties. Moreover the drilling fluids do not generate an abnormal amount of foam, so that any foaming can be controlled by the use of conventional defoamers which are compatible with drilling fluid systems. The drilling fluids show tolerance and stability over wide range in pH and electrolyte concentrations and they perform satisfactorily in the various mud systems (i.e., drilling fluid systems) in general use today, e.g., low and high pH, salt water, also nondispersed and inhibitive mud systems.

Polyethoxylated, Sulfurized Fatty Acids and Polyalkylene Glycols

T.O. Walker; U.S. Patent 4,172,800; October 30, 1979; assigned to Texaco Inc. describes a related process whereby an improved aqueous drilling fluid is provided by incorporating therein a minor amount of a water-soluble or water-dispersible admixture of a polyethoxylated sulfurized fatty acid and a polyalkylene glycol as hereinafter defined.

The expression "an admixture of a water-soluble polyethoxylated sulfurized fatty acid and a polyalkylene glycol" refers to a fatty acid component and a polyethylene glycol component.

Fatty Acid Component: The fatty acid component is an unsaturated fatty acid containing about 14 to 18 carbon atoms therein such as myristoleic, the various hexadecenoic acids including palmitoleic, the various octadecenoic acids including oleic, elaidic, isoleic, linoleic, linolenic, elaeostearic, ricinoleic and the like; and wherein the polyethoxy content of each unsaturated fatty acid can vary from about 10 to 13 mols of ethylene oxide, preferably between 10 and 12 thereof, per mol of the unsaturated fatty acid moiety; and wherein the sulfur content of the unsaturated fatty acid can vary from about 2 to 6% by wt, based on the weight of the unsaturated fatty acid with the polyethylene oxide units and preferably 3 to 5% by wt.

Polyethylene Glycol Component: The polyethylene glycol component is a water-soluble polyalkylene glycol having an average molecular weight in the range of from about 300 to 1,200, preferably between about 400 and 1,000, preferably a polyethylene glycol.

Polyethoxylated fatty acid precursors of the sulfurized water-soluble fatty acid component of the admixture are known as Emerest 2646, 2647, and 2648 (Emery Industries Inc., Cincinnati, Ohio).

The polyalkylene glycol component of the admixture is used in a weight ratio of from about 2 up to about 4 parts per part of the sulfurized polyethoxylated fatty acid component. In general it has been found necessary to maintain the glycol-fatty acid ratio at 2 or more parts by weight of the glycol per part by weight of the fatty acid component so as to maintain the rheology of the drilling fluid within acceptable limits.

The amount of the admixture of the water-soluble polyethoxylated sulfurized fatty acid and polyalkylene glycol added to the well drilling fluid is in the range of from about 0.5 to about 5 lb/bbl of drilling fluid, preferably from about 1 to 2 lb/bbl.

Tall Oil Fraction Lubricating Additive

K.-H. Grodde and A. Schulz; U.S. Patent 4,212,794; July 15, 1980; assigned to Deutsche Texaco AG, Germany describe an improved lubricant for aqueous drilling fluids, which is capable of biological decomposition, is economical, does not foam and which exhibits a lubricating action superior to that of the known prior art lubricants.

The object is met by using as a lubricant for aqueous drilling fluids a tall oil fraction having a high rosin acid content and containing not more than about 25% of fatty acids and fatty oils. A suitable tall oil fraction may be obtained by separating from tall oil some of the fatty acids and fatty oils. The fatty acids and fatty oils are separated until the remaining fatty acids and fatty oils constitute about 10 to 25% by wt, the balance being from about 35 to 85% of rosin acids, the remaining balance being unsaponifiables and neutrals, on a weight basis. These fatty acid and fatty oil constituents are separated to an extent sufficient to ensure that the residual portion thereof present in the tall oil fraction does not exceed about 25% by wt. Preferably fractional distillation is used to effect the separation. The rosin acid content of the lubricant should be in excess of 40% by wt, preferably from 50 to 80%, for effective results.

The following analyses of two tall oil fractions suitable for use as lubricants are given by way of example. These tall oil fractions were obtained by fractional distillation: Initial boiling point: 110° to 130°C at 1.5 mm Hg; and final boiling point: 185° to 195°C at 0.5 mm Hg.

Fraction 1 had a neutralization number of 107; saponification number of 135; iodine number of 215; hydroxyl number of 38; specific gravity of 0.95; and flash point of 160° to 185°C.

Fraction 2 with a neutralization number of 91 and saponification number of 102 has also produced good results.

In the lubricant of the process, it is particularly preferred that the difference between the neutralization number and the saponification number is preferably below about 30.

It has also been found that only very small amounts of the tall oil fraction having a high rosin acid content and substantially free from fatty acids and fatty oils need be added to aqueous drilling fluids in the order of 0.45 to 3% by volume being sufficient to produce a lubricating action superior to that obtained by the known lubricants. The drilling fluid containing the lubricant is advantageously maintained at a pH below 9. The lubricant may, however, also be used in highly alkaline drilling fluids.

It is desirable that from 0.5 to 5% by wt of an aromatic or an aromatic-containing hydrocarbon is preferably added to the tall oil fraction in order to suppress the foaming tendency of highly alkaline fluids. Suitable materials are benzene, toluene and xylene.

Drilling Pipe Release Agent

During drilling operations the drill string may become stuck and cannot be raised, lowered, or rotated. There are a number of mechanisms possible which may contribute to this problem. Namely these are (1) cuttings or slough build-up in the hole, (2) an undergage hole, (3) key-seating, and (4) differential pressures.

Differential sticking may be defined as the sticking of the drill string against a permeable formation containing less pore fluid pressure than the hydrostatic pressure exerted by the drilling fluid column and usually occurs when the drill string remains motionless for a period of time. The mechanism by which this occurs involves the drill string coming into contact with the permeable zone, remaining quiescent for a period of time sufficient for mud cake to build up on each side of the point of contact, thus sealing the pipe against the borehole. The annular pressure exerted by the drilling fluid then holds the pipe against the borehole or the permeable zone.

C.O. Walker; U.S. Patent 4,230,587; October 28, 1980; assigned to Texaco Inc. describes the use of a specific class of additive, namely, a polyethylene glycol having a specified weight or a saturated saltwater solution or a seawater solution of the glycol to effect release of the stuck drill string in the borehole.

The polyethylene glycol must have a molecular weight of at least 106 up to a maximum of about 600, more or less. A polyethylene glycol having a molecular weight less than 106 such as, for example, ethylene glycol, is ineffective. Moreover, polyalkylene glycols having a molecular weight above about 600 are not desirable since the results obtained therewith are not as satisfactory.

A preferred group are the polyethylene glycols having a molecular weight in the range of from about 150 up to about 600, and a particularly preferred class are those having a molecular weight in a range from about 200 to 600. Within these preferred ranges, it has been found that most effective results are obtained in terms of readiness to release the stuck pipe and also to effect release within the shortest period of time. Most effective results have been found when the polyethylene glycol is employed at a 100% concentration in the borehole to release the stuck pipe. However, within certain limits, solutions of the polyethylene glycol in saturated salt water at concentrations of from about 1 part of polyethylene glycol to 0.25 to 3 parts of saturated salt water or seawater can be used.

The additives were evaluated in a laboratory apparatus to determine their effectiveness as a drilling pipe release agent.

The evaluation was carried out in the following apparatus. A 7.0 cm Whatman 42 filter paper is placed in a vacuum funnel measuring 7.859 cm in diameter. A flat metallic plate measuring 4.445 cm in diameter and having a thickness of 1.588 mm is pressed against the filter paper in the funnel. A vacuum is pulled on the suction side of the funnel until a reading of from 0.5 to 1.0 mm of mercury is attained. A standard fluid mud composed of water, clay and a lignosulfonate dispersant is poured into the funnel until the funnel is filled with the mud. The vacuum is maintained for a period of thirty minutes during which time a mud cake forms around the metal disc. Thereafter excess mud is removed from the funnel using a syringe without disturbing the metal plate until only the cake remains in place around the metal plate.

The test fluid is then poured into the vacuum funnel and permitted to remain in contact with the upper surface of the mud cake and plate. A brass rod is attached to the metal plate and extends upwards from the funnel. The other end of the rod is attached to a flexible line that is passed through two pulleys mounted on a horizontal beam to effect an upward vertical force on the metal plate. A 1,000 gram weight is attached to the line to induce separation of the metal plate from the filter cake. At the same time as the 1,000 gram weight is attached to the flexible line a time clock is started. The time for the plate to separate from the filter cake is recorded. The test is terminated after a total time of 7,000 seconds has elapsed. The test is considered a failure if the plate is not freed within this time period.

The test procedure permits a qualitative evaluation of the efficiency of the additive.

Ex. No.	Additive Type	MW	%	Diluent	Release Time (sec)	Remarks
1	Ethylene glycol	62	100	–	7,000	No release
2	Diethylene glycol	106	100	–	227	Released
3	Triethylene glycol	150	100	–	139	Released
4	Polyethylene glycol	200	100	–	180	Released
5	Polyethylene glycol	300	100	–	188	Released
6	Polyethylene glycol	600	100	–	800	Released
7	Diethylene glycol	106	50	Seawater	7,000	No release
8	Triethylene glycol	150	50	Seawater	7,000	No release
9	Polyethylene glycol	300	50	Seawater	830	Released
10	Polyethylene glycol	600	50	Seawater	197	Released
11	Polyethylene glycol	300	25	Seawater	1,120	Released
12	Polyethylene glycol	600	25	Seawater	7,000	No release
13	Diethylene glycol	106	50	Salt water	256	Released
14	Triethylene glycol	150	50	Salt water	215	Released
15	Polyethylene glycol	300	50	Salt water	355	Released
16	Polyethylene glycol	600	50	Salt water	136	Released
17	Diethylene glycol	106	25	Salt water	7,000	No release
18	Triethylene glycol	150	25	Salt water	199	Released
19	Polyethylene glycol	300	25	Salt water	209	Released
20	Polyethylene glycol	600	25	Salt water	166	Released
A	Diesel oil		100	–	7,200	Released
B	Product X*		100	–	7,300	Released
C	Product X*		2.38**	Diesel oil	5,400	Released
D	Product Y*		2.38**	Diesel oil	4,600	Released

*Commercial products.
**1 gal of additive in 1 bbl of diesel oil.

The data in the above table illustrate the operability of the process and demonstrate the unexpected advantages thereof over the commercial products X and Y as well as over diesel oil. In the case of diesel oil and Product X at 100% concentration, the time of the tests was continued beyond the 7,000 second termination time of the other tests.

Examples 1, 7, 12 and 17 illustrate the criticality of the process in terms of molecular weight, concentrations and diluents.

Recovery of Drilling Fluid

J.P. Messines, G. Labat, and B. Tramier; U.S. Patent 4,192,392; March 11, 1980; assigned to Societe Nationale Elf Aquitaine (Production), France describe a process for the recovery of fluids used in ground drilling. It applies particularly to prospecting hydrocarbons, i.e., oil or natural gas. The process relates to wells in which there is no inflow of water into the well and no addition of weighting agents in order to substantially increase the density of circulating mud; it applies to the large number of cases in which it is desirable to have the lowest possible mud density.

The process comprises subjecting the drilling muds to centrifugation in conditions such that the density of the light effluent, i.e., the effluent from the axial region of the centrifuge, is contained between that of the processed mud and that of the initial drilling fluid which is not yet laden with ground material.

The process results from the finding that those elements of the drilled ground which are particularly responsible for the adverse changes in the rheological properties of the mud, i.e., particles of dimensions less than about 100 microns, which are very inadequately eliminated by the hydrocyclones used in this art, can be separated to a much greater degree by mechanical centrifugation.

When a drilling mud formed from an initial fluid of density D has reached a dispersed particle concentration such that it becomes unusable for drilling to continue, it is subjected to one or more known treatments in order to remove therefrom the particles of dimensions above approximately 100 microns; the mud of density D' thus desanded is then subjected to centrifugation so controlled that the centripetal effluent of the centrifuge has a density D_c contained between D and $D + 0.5(D' - D)$, whereafter this effluent is reintroduced into the well undergoing drilling.

Preferably the centrifugation is controlled to give a light effluent of density from D to $D + 0.33(D' - D)$. In other words, in the preferred form of the process, the maximum density accepted for the light effluent for recovery is equal to that of the fresh drilling fluid (D) plus one-third of the difference between it and that of the mud unsuitable for reuse (D'), which contains only particles of dimensions less than about 100 microns.

Referring to Figure 2.1, reference **1** denotes the well, **2** the hollow shaft of the drilling tool, **3** the pipeline containing a pump (not shown) for withdrawing the mud from the well **1**. Conventional devices for separating the solids dispersed in the mud are provided at **4** and are generally vibratory screens followed by one or more cyclones known in the art as desanders; the device **4** may if required, include a finer separator of the desiltor type, which is capable of removing most of the 100 micron particles and approximately half those above 30 microns; however, these latter devices are not essential because the centrifuge **6** carries out the same work to about 10 to 20 microns.

The solids or heavy muds separated in the device **4** are discharged via **10** while the desanded mud, which generally contains no particles above 100 microns, is recovered via pipe **5**; the latter branches into two branches **5a** and **5b**, the former communicating with the recycling pipe **9** carrying the desanded mud to the intake system for recycling to the drilling tool **2**, while **5b** carries a fraction of this mud to the centrifuge **6**; a suitable valve (no reference shown) in the branch **5b** enables the rate of flow of the mud for centrifugation to be controlled.

Figure 2.1: Recycling of Drilling Fluid

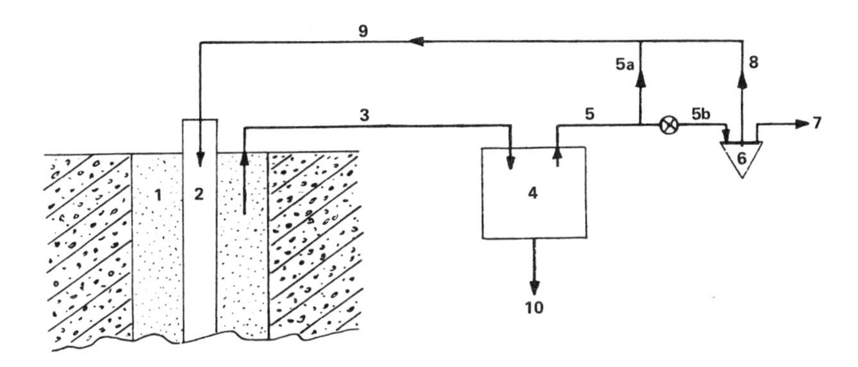

Source: U.S. Patent 4,192,392

The solids separated by centrifugal force at **6** are discarded at **7**, while the centripetal liquid, i.e., a very much lightened mud, passes via pipe **8** to be mixed with the heavy mud to form a mixed mud in the return pipe **9**.

As before, D' will denote the density of the desanded mud leaving device **4** via **5**; D_c is the density of the centripetal light effluent coming from the centrifuge **6** via **8**; the density of the mixed mud formed by dilution of the mud D' with the light effluent D_c is designated D''. Since the most practical criterion for the quality of a drilling mud is its density, the system is controlled on the basis of the densities of the muds.

The continuous process is preferably so carried out that the desanded mud is not yet at its critical utilization limit. In other words, the density D' remains below the critical density after which drilling would be defective. The rate of flow of the muds of density D_c is then controlled so that the density D'' of the mixture reintroduced into the well (pipe **9** in the drawing) is sufficiently low and will give substantially the same density, after the mixture laden with the materials from drilling the ground, has in turn been desanded in the device **4**.

The operation thus takes place with a constant or approximately constant density D', which is very economic because it entails very few fluid losses and applies to the centrifuge only a small fraction of the total flow of the muds.

The volume X (of light effluent D_c) required to produce 100 volumes of mixed mud D'' by mixture with 100 –X volumes of desanded mud D', can be calculated from the conventional dilution equation, which gives:

$$X = 100(D' - D'') \div (D' - D_c).$$

Since the process is applied particularly to drilling with muds whose densities do not exceed 1.25 and the viscosities 80 cp, the following example is based on continuous operation with a mud whose density during operation is less than this value.

Drilling is carried out with bentonite fluid, the initial density D of which is 1.03. The critical density D_L after which operation becomes defective is 1.22 measured of course on desanded mud. The rheological properties of the mud are still very good at a density D' of 1.12 intermediate D and D_L. D' is therefore fixed at that value. The drilling feed and the rate of flow of the mud, and hence the concentration in solids carried by these muds, are such that the density of the desanded mud must be reduced from 1.12 to 1.11 to enable it to be reused, i.e., reintroduced via **9** (see diagram), and to give the value of 1.12 again at the output of group **4**. Also, it is possible to lighten the desanded mud from a density of 1.12 (D') to 1.05 (D_C) by economic centrifugation; the condition D_C = D to 0.33(D' - D) is thus amply satisfied since 1.05 = 0.118(1.12 - 1.03), coefficient 0.118 being contained between 0 and 0.33. Drilling thus operates with the following densities: D' = 1.12; D_C = 1.05; and D'' = 1.11.

Under these conditions, the volume X of centripetal effluent D_C to be mixed with the desanded mud D' to give 100 volumes of recycled mud D'' is:

$$X = 100(D' - D'') \div (D' - D_c) = 100 \times 0.01 \div 0.07 = 14.3.$$

14.3% of centrifuged mud is therefore sufficient to be able to carry out the continuous mud processing with a constant density of 1.12 at the outlet of the separators **4**. Most of the initial bentonite remains in the circulating mud, its losses being reduced to the small proportion of this agent retained by the solids removed at **7** and **10**.

In connection with the above example, it is interesting to note that only a small correction of the density (D' - D'' = 0.01) was sufficient to give perfect drilling operation with a constant density. Without the treatment, the density would increase by 0.01 per 3-hour cycle, and in a short time it would therefore be necessary to remove some of the mud and dilute the remainder with fresh drilling fluid.

AQUEOUS, CLAY-FREE DRILLING FLUIDS

Guar Gum and Magnesia

In the drilling fluid class, clay-based fluids have for years preempted the field, because of the traditional and widely held theory that the viscosity suitable for creating a particle carrying capacity in the drilling fluid could be achieved only with a drilling fluid having thixotropic properties, that is, the viscosity must be supplied by a material that will have sufficient gel strength to prevent the drilled particles from separating from the drilling fluid when agitation of the drilling fluid has ceased, for example, in a holding tank at the surface.

In order to obtain the requisite thixotropy or gel strength, hydratable clay or colloidal clay bodies such as bentonite or fuller's earth have been employed. As a result the drilling fluids are usually referred to as "muds." In other areas where particle carrying capacity may not be as critical, such as completion or workover, brine wellbore fluids are extensively employed. The use of clay-based drilling muds has provided the means of meeting the two basic requirements of drilling fluids, i.e., cooling and particle removal. However, the clay-based drilling muds have created problems for which solutions are needed. For example, since the clays must be hydrated in order to function, it is not possible to employ hydration inhibitors, such as calcium chloride, or if employed, their presence must be at a level which will not interfere with the clay hydration. In certain types of shales generally found in the Gulf Coast area of Texas and Louisiana, there is a tendency for the shale to disintegrate by swelling or cracking upon contact with the water, if hydration is not limited. Thus the uninhibited clay-based or freshwater drilling fluids may be prone to shale disintegration.

The drilled particles and any heaving shale material will be hydrated and taken up by the conventional clay-based drilling fluids. The continued addition of extraneous hydrated solid particles to the drilling fluid will increase the viscosity and necessitated costly and constant thinning and reformulation of the drilling mud to maintain its original properties.

Another serious disadvantage of the clay-based fluids is their susceptibility to the detrimental effect of brines which are often found in drilled formations, particularly Gulf Coast formations. Such brines can have a hydration-inhibiting effect, detrimental to the hydration requirement for the clays.

A third serious disadvantage of clay-based drilling fluids arises out of the thixotropic nature of the fluid. The separation of drilled particles is inhibited by the gel strength of the drilling mud. Settling of the drilled particles can require rather long periods of time and require settling ponds of large size.

Other disadvantages of clay-based drilling fluids are their (1) tendency to prevent the escape of gas bubbles, when the viscosity of the mud raises too high by the incidental addition of hydratable material, which can result in blowouts; (2) the need for constant human control and supervision of the clay-based fluids because of the expectable, yet unpredictable, variations in properties; and (3) the formation of a thick cake on the internal surfaces of the wellbore.

Freshwater wellbore fluids avoid many of the clay-based fluid problems, but may cause hydration of the formation. The brines have the advantage of containing hydration-inhibiting materials such as potassium chloride, calcium chloride or the like. Quite apparently any solid particulate material would be easily separated from the brine solution since it is not hydrated. Thus, the properties of the brine are not changed by solid particulate matter from the wellbore. Similarly, since there is no opportunity for gas bubbles to become entrapped, blowouts are less likely in a clay-free brine-type wellbore fluid.

J.M. Jackson; U.S. Patent 4,151,096; April 24, 1979; assigned to Brinadd Company describes an additive composition for use in clay-free, nonargillaceous, wellbore fluids which comprises guar gum and at least 10 wt % magnesia or magnesium hydroxide based on the weight of guar gum. Preferably the composition will contain about 28 to 50 wt % of magnesia or magnesium hydroxide on this same basis.

The compositions are substantially free or devoid of dolomite (CaMg carbonate). The only solids present, other than drilled particles or other materials coming into the fluid from its use in the wellbore operation results from magnesia present in excess over that portion soluble in the fluid.

One theory for the effectiveness of the process is that the very slightly soluble magnesia which is present in excess of its solubility in the wellbore fluid provides a reservoir of basicity of just the correct amount to maintain the pH of the fluid in the range at which the guar gum is most stable. Magnesium hydroxide may be similarly viewed.

The amount of the composition employed in the wellbore fluid is not critical and may vary for different applications of the fluid. Generally at least 0.5 lb up to about 5 or 10 lb of guar gum per U.S. barrel (42 U.S. gal) will be employed. There will be at least 0.05 ppb of magnesia or magnesium hydroxide present in the wellbore fluid based on the weight of guar gum. This minimum amount of magnesia and magnesium hydroxide represents an excess of magnesia beyond that soluble in the wellbore fluid, e.g., greater than 0.0035 lb of magnesia per barrel of water.

In addition to the guar gum and magnesia or magnesium hydroxide other conventional wellbore additives can be present, serving their usual functions.

It has been found that the pH of the drilling fluid after combining it with the additive composition should be highly alkaline, i.e., preferably about 8.5 or more preferably 8.5 to about 11 to obtain yield.

One embodiment is a wellbore fluid consisting essentially of water and an electrolyte inhibitor for preventing hydration, selected from the group consisting of at least 600 ppm calcium ion, at least 200 ppm aluminum ion or chromium ion, at least 1,500 ppm potassium chloride, at least 5,000 ppm sodium chloride and combinations thereof.

Brines provide a preferred wellbore fluid, generally containing at least 1.0% by wt of a soluble salt of potassium, sodium or calcium in water. In addition, the brine may contain other soluble salts, for example, zinc, chromium, iron, copper and the like. Generally, the chlorides are employed because of availability, but other salts such as the bromides, sulfates and the like may be used.

It is an advantage of the additive composition that it has extended stability and effectiveness over a higher temperature range. A particular feature is that faster yields are obtained by using the additive composition in wellbore fluids. A particular advantage of the additive composition is that the water loss effectiveness is greater and is extended beyond that normally achieved with guar gum.

Guar Gum and Lignosulfonate Salt

In a similar process, *J.M. Jackson; U.S. Patents 4,172,801; October 30, 1979; and 4,140,639; February 20, 1979; both assigned to Brinadd Company* describes an additive composition for use in brine well bore fluids consisting essentially of from 15 to 95 wt % guar gum or hydroxyalkyl guar gum and from 5 to 85 wt % of a salt of lignosulfonate based on the total of guar or hydroxyalkyl guar and lignosulfonate salt. Preferably the additive composition will consist essentially of from 25 to 75 wt % guar gum or hydroxyalkyl guar gum and 25 to 75 wt % of a lignosulfonate salt on the same basis.

The amount of the defined additive composition which will reduce the water loss of the brine well bore over an untreated well bore fluid is preferably from about 0.1 to 3 wt % of the additive composition, based on the weight of the well bore fluid. It has been found that the combination of guar gum or hydroxyalkyl guar gum and lignosulfonate salt produce an unexpected improvement in fluid loss over either component alone. Some suitable water-soluble lignosulfonate salts are alkali and alkaline earth metal salts, chromium salts, iron salts, lead salts, ammonium salts and the like. Particularly preferred are sodium and calcium salts.

Hydroxyalkyl Cellulose and Magnesia

J.M. Jackson; U.S. Patent 4,155,410; May 22, 1979; assigned to Brinadd Company describes a method for preventing or minimizing the loss of drilling fluid from a well bore into porous and permeable formations and for correcting so-called lost circulation of drilling fluids in the well by the use of specific aerated clay-free, nonthixotropic, drilling fluids, which consist essentially of a viscosifying amount of water-soluble nonionic hydroxyalkylcellulose, preferably hydroxyethylcellulose therein. Preferably a stabilizing amount of magnesia or magnesium hydroxide is also present in the drilling fluid to stabilize the viscosity-increasing effect of the hydroxyalkylcellulose, magnesia being more preferred. The process will not eliminate lost circulation in all situations, it will however, provide a means for correcting or reducing lost circulation to allow successful completion of the drilling.

The minimum amount of magnesia employed is at least an excess beyond that magnesia which is soluble in well bore fluid or >0.001% by wt based on the aqueous component, which is about 0.0035 lb of magnesia per bbl. The amount of hydroxyalkylcellulose will vary depending on drilling conditions for each site and the viscosity desired. The amount of hydroxyalkylcellulose in relation to the magnesia is not critical. Preferably at least 25 wt % of magnesia or magnesium hydroxide based on the combined weight with hydroxyalkylcellulose is present.

Reaction Product of Polycyclic, Polycarboxylic Acids and a Base

J.G.D. Schulz and J. Zajac; U.S. Patent 4,235,728; November 25, 1980; assigned to Gulf Research & Development Company describe compositions of matter resulting from the reaction of (1) a mixture of substantially water-insoluble polycyclic, polycarboxylic acids obtained as a result of the oxidation of coal with (2) a base. It was found that these compositions can be added to fresh or salt water, and the resulting compositions obtained from these components are heat-stable drilling fluids possessing satisfactory range of plastic viscosity, yield value and gel strength, the ability to substantially reduce or minimize drilling fluid leakage in a formation and the additional ability to emulsify oil in the drilling fluid when, for example, an oil is present as a lubricant.

The substantially water-insoluble polycyclic, polycarboxylic acids employed in the reaction with a base herein can be obtained by any conventional or suitable procedure for the oxidation of coal. Bituminous and subbituminous coals, lignitic materials and other types of coal products are exemplary of coals that are suitable herein. Some of these coals in their raw state will contain relatively large amounts of water. These can be dried prior to use and preferably can be ground in a suitable attrition machine, such as a hammer mill, to a size such that at least about 50% of the coal will pass through a 40 mesh (U.S. Series) sieve. The carbon and hydrogen content of the coal are believed to reside primarily in multiring aromatic and nonaromatic compounds (condensed and/or uncondensed), heterocyclic compounds, etc. On a moisture-free, ash-free basis the coal can have the following composition:

	Broad Range	Preferred Range
	............(wt %)............	
Carbon	45-95	60-85
Hydrogen	2.2-8	5-7
Oxygen	2-46	8-40
Nitrogen	0.7-3	1-2
Sulfur	0.1-10	0.2-5

Any conventional or suitable oxidation procedure can be used to convert the coal to the desired substantially water-insoluble polycyclic, polycarboxylic acids. For example, a stirred aqueous slurry containing coal in particulate form, with or without a catalyst, such as cobalt, manganese, vanadium, or their compounds, can be subjected to a temperature of about 60° to about 225°C and an oxygen pressure of about atmospheric (ambient) to about 2,000 psi gauge (about atmospheric to about 13.8 MPa) for about one to about 20 hours. The product so obtained can then be subjected to mechanical separation, for example, filtration, and solid residue can be washed with water, if desired, and dried. The solid product remaining will be the desired mixture of substantially water-insoluble polycyclic, polycarboxylic acids, hereinafter referred to as coal carboxylate.

Any base or basic salt, organic or inorganic, that can react with an acid can be used herein to react with the coal carboxylate. Thus, hydroxides of the elements of Group I-A and Group II-A of the Periodic Table can be used. Of these, it is preferred to use potassium, sodium or calcium hydroxide. In addition ammonium hydroxide can also be used. Among the organic bases that can be used are aliphatic amines having from 1 to 12 carbon atoms, preferably from 1 to 6 carbon atoms, such as methylamine, ethylamine, ethanolamine and hexamethylenediamine, aromatic amines having from 6 to 60 carbon atoms, preferably from 6 to 30 carbon atoms, such as aniline and naphthylamine, aromatic structures carrying nitrogen as a ring constituent, such as pyridine and quinoline, etc. By "basic salt" is meant to include salts whose aqueous solutions exhibit a pH in the basic region, such as potassium carbonate, sodium metasilicate, calcium acetate, barium formate, etc.

The reaction between the coal carboxylate and the base is easily effected. The amounts of reactants are so correlated that the amount of base used is at least that amount stoichiometrically required to react with all, or a portion (for example, at least about 10%, preferably at least about 50%), of the carboxyl groups present in the coal carboxylate. This can be done, for example, by dispersing the coal carboxylate in an aqueous medium, such as water, noting the initial pH thereof, adding base thereto while stirring and continuing such addition while noting the pH of the resulting mixture. Such addition can be stopped anytime. In the preferred embodiment wherein a large portion or substantially all of the carboxyl groups are desirably reacted with the base, addition of base is continued until a stable pH reading is obtained.

The reactions can be varied over a wide range, for example, using a temperature of about 5° to 150°C, preferably about 15° to about 90°C, and a pressure of about atmospheric to 75 psig (about atmospheric to about 0.5 MPa), preferably about atmospheric (about 0.1 MPa). The resulting product can then be subjected, for example, to a temperature of about 20° to about 200°C under vacuum to about 100 psi gauge (under vacuum to about 0.69 MPa) for the removal of water therefrom. However, if desired the water need not be removed from the total reaction product, and the total reaction product, or after removal of a portion of the water therefrom, can be used to prepare a drilling fluid as taught herein.

These drilling fluids are easily prepared. Thus they can be admixed with water using any suitable means, for example stirring, until the final drilling fluid contains from about 1 to about 25 wt %, preferably about 5 to about 15 wt % of the reaction product defined above. Although the drilling fluid can be prepared at any suitable temperature and pressure, for example, a temperature of about 5° to 90°C and atmospheric pressure over a period of about 12 to 24 hours, it is preferred that they be prepared at ambient temperature and ambient pressure. If desired, additives that are often incorporated into drilling fluids for a particular purpose, for example, electrolytes, such as potassium chloride or sodium chloride to enhance the inhibitive property of the drilling fluid toward clay cuttings (for example, about 1 to 30 wt % of the drilling fluid), weighting agents, such as barite (for example, amounts sufficient to obtain a density in the drilling fluid up to about 2.16 g/cm^3), oil lubricants (for example, about 3 to 20 vol % of the drilling fluid), pH adjustment compounds, such as sodium hydroxide, potassium hydroxide or calcium hydroxide, can also be added to the drilling fluid at the time of its preparation or at any time thereafter.

Aluminum Crosslinking Agent for Polysaccharide Derivatives

C.H. Kucera and D.N. DeMott; U.S. Patent 4,257,903; March 24, 1981; assigned to The Dow Chemical Company describe a drilling fluid comprising an aqueous solution of a hydroxyalkyl polysaccharide derivative crosslinked with a water-soluble ionic aluminum compound. A most preferred ionic aluminum crosslinking agent is sodium aluminate. Most preferably, the unweighted drilling fluid comprises from about 0.5 to 1.5 lb polymer per barrel (1.43 to 4.28 kg/m^3), and from about 1.5 to 3.5 lb crosslinker per barrel (4.28 to 9.99 kg/m^3). The subject drilling fluid is made by admixing the hydroxyalkyl polysaccharide derivative with water, circulating the admixture thus formed for a period effective to permit hydration of the polymer, and thereafter adding the ionic aluminum crosslinking agent.

The most preferred hydroxypropyl polysaccharide derivative, hydroxypropyl guar, has the following generalized molecular structure:

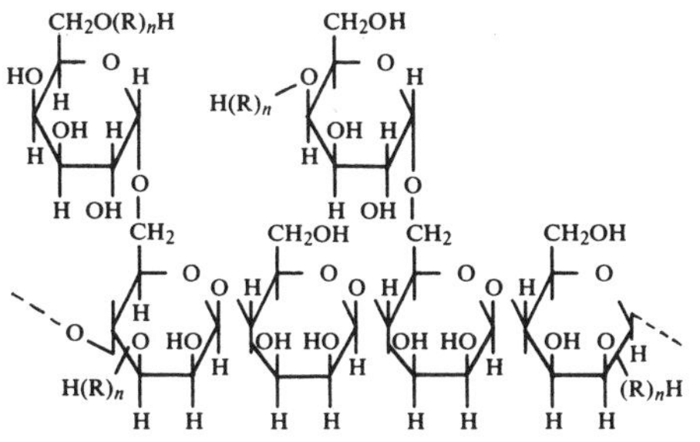

wherein R = $CH_2CH(CH_3)O$, although it is understood that both the degree of substitution and the position of the derivative groups may vary from that shown above. The molecular substitution of the hydroxypropyl polysaccharide derivative will preferably range from about 0.35 to 0.50%. A satisfactory hydroxypropyl polysaccharide derivative is a hydroxypropyl guar gum derivative (galactomannan ether) known as Jaguar HP-8 (Stein, Hall and Co. Inc.). Jaguar HP-8 disperses readily into water, hydrates slowly to form a solution, and will eventually gel if left at its inherent alkaline pH. However, if an acid diluent is added immediately after initial dispersion to reduce the pH to neutral or slightly acidic conditions, the product will hydrate in about 15 minutes to a high viscosity solution having a smooth appearance. Jaguar HP-11, a hydroxypropyl guar gum derivative with the same generalized molecular structure as HP-8, but with an added buffer salt, is capable of achieving complete hydration within about eight minutes, and can also be used.

According to a preferred embodiment of the process the density of the subject drilling fluid ranges from about 8.34 to about 20 lb/gal (1.0 to 2.4 g/ml), or greater, depending upon whether or not an additional weighting agent such as barite is combined therewith. The Fann viscosity of the drilling fluid preferably ranges from about 5 to 60 cp at 300 rpm at ambient temperature, and the pH preferably ranges from about 8 to 11, depending upon the formation, formation water pH, optimum crosslinking pH for the particular combination of polymer and crosslinker employed, and the like. Furthermore, the preferred drilling fluid remains shear stable at static temperatures ranging up to about 180°F (82°C) and circulating temperatures ranging up to about 250°F (121°C).

Through the manufacture and use of the drilling fluid disclosed herein, those working in the drilling industry are provided with a drilling fluid which significantly reduces pressure loss in the drill string, thereby increasing the bit penetration rate; which improves lubricity, reduces regrinding of cuttings, and significantly prolongs bit life; and which exhibits unexpected improvements in lubricating and stabilizing the borehole.

Shale-Stabilizing Drilling Fluid

D.B. Anderson; U.S. Patent 4,142,595; March 6, 1979; assigned to Standard Oil Company (Indiana) describe non-clay-based aqueous drilling fluid and the use thereof in a shale environment. Typically, the fluid contains 0.2 to 1.5 lb/bbl of flaxseed gum and at least one salt having a cation selected from the group consisting of potassium and ammonium, with the cation concentration of potassium and/or ammonium being at least 10,000 ppm. A nonclay viscosifier is also used.

Preferably, the cation concentration of the salt additive in parts per million is at least 10,000 x (1 + 0.025 x the percent of saturation of sodium chloride in the drilling fluid). For seawater, the cation concentration is desirably between 20,000 and 30,000 ppm. Preferably gilsonite is also added to the drilling fluids when formation evaluation considerations do not make the use of gilsonite impractical. The use of this combination provides long-lasting stabilization of the shale and the drilling fluid can be used (in contact with the shale formation) to drill formations beneath the shale without additions of further shale-stabilizing agents.

The flaxseed gum is a component of flax meal and while the flaxseed gum can be extracted and used in the amounts specified herein, it is generally convenient to add the flaxseed gum to the drilling fluid in the form of flax meal. As only approximately 10% by wt of the flax meal is flaxseed gum, a correspondingly greater amount of the meal should be added. As used herein, the flaxseed gum (or flax meal) is not a viscosifier. The desired viscosity is to be obtained by adding nonclay viscosifiers, such as hydroxyethylcellulose, carboxymethylcellulose or hydroxyethylcellulose plus biopolymer. The salts used herein tend to precipitate clays and, thus, clay viscosifiers are not appropriate.

Generally, the drilling fluid can be prepared by dissolving about 20 lb/bbl of potassium chloride (about 29,470 ppm of potassium ion) in seawater and adding about 10 lb/bbl of flax meal. The non-clay-based viscosifier (such as carboxymethylcellulose, hydroxyethylcellulose, or hydroxyethylcellulose plus biopolymer) is then added to obtain a desired viscosity, as known in the art. Any of these materials may be used as viscosifiers, typically, in a concentration of about 1 lb/bbl. The preferred viscosifier is hydroxyethylcellulose plus a biopolymer which is a heteropolysaccharide produced by the bacterium *Xanthomonas campestris* NRRL B-1459. Such hydroxyethylcellulose/polysaccharide viscosifiers are well-known in the drilling art, e.g., see U.S. Patent 3,953,336.

Preferably, the amount of potassium chloride is varied depending on the salinity of the water. Testing has shown that in a seawater drilling fluid approximately 2½ times the amount of potassium chloride is required, as is necessary for fresh-water fluids. In saturated salt water, the ratio should be increased to 3½ times.

Further, it has been discovered that there is a relationship between the MBT value of the shale being drilled and the amount of potassium chloride required to maintain the necessary concentration. The MBT values are the Methylene Blue Test results as determined by the API Standard Procedure for Testing Drilling Fluids using 100 lb of shale per barrel of water, and are a measure of the cation capacity of the shale. For every 100 lb of shale drilled, the amount of potassium chloride (in pounds) which should be added is equal to from about one-tenth to one-twentieth the MBT value of the shale.

Generally, of course, a well is drilled using conventional (nonstabilizing) drilling fluids until a shale formation is encountered. Samples of the shale should then be taken to be tested to determine its MBT value. The shale-stabilizing drilling mud is then prepared using the composition as abovedescribed, also including about 1 to 5 lb/bbl of gilsonite if formation evaluation considerations permit. As the drilling progresses, additional potassium chloride is added. The weight of potassium chloride to be added is preferably equal to the MBT value of the shale times the weight of shale drilled divided by 1,500. Thus, if the MBT value of the shale is 15 and the shale is being drilled at a rate of 100 lb/hr, KCl should be added at a rate of 1 lb/hr. While the KCl can be added periodically, it is preferably added continuously. Flax meal is preferably also added generally at a rate of about 1 to 3 lb/ft of hole drilled.

Once the shale formation has been penetrated, drilling could be done using the conventional nonstabilizing aqueous drilling fluid, but it is generally more convenient to continue drilling with the same fluid and just stop the KCl and flax meal additions. An even more conservative approach is to continue the additions but at about one-fifth the rate.

OIL-BASED DRILLING FLUIDS

Asbestos Composition Having Organo-Silane Coating

S. Ramachandran; U.S. Patent 4,183,814; January 15, 1980; assigned to Union Carbide Corporation describes a chrysotile asbestos base material which can be used to enhance the rheological properties of fluids such as oil drilling muds.

An asbestos base material in accordance with the process comprises opened chrysotile asbestos having a precipitation deposited siliceous layer and an organosilane coating bonded to and overlying the siliceous layer.

The siliceous layer on the chrysotile asbestos can be provided by the method of U.S. Patent 3,471,438. In a particular embodiment of the method of this patent opened particulate chrysotile asbestos is provided, such as that available as High Purity Grade Asbestos from Union Carbide Corporation. This asbestos is slurried with water, the slurry conveniently containing from about 0.5 to 4% by wt asbestos and more suitably about 1 to 2% asbestos by weight. A predetermined amount of concentrated sodium silicate solution, suitably an amount which provides about 12 parts by weight of SiO_2/100 parts of asbestos, is added to the slurry followed by neutralization to provide in the slurry a pH of about 9.5 or less. Acetic acid is preferably used as the neutralizing agent. Under these circumstances, i.e., a pH of 9.5 or less, a siliceous gel is precipitated and this gel adheres to and coats the slurried asbestos and provides a siliceous layer thereon.

Opened particulate asbestos refers to particulate asbestos in which the naturally-occurring fiber bundles have been separated into their ultimate individual fibers to the extent that most of the constituent particles are in the form of individual fibrils.

Following the precipitation coating of the asbestos as abovedescribed, the slurry is adjusted to the extent necessary to provide a pH in the range of about 6 to 9.0, preferably 7 to 8, and an organosilane is added to the slurry in an amount of about 0.5 to 10% by wt of the asbestos base material in the slurry. The thus treated solids are recovered and dried by conventional techniques. The resulting asbestos material has a coating of organosilane overlying the siliceous layer of the asbestos base material.

The properties of the organosilane coated material are influenced by the particular organosilane employed. For example, octyl thiethoxy silane provides an oleophilic coating which has a positive interaction with oil base fluids, such as drilling muds, and improves the rheological properties of these fluids. The organosilanes used in the process are characterized by one of the two following structures:

$$G-\overset{\displaystyle R}{\underset{\displaystyle R'}{\overset{|}{\underset{|}{Si}}}}-Y \qquad (1)$$

where G is a hydroxyl group or a group hydroxyzable to hydroxyl such as, for example, alkoxy or halogen; Y is an alkyl group containing from 1 to 20 carbon atoms, a phenyl group, an alkyl substituted phenyl group where the alkyl groups can contain a total of from 1 to 12 carbon atoms or a polyoxyalkylene radical having up to 25 carbon atoms bonded to the silicon atom by a silicon to carbon bond, R and R' are selected from the group described by G and Y or hydrogen; or:

(2)

$$G{-}\overset{\displaystyle R}{\underset{\displaystyle R'}{\overset{|}{\underset{|}{Si}}}}{-}Z$$

where G is a hydroxyl group or a group hydroxyzable to a hydroxyl such as, for example, alkoxyl or halogen; Z is an alkyl group containing from 1 to 20 carbon atoms bearing a functional group such as, for example, amino, oxirane, mercapto or acryloxy; R and R' are selected from the groups described by G and Z, hydrogen, an alkyl group containing from 1 to 20 carbon atoms, phenyl, or alkyl substituted phenyl where the alkyl groups can contain a total of from 1 to 12 carbon atoms.

Example: Short fiber chrysotile asbestos from the Coalinga, California deposit, (Union Carbide Corporation, High Purity Grade) and having the following properties was added to water at 31°C in an amount of 1.4% by wt: specific surface area, 60 to 80 m^2/g; magnetite content, 0.04 to 0.5%; and reflectance, 72 to 78%.

Specific surface area is calculated from adsorption data using the BET (Brunauer, Emmet, Teller) method as described in Brunauer, *The Adsorption of Gases and Vapors,* Princeton University Press (1945).

Magnetite content is measured by permeametric device patterned after ASTM standard method D-1118-57. In order to obtain greater range and improved sensitivity, the ASTM method has been modified to detect the phase changes of the current generated when magnetic materials are placed in a transformer core rather than the voltage changes generated.

Reflectance is measured on a sample prepared according to TAPPI (Technical Association of the Pulp and Paper Industry) standard T-452-m-58 and reported as the percent of ultimate reflectance based on magnesium oxide as 100% reflectance.

The water-asbestos mixture (35 g of asbestos in 2.5 liters of water) was introduced into a large Waring blender (Model No. CB5) and the blender was run at its highest speed for about 3 minutes. Following this treatment, sodium silicate (1 M solution) in the amount of 70 ml was gradually added to the asbestos-water slurry while slowly stirring the slurry with a mechanical stirrer. This provided in the slurry the equivalent of about 4.2 g of SiO_2. This mixture was then treated by slowly adding acetic acid (1 M) to neutralize the solution and obtain a pH of about 9.5. At this pH, precipitation of siliceous material occurred which was substantially all adsorbed by the slurried asbestos particles and such particles when added to water exhibit a negative charge as can be demonstrated by standard electrophoresis techniques. On the other hand, untreated asbestos exhibits a strong positive charge under the same circumstances.

After this step, additional acetic acid (1 M) was added to bring the pH to about 7.6 and the slurry was slowly stirred by a mechanical mixer for about 3 minutes. The total amount of acetic acid used was about 40 ml. 2.1 g of octyl triethoxy silane was added to the slurry which was mixed further for about 5 minutes.

The solids were removed by filtration and dried at about 110°C for about 3 hr. The resulting product had an oleophilic organosilane coating overlying and chemically bonded to the siliceous layer on the opened chrysotile asbestos.

A particular embodiment of the process is the use of the material as an additive to conventional and well known drilling fluids used as drilling muds in oil and gas well drilling operations. In connection with this embodiment material prepared as above was opened in a Waring blender (Model 91-264) at high speed for about 30 seconds. It was then employed, in the proportions shown in the table below as an additive in a standard oil base fluid (drilling mud) having the following composition: 332.5 ml of No. 2 diesel oil and 17.5 ml of water.

The actual amounts of the additions were 2, 4 and 7 g.

The testing procedure for viscosity evaluation of the drilling mud was as follows: The oil and water were mixed at high speed in a Waring blender (Model No. 91-264) for 2 minutes prior to the addition of the prepared asbestos base material. Following the addition, stirring was continued in the blender, also at high speed, for 10 minutes, after which time, the sample was removed to a Fann viscometer (Model No. 35A), cooled to 115°F (46°C) and the viscosity and gel strengths determined using standard procedures as described in *American Petroleum Institute Publication No. API RP 13B,* fourth edition, November 1972. The results are shown below.

Effect of Asbestos Base Material on Mud Properties

		Fann Viscosity Results			
No.	Additive Loading (lb/bbl)	Apparent Viscosity (cp)	Plastic Viscosity (cp)	Yield Point (lb/100 ft²)	Gel Strength Initial (lb/100 ft²)
1	2	8	4	8	3
2	4	17.5	6	23	9
3	7	70.0	16	108	39

WORKOVER AND COMPLETION FLUIDS

Sized Salt Bridging Material

Various types of work-over and completion fluids with bridging agents are available and in use. Some are termed oil-soluble and others are termed acid-soluble, depending upon whether or not the bridging agents in the completion and work-over fluids are soluble in oil or in acid. An important function of a completion and work-over fluid is to seal off or temporarily plug the face of the producing formation in the well bore so that during the completion and work-over operations fluid and solids in the fluid are not lost to the producing formation. Theoretically and ideally it is desired to accomplish this by depositing a thin film of solids over the surface of the producing formation without any loss of solids to the formation. Solids which are coated or deposited on the formation are generally termed bridging agents and temporarily bridge over the formation pores rather than permanently plugging the pores. The coating of bridging agent is then dissolved when the work-over or completion operation is completed so that oil or gas may then be produced through the formation and into the well bore.

A nondamaging work-over and completion fluid is so termed in that it causes a minimum of permanent plugging of the formation pores by loss of solids, or particles in the completion and work-over fluid to the producing formation.

T.C. Mondshine; U.S. Patents 4,186,803; Feb. 5, 1980; and 4,175,042; Nov. 20, 1979; both assigned to Texas Brine Corporation describes a substantially nondamaging work-over and completion fluid including saturated brine solution having water-soluble salts which are insoluble therein to provide water-soluble bridging agents with the water-soluble salts having particle size range from 5 to 800 μ in size, wherein greater than about 5% of the particles are coarser than 44 μ, and in the range of approximately 4 to 50 lb/bbl of saturated brine solution along with an additive for viscosity and suspension in the range of about 0.2 to about 5 lb/bbl of saturated brine solution. For additional fluid loss reduction a fluid loss additive in the range of about 0.2 to 10 lb/bbl of saturated brine solution may be added.

The saturated brine solution is formed by dissolving a salt or mixture of salts in water and normally the minimum density of the saturated brine solution is approximately 10 lb/gal. In those situations where it is desirable to employ the process with a density less than 10 lb/gal, the saturated brine solution can be diluted with some suitable substance such as diesel oil. In addition, the density of the saturated brine solution can be increased by the addition of iron carbonate or barites to provide a completion and work-over fluid with bridging agents having a density of approximately 19 lb/gal when desired. Additional quantities of sized salt particles may be added to increase the density as may be desired.

Some salts that are generally available and which may be used include potassium chloride, sodium chloride, calcium chloride, sodium sulfate, sodium carbonate, sodium bicarbonate, calcium bromide and potassium carbonate. When the process is employed in well bores which have increased temperatures, the sized salt which is employed as the bridging agent in the completion fluid is added in a sufficient quantity so that even though some of it may dissolve at higher temperatures, the amount dissolved will not materially affect the action of the sized salt particles suspended in the saturated brine solution in functioning as a water-soluble bridging agent for temporarily plugging the producing formation pores during the completion and work-over procedure.

The composition and properties of various work-over and completion fluids using specially sized salt bridging material are given in the table below.

	Example Number							
	1	2	3	4	5	6	7	8
Composition, lb/bbl saturated brine								
Hydroxyethylcellulose*	2	2	–	–	–	–	–	–
Salt**	19	19	19	19	19	19	19	31
Guar gum	–	–	–	–	2	2	–	–
Calcium lignosulfonate	–	–	–	6	–	–	–	7
Xanthum gum	–	–	1	1	–	–	–	1.5
Pregelatinized starch	–	–	–	–	–	6	–	–
Carboxymethylcellulose (CMC-9)	–	–	–	–	–	–	2	–
Carboxymethylcellulose (CMC-7)	–	4	–	–	–	–	–	–

(continued)

	Example Number							
	1	2	3	4	5	6	7	8
Properties								
Density, lb/gal	10.2	10.2	10.2	10.2	10.2	10.2	10.2	10.2
Plastic viscosity, cp	33	80	12	27	14	25	22	12
Yield point, lb/100 ft^2	58	70	11	31	23	32	18	16
API filtrate, ml	29.2	12.0	14.0	9.4	20.0	8.0	9.0	5.8
Seal on 1,000 millidarcy sand bed at 100 psi	Partial	Good	Good	Good	Partial	Partial	Good	Good

*Cellosize QP 100 MH.
**Ranging from 5 to 100 μ, averaging 15 to 30 μ with about 5% coarser than 44 μ.

Low Fluid Loss Composition

D.S. Pye, J.P. Gallus, and P.W. Fischer; U.S. Patent 4,192,753; March 11, 1980; assigned to Union Oil Company of California describe an aqueous well drilling, completion or work-over fluid comprising a mixture of:

(1) about 8 to 60 lb/bbl of a microemulsion comprising:

- (a) about 45 to 85 parts by weight of water;
- (b) about 15 to 55 parts by weight of solids comprising
  - (i) about 12 to 30% by wt of an oil-soluble surface active agent;
  - (ii) about 20 to 38% by wt of a water-dispersible surface active agent;
  - (iii) about 8 to 30% by wt of wax;
  - (iv) about 20 to 50 % by wt of a relatively low softening point resin having a softening point of about 90° to 134°C, selected from the group consisting of rosin acids, rosin esters, coumarone-indene resins, petroleum resins, polymers derived from one or more terpenes and condensation products of aromatic hydrocarbons with formaldehyde,

(2) about 270 to 400 lb/bbl of water;

(3) about 14 to 91 lb/bbl of a water-soluble inorganic salt; and

(4) about 1 to 8 lb/bbl of a particulate relatively high softening point resin having a softening point of 135°C or above, selected from the group as described in (1-iv) above.

The water-soluble inorganic salt component of the composition both prevents or inhibits hydration and swelling of the water-sensitive components of the reservoir and increases the density of the fluid composition. The fluid contains at least about 1% by wt dissolved salts to inhibit hydration of the water-sensitive clays contained in the reservoir. Also, the density (mud weight) can be increased by the addition of these dissolved salts, the maximum density being limited by the solubility of the salts. The concentration of salts must be maintained below the saturation concentration at the temperatures to which the fluid will be subjected to prevent salting out, i.e., solids precipitation. While the maximum mud weight is dependent upon the particular salt or salts dissolved in the fluid and the temperature, mud weight can generally be increased from the minimum of about 63 to 85 lb/ft^3 by the control of salt concentration.

Inorganic salts which can be dissolved in the fluid include alkali metal and ammonium halides and nitrates. The preferred salts include sodium, potassium, and ammonium halides, and particularly sodium chloride, potassium chloride, and ammonium chloride. Either a single salt may be dissolved in the fluid, or a mixture of salts can be employed. The preferred concentration ranges for various salts are as follows in percent by weight: 1 to 10% sodium chloride, 1 to 20% ammonium chloride, 1 to 22% potassium chloride, 1 to 20% sodium nitrate, 1 to 40% ammonium nitrate, and 1 to 30% potassium nitrate.

The solids component of the microemulsion component of the completion fluid preferably contains the following ingredients in percent by weight: 16 to 22% oil-soluble surfactant, 24 to 33% water-dispersible surfactant, 10 to 25% wax, 25 to 40% low softening point resin, 1 to 3% agent to control phase separation (optional) and 0.05 to 0.2% bactericide (optional).

The oil-soluble surface active agent employed in the microemulsion is a glyceryl or sorbitan higher fatty acid partial ester, exemplary of which are the glyceryl and sorbitan mono- and diesters of saturated fatty acids containing between 12 and 20 carbon atoms. Specific esters that can be employed in these compositions include glyceryl monolaurate, glyceryl monomyristate, glyceryl monopalmitate, glyceryl monostearate, glyceryl 1,3-distearate, sorbitan monolaurate, sorbitan monomyristate, sorbitan monopalmitate, sorbitan monostearate, and sorbitan 1,3-distearate. Preferred oil-soluble surface active agents are glyceryl and sorbitan monostearate.

The water-dispersible surface active agent is a water-dispersible polyethylene glycol higher fatty acid ester. Exemplary polyethylene glycol higher fatty acid esters include esters of polyethylene glycols having molecular weights of about 300 to 10,000 and saturated fatty acids containing between 12 and 20 carbon atoms. Specific polyethylene glycol fatty acid esters include polyethylene glycol monolaurate, polyethylene glycol monomyristate, polyethylene glycol monopalmitate, and polyethylene glycol monostearate esters of polyethylene glycols having molecular weights of about 300 to 10,000.

Preferred surface active compounds include polyethylene glycol 300 monolaurate, polyethylene glycol 900 monostearate, polyethylene glycol 1000 monostearate, polyethylene glycol 4000 monostearate, and polyoxyethylene glycol 6000 monostearate. The designations 300, 900,1000, 4000, and 6000 indicate the approximate molecular weight of the polyethylene glycol employed to form the ester. A particularly preferred water-dispersible surface active agent is polyethylene glycol 6000 monostearate.

Any of a wide variety of waxes can be used in the microemulsion. Suitable waxes include crystalline and microcrystalline petroleum waxes, beeswax, carnauba wax, candellia wax, montan wax, and the like. One preferred class of waxes includes the fully and partially refined paraffin waxes melting between about 52° and 77°C. Another class of waxes and wax-like substances that can be employed, particularly in higher temperature applications, are long chain aliphatic hydrocarbon and oxidized hydrocarbon waxes melting above about 77°C, such as the synthetic Fischer-Tropsch waxes. These waxes are characteristically straight or branched chain aliphatic hydrocarbons and oxygenated aliphatic hydrocarbons such as aliphatic carboxylic acids, esters and amides having molecular weights higher than the paraffin waxes, and particularly having molecular weights of about 500 to 2,500. A preferred class of synthetic waxes are long chain aliphatic hydrocarbons and oxidized hydrocarbon waxes melting between about 90° and 121°C.

The microemulsion component of the composition can be prepared readily by melting in proper proportions the individual components of the oil-soluble surface active agent, the water-dispersible surface active agent, at least a portion of the second low softening point resin, and, if employed, the water-dispersible thickening agent, the additive to prevent phase separation and/or the bactericide. Alternatively these ingredients can be combined in the proper proportions and then melted and agitated to form a homogeneous mixture. The melted mixture is then stirred into hot water of approximately the same temperature as the melted mixture and the resulting composition allowed to slowly cool down while agitation continues until the solids reform to produce an oil-in-water microemulsion.

The well completion and work-over fluid is then prepared by mixing together in any order desired, the microemulsion, water, the water-insoluble inorganic salt, the first high softening point resin, about 1.5 to 10 lb/bbl of a water-dispersible thickening agent, if used, and about 1 to 20 lb/bbl of a pH control agent, if used. It is preferred that the microemulsion be stirred into the water followed by the first resin, the water-dispersible thickening agent, the water-insoluble inorganic salt and the pH control agent.

The compositions of this process are useful in treating permeable subterranean reservoirs. In its broadest application, the process comprises contacting a subterranean reservoir penetrated by a well with the aforesaid composition injected through the well. This treatment can comprise a single temporary and selective plugging step, or it can be an integral part of a comprehensive fracturing, well drilling, acidizing or solvent-treating process. Also the fluid composition of this process can be effective used as a low fluid loss drilling fluid employed in the drilling of oil and gas wells and as a work-over fluid employed in recompleting oil and gas wells. In the drilling and work-over applications, the drilling fluid is circulated from the surface to a drilling zone in a reservoir during a rotary drilling operation, and at least a portion of the fluid is returned to the surface.

The compositions can be used as a fracturing fluid employed in hydraulically fracturing the reservoir surrounding a well, wherein the fracturing fluid is injected through the well and into contact with the reservoir at a pressure and volume flow rate sufficient to fracture the reservoir. The composition also can be employed in chemically treating, acidizing and other well treating operations wherein it is desired to control fluid loss to permeable underground structures.

Examples 1 through 4: The microemulsion component of a slowly oil-soluble, water-insoluble low fluid loss fluid well composition is prepared by first melting and combining 18.3% by wt sorbitan monostearate, 27.4% by wt polyethylene glycol monostearate, 18.3% by wt paraffin wax having a melting point of 60°C, 34% by wt Pexalyn 600 hydrocarbon resin (Hercules Inc.) and having a softening point of 104°C, 1.9% by wt lauryl alcohol and 0.1% by wt paraformaldehyde. Next 28 parts by weight of the melted composition is added to 72 parts by weight water heated to about 66°C. The mixture is stirred and allowed to cool slowly to room temperature. Upon cooling, the solids reform and a microemulsion forms. The fluid well composition is prepared by mixing together 24.5 lb/bbl of the abovedescribed microemulsion, 327 lb/bbl water, 21 lb/bbl of potassium chloride, 2.5 lb/bbl of Picco 6140 petroleum hydrocarbon resin (Hercules Inc.), having a softening point of 140°C and having various particle size distributions or substitutes therefor (see table below), 1 lb/bbl of Natrosol 250 HR hydroxyethylcellulose (Hercules, Inc.), 1 lb/bbl of Kelzan XC-AL biopolymer (Kelco Company), a xanthan gum, and 2 lb/bbl of potassium hydroxide. The static fluid loss and return permeability using this composition is determined as follows:

A 1" diameter, 1" long Berea sandstone core is sealed in a 1" diameter brass tubing jacket with the core end faces remaining unsealed. The core is placed under vacuum to remove the air therefrom, soaked in a 3% by wt aqueous solution of sodium chloride for 24 hours and stored under kerosine. The flow rate of kerosine through the core at 38°C is determined. The core is placed in a static fluid loss apparatus, 150 ml of the composition poured into the apparatus, the temperature of the system raised to 38°C, the fluid forced against the core by applying 500 psig nitrogen pressure to the system, and the fluid loss through the core determined at various time intervals up to 30 minutes. The core is then removed from the apparatus and the flow rate of kerosine through the core at 38°C determined in the direction opposite that in which the fluid was forced against the core in the fluid loss test. The return permeability (percent) is 100 times the permeability to kerosine after the fluid loss test divided by the permeability to kerosine before the fluid loss test. The results of these tests are reported in the following table:

Fluid Loss Control Properties and Return Permeability of Various Completion Fluids

Ex. No.	Low SP Resin Component or Equivalent Material	Particle Size (%) 5 μ	5-30 μ	30 μ	Static Fluid Loss* (ml)	Return Permeability (%)
1	XP-20 chrome lignite	3	17	80	48	44
2	Picco 6140 resin	1	14	85	48	84
3	Picco 6140 resin	5	35	60	12	90
4	Picco 6140 resin	75	25	–	34	90

*In 30 minutes.

The results of Example 1 show that when using XP-20 chrome lignite, a known non-oil-soluble fluid loss control additive of the prior art in the well completion fluid, adequate control of fluid loss is achieved but the low return permeability indicates considerable plugging of the core by the completion fluid. The results of Examples 2 through 4 show that, when using the low softening point resin component of this process as a fluid loss control additive in the well completion fluid, adequate control of fluid loss is achieved and the high return permeability indicates little undesirable plugging of the core by the completion fluid.

Foam Drilling and Work-Over Composition

D.S. Pye and P.W. Fischer; U.S. Patent 4,201,678; May 6, 1980; assigned to

Union Oil Company of California describe a method for conducting foam drilling and work-over operations in subterranean reservoirs which employs a foam having stability at high temperatures. The foam is formed by contacting a gas at elevated pressure with an aqueous foaming solution containing about 0.08 to 1.0% by wt of a first foaming agent, an amphoteric betaine having the formula:

$$\begin{array}{c} R_2 \\ | \\ R_1-\overset{+}{N}-R_4X^- \\ | \\ R_3 \end{array}$$

wherein R_1 is a high molecular weight alkyl radical having from 10 to 18 carbon atoms or the amide radical, $RCONH(CH_2)_3$, wherein R is a higher alkyl radical having from 10 to 18 carbon atoms, R_2 and R_3 are each alkyl radicals having from about 1 to 3 carbon atoms, R_4 is an alkylene or hydroxyalkylene radical having from 1 to 4 carbon atoms, and X is an anion selected from the group consisting of SO_3 and COO radicals; about 0.02 to 0.4% by wt of a second foaming agent, a salt of a linear aliphatic or alkyl aryl hydrocarbon sulfonate having the formula:

$$R_5-S(=O)(=O)-OM$$

wherein R_5 is an oleophilic group having from 10 to 18 carbon atoms, and M is an alkali metal or ammonium cation; and optionally, from 0.03 to 3.5% by wt of unneutralized ammonia. While the method is useful in low-temperature subterranean reservoirs, it has particular utility in high-temperature reservoirs, such as those having a temperature of above 200°F, and particularly above 400°F. The method is particularly suited for use in geothermal reservoirs.

Examples of amphoteric betaines useful herein include the high alkyl betaines such as cocodimethylcarboxymethylbetaine, lauryldimethylcarboxymethylbetaine, lauryldimethyl-α-carboxyethylbetaine, cetyldimethylcarboxymethylbetaine, laurylbis-(2-hydroxyethyl)-carboxymethylbetaine, oleyl dimethyl-γ-carboxypropylbetaine, laurylbis-(2-hydroxylpropyl)-α-carboxyethylbetaine, ammonium salts of the foregoing and the like. Specific sulfobetaines include cocodimethylsulfopropylbetaine, stearyldimethylsulfopropylbetaine, laurylbis-(2-hydroxyethyl)-sulfopropylbetaine and the like, amidobetaines and amidosulfobetaines wherein the $RCONH(CH_2)_3$ radical is attached to the nitrogen atom of the betaine.

Preferred classes of the second foaming agents include the alkylbenzene sulfonates, the paraffin sulfonates, the α-olefin sulfonates and the internal olefin sulfonates. Of these foaming agents, the α-olefin sulfonates are particularly preferred.

The foaming agent solution and gas can be injected at an elevated pressure through the drill pipe penetrating the subterranean reservoir and the foam generated by contact of the foaming agent solution and the gas caused to travel down the drill pipe and then up the borehole annulus so that the foam carries the drill cuttings, liquids and other debris from the bottom of the borehole to the surface of the earth. In a modification of this conventional mode, the foam can be preformed at the surface before injection into the borehole. Alternatively, the two fluids can be injected simultaneously, but separately, down separate conduits and allowed to mix at the bottom of the hole. For example, in the drilling of a borehole, the gas can be injected down a separate central pipe within the drill pipe, and the foaming agent solution can be injected down the annulus between the central pipe and the drill pipe. In some cases the gas can be injected down the annulus between the central pipe and the drill pipe, while the foaming agent solution is injected down the central pipe. The fluid injected down the central pipe will emerge from the drill pipe via ports in the drill bit. The fluid injected down the annulus will exit the drill pipe through perforations near the drill bit. It may be desired in conducting a foam drilling operation to inject the gas down the central pipe and the foaming agent solution down the annulus. This method reduces the loss of lubrication in the drill bit as a result of the detergent action of the foaming agent solution and also prevents the corrosion of the drill bit as a result of the direct contact of the foaming agent solution on the lubricant-free drill bit surfaces.

When the foam is used in high-temperature reservoirs, sometimes it is desired to maintain sufficient pressure in the borehole to prevent the liquid portion of the foam from flashing. For this reason, the top of the borehole may be sealed so that the borehole annulus does not directly communicate with the atmosphere; and the conduit at the surface for discharging the foam may be equipped with a valve for supplying the necessary back pressure to the borehole.

Phenol-Aldehyde Resin and Polyalkylene Oxide Polymer

B.L. Swanson; U.S. Patent 4,212,747; July 15, 1980; assigned to Phillips Petroleum describes shear-thickening fluid compositions prepared from a high molecular weight polyalkylene oxide polymer and a synthetic resin of an aldehyde and a phenol which composition is prepared in an alkaline environment. The shear-thickening aqueous compositions are useful as work-over fluids in producing or injection wells, as water diversion agents and mobility control fluids, and as hydraulic fracturing fluids.

The synthetic resins that can be used can be prepared in accordance with known methods and generally comprise the reaction of a suitable aldehyde with a suitable phenol compound in the presence of an alkaline catalyst. The molar ratio of aldehyde to phenol is generally in the range of 0.1:1 to 2:1 with a preferred range of about 0.3:1 to 1:1. The pH maintained during the preparation of the synthetic resin is in the range of about 7 to 12, preferably in the range of about 9 to 11.

Suitable aldehydes that can be employed include those having from 1 to 4 carbon atoms such as formaldehyde, acetaldehyde, butyraldehyde, and the like. Suitable phenols include phenol, catechol, resorcinol, the cresols, and the like, with the proviso that catechols are operable only in brine-containing systems. Suitable alkaline catalysts for the phenol-aldehyde condensation are selected from the alkali metal and ammonium hydroxides and carbonates with sodium hydroxide and sodium carbonate being preferred because of cost and availability.

The shear-thickening composition can also be prepared by contacting the phenol, aldehyde, and polyalkylene oxide under suitable conditions in an aqueous reaction medium.

The synthetic resin-forming reaction mass is maintained in the broad temperature range of 100° to 212°F, preferably 120° to 180°F, for a time in the range of 5 minutes to 48 hours, preferably 10 minutes to 24 hours, and most preferably, 10 minutes to 1 hour.

Alternatively, the shear-thickening compositions can be prepared by reacting the phenolic component and aldehyde component in a thickened aqueous solution of the polyethylene oxide over a pH range of about 7 to 9. The ranges of component concentrations for such a procedure are shown below:

Components	Broad Range	Preferred Range
	.... (ppm).	....
Polyalkylene oxide	1,000-30,000	5,000-20,000
Phenolic compound	200-20,000	500-10,000
$NaHCO_3$ (optional)	0-5,000	1,000-3,000
HCHO (dry basis)	100-5,000	1,000-3,000

The term polyalkylene oxide polymer as used herein is meant to include high molecular weight polyethylene oxides, polypropylene oxides, and polybutylene oxides, etc. These polyalkylene oxides can be chemically modified to obtain improved results. The molecular weight of the polyalkylene oxide polymers is about 10,000 to 10,000,000 or more, and preferably about 600,000 to about 7,500,000, and more preferably about 4,000,000 to about 6,000,000. Lower or higher molecular weight polymers may be desired for particular purposes; but the above ranges should be suited for most applications. The preferred polyalkylene oxide is known as Polyox Coagulant and Polyox WSR 301 (Union Carbide Company).

The amount of synthetic phenol-aldehyde resin and polyalkylene oxide polymer used will vary somewhat but will be sufficient to provide a shear-thickening composition. In general, the amount of synthetic resin will range from 0.025 to 2.5 wt %, and the amount of polyalkylene oxide polymer will range from 0.1 to 3.0 wt %, with the balance being water.

Any suitable method can be employed for preparing the shear-thickening compositions. It is ordinarily preferred to first dissolve or disperse the polymer in water before contacting the polymer solution with the alkaline synthetic phenol-formaldehyde resin reaction mixture. Alternatively, the phenol and formaldehyde can be reacted to form the synthetic resin in an aqueous solution containing the polyalkylene oxide, the phenol, and the aldehyde.

Example: A shear-thickening composition was prepared in aqueous solution at a pH of 7.6 by combining the following components in the indicated quantities: polyethylene oxide (WSR 301), 15,000 ppm; resorcinol, 1,000 ppm; $NaHCO_3$, 2,000 ppm; and HCHO (as formalin), 2,000 ppm; and then aging this mixture for 24 hours at 125°F. The shear-thickening character of this composition was verified by testing in a Stormer viscometer. Such a composition is potentially useful as an oil well treatment fluid.

Gel from CMHEC and Dichromate Compound

C.A. Sauber; U.S. Patent 4,239,629; December 16, 1980; assigned to Phillips Petroleum Company describes a process relating to drilling work-over and completion fluids comprising a water-soluble carboxymethylhydroxyethyl mixed cellulose ether (CMHEC) having a specified carboxymethyl degree of substitution and a specified hydroxyethyl molar substitution and sodium dichromate dihydrate as a gelling agent.

According to the process, an increase in gel strength of CMHEC solutions is realized by adding only a minor amount of dichromate compound such as sodium dichromate dihydrate, $Na_2Cr_2O_7{\cdot}2H_2O$, at about 0.5 to 2 lb/bbl of solution.

Type 420H CMHEC is a preferred material and is used at about 2 lb/bbl in solution. The gel is formed with sodium dichromate dihydrate without use of a reducing agent. A pH of the fluid of about 5 or lower is necessary because if the pH is above 5, gelation occurs more slowly, if at all. Dichromate ion is needed and does not exist except at an acid pH.

The carboxymethyl DS (degree of substitution) of the CMHEC can be from about 0.1 to at least about 0.9 and the hydroxyethyl MS (molar substitution) can be from about 0.2 to at least about 2.7. High viscosity grade CMHEC is preferable because gelling occurs more rapidly when the solution viscosity is high. If solution viscosity is too low the gel appears as discrete gel particles throughout the fluid and gel strength does not develop readily.

Anticaking Agents for NaCl and KCl

In some instances sodium chloride or potassium chloride is packaged and stored for substantial periods of time, thus subjecting the potassium and sodium chloride to varying temperatures and humidities. In those instances where it is desired to use the sodium chloride or potassium chloride as an additive in well completion fluids, it is generally desirable that the sodium and potassium chloride be in the range of 20 mesh and finer.

When the salt is in this particle size range, it is subject to severe caking problems when exposed to varying temperature and humidity conditions so that it cannot be readily employed in the desired environment prior to further treatment.

T.C. Mondshine; U.S. Patent 4,192,756; March 11, 1980; assigned to Texas Brine Corporation has found that when one or more of the metal salts of lignosulfonate formed from the group consisting of alkali earth metals or other multivalent metals such as calcium, iron, chromium, ferrochrome, zinc, copper and magnesium is added to the sodium chloride or potassium chloride, the tendency to cake is substantially reduced, even over extended periods of storage and warehousing in cartons, bags or bulk. The metal salts of lignosulfonate may be added in a range of about 1 to 20% by wt to the sodium chloride or potassium chloride to substantially reduce, if not completely eliminate the problems attendant with caking of these substances under severe humidity and temperature conditions that may be encountered.

Specifically, the anticaking agent is a metallic sulfonated salt of lignin which is the binding agent in wood. Lignin is removed from wood by rendering it soluble by the process of sulfonating it with sulfurous acid and one of its soluble salts. This soluble fraction is then separated from the insoluble cellulose fibers and processed independently to produce the desired metallic salt.

Examples of commercially available lignosulfonates suitable for use include Toranil B (St. Regis Paper Company) and Lignosite A (Georgia-Pacific Corporation).

FRACTURING AND STIMULATION

ACIDIZING COMPOSITIONS

Polyvalent Metal Crosslinking Agent

Acid treating or acidizing of porous subterranean formations penetrated by a well bore has been widely employed for increasing the production of fluids, e.g., crude oil, natural gas, etc., from the formations. The usual technique of acidizing a formation comprises introducing a nonoxidizing acid into the well under sufficient pressure to force the acid out into the formation where it reacts with the acid-soluble components of the formation. The technique is not limited to formations of high acid solubility such as limestone, dolomite, etc. The technique is also applicable to other types of formations such as a sandstone containing streaks or striations of acid-soluble components such as the various carbonates.

During the acid treating operation, passageways for fluid flow are created in the formation, or existing passageways therein are enlarged, thus stimulating the production of fluids from the formation. This action of the acid on the formation is often called etching. Acid treating or acidizing operations wherein the acid is injected into the formation at a pressure or rate insufficient to create cracks or fractures in the formation is usually referred to as matrix-acidizing.

R.L. Clampitt and J.E. Hessert; U.S. Patent 4,172,041; October 23, 1979; assigned to Phillips Petroleum Company describe a gelled acidic composition, suitable for matrix-acidizing or fracture-acidizing of a subterranean formation. The composition comprises water; a water-thickening amount of a water-dispersible polymer of acrylamide; an amount of a water-soluble compound of a polyvalent metal wherein the metal present is capable of being reduced to a lower polyvalent valence state and which is sufficient to cause gelation of an aqueous dispersion of the components of the composition when the valence of at least a portion of the metal is reduced to the lower valence state; an amount of a water-soluble reducing agent which is effective to reduce at least a portion of the metal to the lower valence state and cause gelation; and an amount of a nonoxidizing acid which is capable of reacting with a significant amount of the acid-soluble components of the formation. The polymer, the polyvalent metal compound, the reducing agent, and the acid, in the amounts used, are sufficiently compatible with each other in an aqueous dispersion thereof to permit gelation and thus form a composition having sufficient stability to degeneration by the heat of the formation to permit good penetration of the composition into the formation and the maintenance of the composition in the formation in contact therewith for a period of time sufficient for the acid in the composition to significantly react with the acid-soluble components of the formation and stimulate the production of fluids therefrom.

Any suitable polymer of acrylamide meeting the above-stated compatibility requirements can be used. Thus, under proper conditions of use, such polymers can include various polyacrylamides and related polymers which are water-dispersible and which can be used in an aqueous medium, with the gelling agents described herein, to give an aqueous gel. These can include the various substantially linear homopolymers and copolymers of acrylamide and methacrylamide. By substantially linear it is meant that the polymers are substantially free of crosslinking between the polymer chains. The polymers can have up to about 75%, preferably up to about 45% of the carboxamide groups hydrolyzed to carboxyl groups. One preferred group of polymers includes those wherein from about 20 to 25% of the carboxamide groups are hydrolyzed.

Crosslinked polyacrylamides and crosslinked polymethacrylamides, at various stages of hydrolysis can also be used.

It is preferred that the polymer have a molecular weight of at least 500,000, more preferably at least about 2,000,000.

Generally speaking, amounts of the abovedescribed polymers in the range preferably from 0.1 to 1.5 wt %, based on the total weight of the composition, can be used in preparing gelled acidic compositions.

Metal compounds which can be used can include potassium permanganate, sodium permanganate, ammonium chromate, ammonium dichromate, the alkali metal chromates, the alkali metal dichromates, and chromium trioxide. Sodium dichromate and potassium dichromate, because of low cost and ready availability, are the preferred metal-containing compounds.

As a general guide, the amount of the starting polyvalent metal-containing compound used in preparing the gelled acidic compositions will be in the range of from 0.05 to 30 wt %, preferably 0.5 to 20 wt %, of the amount of the polymer used.

Suitable water-soluble reducing agents which can be used can include sulfur-containing compounds such as sodium sulfite, potassium sulfite, sodium hydrosulfite, potassium hydrosulfite, sodium metabisulfite, potassium metabisulfite, sodium bisulfite, potassium bisulfite, sodium sulfide, potassium sulfide, sodium thiosulfate, potassium thiosulfate, ferrous sulfate, thioacetamide, hydrogen sulfide, and others; and non-sulfur-containing compounds such as hydroquinone, ferrous chloride, p-hydrazinobenzoic acid, hydrazine phosphite, hydrazine dichloride, and others.

As a general guide, the amount of reducing agent used will generally be within the range of from 0.1 to at least 150 wt %, preferably at least about 200 wt %, of the stoichiometric amount required to reduce the metal in the starting polyvalent metal compound to the lower polyvalent valence state, e.g., +6 Cr to +3 Cr. In most instances, it will be preferred to use at least a stoichiometric amount.

Examples of useful acids can include inorganic acids such as hydrochloric acid and sulfuric acid; C_{1-3} organic acids such as formic acid, acetic acid, propionic acid, and mixtures thereof, and combinations of inorganic and organic acids. The concentration or strength of the acid can vary depending upon the type of acid, the type of formation being treated, the above-stated compatibility requirements, and the results desired in the particular treating operation. The concentration can vary from about 1 to 60 wt %, with concentrations within the range of 5 to 50 wt % usually preferred, based upon the total weight of the gelled acidic composition. When an inorganic acid such as hydrochloric acid is used it is presently preferred to use an amount which is sufficient to provide an amount of HCl within the range of from 1 to 12 wt %, preferably up to about 10 wt %, based on the total weight of the composition. The acids used in the process can contain any of the known corrosion inhibitors, deemulsifying agents, sequestering agents, surfactants, fraction reducers, etc., known in the art, and which meet the above-stated compatibility requirements.

J.E. Hessert and B.J. Bertus; U.S. Patent 4,146,486; March 27, 1979; assigned to Phillips Petroleum Company describe a similar process where the polyacrylamide is replaced by a water-dispersible biopolysaccharide produced by the action of bacteria of the genus *Xanthomonas* on a carbohydrate. Preferred species include *Xanthomonas begoniae*, *Xanthomonas campestris*, *Xanthomonas incanae*, and *Xanthomonas pisi*.

Polysaccharide B-1459 is an example of a biopolysaccharide produced by the action of *Xanthomonas campestris* bacteria, and which is commercially available in various grades as Kelzan (Kelco Company).

Generally speaking, amounts of the abovedescribed biopolysaccharides in the range of from 0.01 to 5 wt %, preferably 0.1 to 1.5 wt %, based on the total weight of the composition, can be used in preparing gelled acidic compositions. Only organic acids may be used in these compositions.

C.C. Johnston, Jr., B.J. Bertus and J.E. Hessert; U.S. Patent 4,169,797; October 2, 1979; assigned to Phillips Petroleum Company describe another similar process where the thickening agent is a water-soluble cellulose ether.

Cellulose ethers which can be used can include: the various carboxyalkyl cellulose ethers, e.g., carboxyethyl cellulose and carboxymethyl cellulose (CMC); mixed ethers such as carboxyalkyl hydroxyalkyl ethers, e.g., carboxymethyl hydroxyethyl cellulose (CMHEC); hydroxyalkyl celluloses such as hydroxyethyl cellulose, and hydroxypropyl cellulose; alkylhydroxyalkyl celluloses such as methylhydroxypropyl cellulose; alkyl celluloses such as methyl cellulose, ethyl cellulose, and propyl cellulose; alkylcarboxyalkyl celluloses such as ethylcarboxymethyl cellulose; alkylalkyl celluloses such as methylethyl cellulose; and hydroxyalkylalkyl celluloses such as hydroxypropylmethyl cellulose; and the like.

The amount of cellulose ether used in preparing the gelled acidic compositions can vary widely depending upon the viscosity grade and purity of the ether, and properties desired in the compositions. In general, the amount of cellulose ether used will be a water-thickening amount, i.e., at least an amount which will significantly thicken the water to which it is added. For example, amounts in the order of 25 to 100 parts per million weight (0.0025 to 0.01 weight percent) have been found to significantly thicken water.

Iodine-Containing Redox Couple System

B.L. Swanson and L.E. Roper; U.S. Patent 4,205,724; June 3, 1980; assigned to Phillips Petroleum Company describe gelled acidic compositions comprising water, a water-dispersible polymeric viscosifier selected from polyoxyalkylated cellulose ethers and acrylamide-derived cationic polymers, an acid such as aqueous HCl, and an iodine-containing redox couple comprising a polyvalent nonmetal-containing compound and a reducing agent.

A preferred group of copolymers for use in the process comprises the copolymers of acrylamide or methacrylamide with a monomer of the formula:

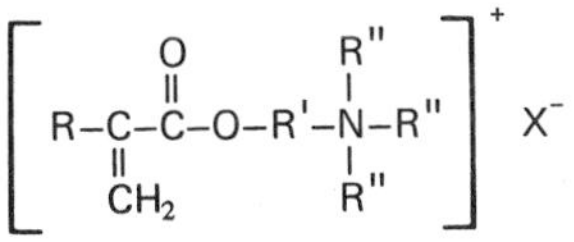

$$\left[R{-}\underset{\underset{CH_2}{\|}}{C}{-}\overset{\overset{O}{\|}}{C}{-}O{-}R'{-}\overset{\overset{R''}{|}}{\underset{\underset{R''}{|}}{N}}{-}R'' \right]^{+} X^{-}$$

wherein R is hydrogen or a lower alkyl radical containing from 1 to 6 carbon atoms, R preferably being hydrogen or a methyl radical; R' is an alkylene radical containing from 1 to 24 carbon atoms or an arylene radical containing from 6 to 10 carbon atoms, R' preferably being an alkylene radical containing from 2 to about 10 carbon atoms; each R" is an alkyl radical containing from 1 to 6 carbon atoms, preferably from 1 to 4 carbon atoms; X is any suitable anion such as methylsulfate, ethylsulfate, chloride, bromide, acetate, nitrate, and the like; and wherein the number of repeating units from the formula monomer is within the range of from 1 to 90, preferably 5 to 70, more preferably 10 to 60 mol %.

Monomers of the above formula and methods for their preparation are known in the art. For example, see U.S. Patent 3,573,263.

Suitable water-soluble iodine-containing redox couple systems that can be used include those containing iodine in two different valent states such as a mixture of water-soluble metal or ammonium iodide and iodate, e.g., KI and KIO_3. The iodine-containing materials can also be selected from compounds containing polyhalide anions such as tetra-n-butylammonium triiodide and the like prepared as described in *Inorg. Syn.*, 5, 166. Other water-soluble iodine-containing redox couples suitable for use include NO_3^-/I^-, O^{2-}/I^-, and the like.

The parameters suitable for use in the process are summarized below:

	Broad Range	Preferred Range
Weight ratio of reductant:oxidant in the redox couple ($I^-:IO_3^-$)	2:1-15:1	4:1-8:1
Weight percent acid in final composition (HCl)	0.5-15	2-5
Weight percent polymer in final composition	0.1-3	0.5-2

(continued)

	Broad Range	Preferred Range
Weight ratio of polymer:total redox couple	3.75:1-0.5:1	2.5:1-1.5:1

Propping agents can be included in the gelled acidic compositions, if desired.

Mixture of Polymer, Acid, Aldehyde and Phenol

B.L. Swanson; U.S. Patents 4,244,826; January 13, 1981; and 4,191,657; Mar. 4, 1980; both assigned to Phillips Petroleum Company describes a method for acid-treating a porous subterranean formation susceptible to attack by an acid and penetrated by a well bore which comprises injecting into the formation a gelled acidic composition comprising water, a water-dispersible polymer, an acid, one or more water-dispersible aldehydes, and one or more phenolic compounds and maintaining the gelled acidic composition in the formation for a period of time sufficient for the acid in the composition to react significantly with the acid-soluble components in the formation and stimulate the production of fluids therefrom.

The polymers which may be used are the same as described in the previous four processes.

Any suitable water-dispersible aldehyde meeting the compatibility requirements can be used. Thus, under proper conditions of use, both aliphatic and aromatic monoaldehydes, and also dialdehydes, can be used. The aliphatic monoaldehydes containing from 1 to about 10 carbon atoms per molecule are preferred. Representative examples of such aldehydes include formaldehyde, paraformaldehyde, acetaldehyde, propionaldehyde, butyraldehyde, isobutyraldehyde, valeraldehyde, heptaldehyde, decanal, and the like. Representative examples of dialdehydes include glyoxal, glutaraldehyde, terephthaldehyde, and the like. Various mixtures of the aldehydes can also be used.

Any suitable water-dispersible phenol or naphthol meeting the compatibility requirements can be used. Suitable phenols include monohydroxy, as well as polyhydroxy, compounds including monohydroxy and polyhydroxy naphthols.

Phenolic compounds suitable for use include phenol, catechol, resorcinol, hydroquinone, phloroglucinol, pyrogallol, 4,4'-diphenol, 1,3-dihydroxynaphthalene, and the like. Resorcinol is the preferred phenolic compound for use in the process.

Any suitable amount of aldehydes and phenolic compounds can be used. In all instances the amounts of aldehyde and phenolic compound used will be a small but effective amount which is sufficient to cause gelation of an aqueous dispersion of the polymer, the acid, the aldehyde, and the phenolic compound. As a general guide, the amount of aldehyde used in preparing the gelled acidic compositions will be in the range of from about 0.003 to 1.2 wt %, preferably from 0.04 to about 1 wt % based on the total weight of the composition. The amount of phenolic compound used will be in the range of from about 0.001 to 2 wt %, preferably from about 0.04 to 1 wt % based on the total weight of the composition. The molar ratio of aldehyde to phenolic compound will be in the broad range of from about 0.1:1 to 25:1, preferably from 1:1 to 2:1.

Acids useful in the process include any acid meeting the compatibility requirements and which is effective in increasing the flow of fluids, e.g., hydrocarbons, through the formation and into the well. Thus, under proper conditions of use, examples of such acids can include inorganic acids such as hydrochloric acid, phosphoric acid, nitric acid, and sulfuric acid; C_{1-4} organic acids as formic acid, acetic acid, propionic acid, butyric acid, and mixtures thereof, and combinations of inorganic and organic acids. The nonoxidizing acids are preferred.

Propping agents can be included in the gelled acidic compositions, if desired.

Acid and Vinylpyrrolidone Terpolymer Solution

J.C. Allen and J.F. Tate; U.S. Patents 4,210,205; July 1, 1980; and 4,219,429; August 26, 1980; both assigned to Texaco Inc. describe a process for the improved recovery of fluids and especially hydrocarbons from subterranean fluid-bearing formations by providing a process wherein a composition comprising an acidic aqueous solution of a vinylpyrrolidone polymer is injected into a formation communicating between a producing well and an adjacent injection well, the formation containing acid-soluble components and in some instances also containing water-sensitive clays or shales, and whereafter the acid contained in the composition reacts with the acid-soluble components of the formation to increase permeability and porosity of the formation thereby facilitating the flow of fluids therethrough.

A number of advantages result in treating subterranean hydrocarbon-bearing formations having acid-soluble components therein with the acidic aqueous polymer-containing compositions of this process, namely:

(1) The reaction rate of the acid with the formation acid-solubles, such as carbonates or dolomites, is greatly lessened. One of the most serious problems encountered in the use of mineral acids as acidizing agents is the very rapid rate with which they react with such acid-solubles in the formation with the result that the acid necessarily spends itself in the formation immediately adjacent the injection well bore so that little beneficial effect is realized at any great distance from the bore within the formation under treatment.

(2) The viscosity of the displacing fluid is increased. The viscosities of oil present in subsurface geologic formations and its displacing fluid are important factors in the determination of the effectiveness with which oil is pushed through the pore space of the oil-bearing formation and the degree to which the oil is permitted to stick to formation surfaces (such as sand grains) to remain as residual oil. When the displacing fluid is lower in viscosity than the oil to be displaced, the high viscosity oil preferentially sticks to the walls of pore channels and permits the low viscosity displacing fluid to move ahead. Thus, low recoveries generally are obtained from reservoirs where oil viscosity is high.

(3) The injection rate of the "polymer flood" is increased. The injection rate of a viscous polymer solution, though accomplishing the advantage cited in (2) is often greater than that of water flood, at the same pressure. Acidization in-depth accomplished during flooding greatly enhances the rate of injection.

(4) The cited polymer is effective in preventing swelling of water-sensitive clays or shales and thus formation damage of this type during the flooding.

In the first step of preparing the acidic aqueous polymer solution, a solution containing from about 3 to 30% by weight of a nonoxidizing mineral acid, such as hydrochloric acid, in water is prepared. An inhibitor to prevent corrosion on the metal equipment associated with the wells is usually added with mixing in the next step. The required amount of the polymer is then admixed with the aqueous acid solution employing a blender whereupon the polymer dissolves at a rather rapid rate.

In conducting the process, the acidic aqueous polymer solution prepared as described above is forced, usually via a suitable pumping system, down the well bore of an injection well and into the producing formation through which it is then displaced together with hydrocarbons of the formation in the direction of a production well.

It should be understood that the concentration of the polymer and the acid may be chosen to provide a displacing fluid of the desired rheological properties. Similarly, the appropriate molecular weight polymer is selected on the basis of the formation being treated as well as other operating conditions employed.

Water-soluble vinylpyrrolidone polymers useful in preparing the acidizing compositions include the following terpolymers:

(A) Vinylpyrrolidone-vinyl acetate-2-acrylamido-2-methylpropanesulfonic acid,

(B) Vinylpyrrolidone-hydroxyethyl acrylate-2-acrylamido-2-methylpropanesulfonic acid, and

(C) Terpolymer A or B alkoxylated with from 2 to about 100 wt % of ethylene oxide.

Terpolymer A comprises repeating units of

$$\left[\begin{array}{c} N \quad = O \\ | \\ -CH-CH_2- \end{array}\right], \qquad \left[\begin{array}{c} -CH_2-CH- \\ | \\ O \\ | \\ O=C-CH_3 \end{array}\right] \quad \text{and} \quad \left[\begin{array}{c} -CH_2-CH- \\ | \\ C=O \\ | \\ N-H \\ | \\ CH_3-C-CH_2SO_3H \\ | \\ CH_3 \end{array}\right]$$

Terpolymer B comprises repeating units of

$$\left[\begin{array}{c} N \quad = O \\ | \\ -CH-CH_2- \end{array}\right], \qquad \left[\begin{array}{c} -CH_2-CH- \\ | \\ C=O \\ | \\ O-CH_2CH_2OH \end{array}\right] \quad \text{and} \quad \left[\begin{array}{c} -CH_2-CH- \\ | \\ C=O \\ | \\ N-H \\ | \\ CH_3-C-CH_2SO_3H \\ | \\ CH_3 \end{array}\right]$$

The terpolymers can be conveniently prepared by the usual vinyl compound polymerization methods at temperatures of about 30° to 100°C employing a suitable polymerization catalyst such as azobis(isobutyronitrile), ammonium persulfate, etc. The preparation of such vinyl-type polymers is described in detail in numerous patents including U.S. Patents 3,264,272; 3,779,917; 3,405,003; etc.

In Terpolymer A the weight percent of vinylpyrrolidone units will vary from about 65 to 80, the weight percent of vinyl acetate units from about 8 to 15 and with the balance being 2-acrylamido-2-methylpropanesulfonic acid units. Likewise in Terpolymer B the weight percent of vinylpyrrolidone units will vary from about 65 to 80, the weight percent of hydroxyethyl acrylate units from 8 to about 15 and with the balance being 2-acrylamido-2-methylpropanesulfonic acid. Generally, the number average molecular weight of Terpolymers A and B and their alkoxylated derivatives useful in preparing the acidizing compositions will range from about 10,000 to 2,000,000 or more and preferably will be from about 100,000 to 400,000.

Example 1: A total of 400 cc of xylene, 1.5 g of powdered potassium hydroxide and 65 g of terpolymer consisting of repeating units of vinylpyrrolidone, vinyl acetate and 2-acrylamido-2-methylpropanesulfonic acid (number average molecular weight of about 210,000) in particulate form are added to an autoclave and stirring is commenced in order to form a slurry or dispersion of the terpolymer and catalyst in the xylene. The autoclave and contents are then heated to a temperature of 110°C. In the terpolymer the weight percent of vinylpyrrolidone units is about 68, the weight percent of vinyl acetate units is about 14 and the balance is 2-acrylamido-2-methylpropanesulfonic acid units. Ethylene oxide in the amount of 40 g is added to the autoclave under nitrogen pressure over a 1.2-hour period during which time the temperature of the autoclave is maintained at 110°C.

Next, the autoclave and contents are allowed to cool to room temperature after which the autoclave is vented. The reaction mixture is then stripped of volatiles using a nitrogen purge. The resulting water-soluble product is the vinylpyrrolidone-vinyl acetate-2-acrylamido-2-methylpropane sulfonic acid terpolymer alkoxylated with about 37 wt % of ethylene oxide.

Example 2: Through a water injection well drilled into a limestone formation there is displaced under pressure down the tubing and into the formation an acidic aqueous polymer solution containing 1% by wt based on the total weight of a terpolymer consisting of 70 wt % of vinylpyrrolidone, 18 wt % of hydroxyethyl acrylate and with the balance being 2-acrylamido-2-methylpropanesulfonic acid alkoxylated with about 9 wt % of ethylene oxide and having an average molecular weight of 200,000 dissolved in a 7.6 wt % of an aqueous solution of hydrochloric acid. After about 90 days the production of hydrocarbons from an adjacent producing well is substantially increased over that obtained utilizing water as the drive fluid.

Oxyalkylated Acrylamido Alkanesulfonic Acid Polymer

J.F. Tate; U.S. Patents 4,206,058; June 3, 1980; and 4,200,151; April 29, 1980; both assigned to Texaco Inc. describes a process which comprises introducing into a subsurface calcareous formation an acid solution of a water-soluble, oxyalkylated acrylamido alkanesulfonic acid polymer wherein the solution is maintained in contact with the formation for a time sufficient to chemically react with the formation so as to increase substantially the flow capabilities of the formation and to release carbon dioxide concomitantly whereby a beneficial effect due to the mutual miscibility of carbon dioxide in the fluid phase is realized as a reduction in viscosity and retentive capillary forces. Another beneficial effect is realized in the form of increased formation energy, due to the pressure generated by the released carbon dioxide.

The average molecular weight of the oxyalkylated acrylamido alkanesulfonic acid polymers and copolymers utilized in the method generally will be from about 1,000 to 1,000,000 or more and preferably will be from about 1,000 to 400,000.

Highly advantageous results are realized with the method when the water-soluble oxyalkylated acrylamido alkanesulfonic acid polymer employed comprises recurring units of the formula:

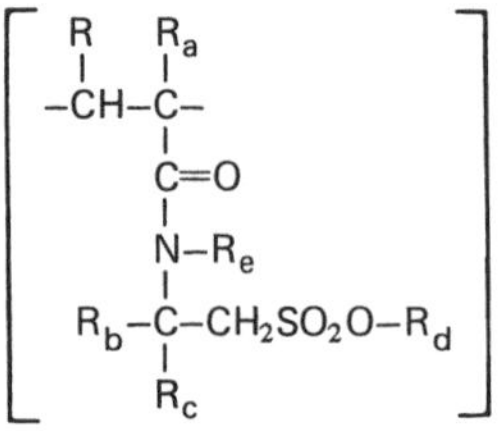

$$\left[\begin{array}{l} \quad R \quad R_a \\ \quad | \quad\ \ | \\ -CH-C- \\ \qquad\ \ | \\ \qquad\ C{=}O \\ \qquad\ \ | \\ \qquad\ N-R_e \\ \qquad\ \ | \\ R_b-C-CH_2SO_2O-R_d \\ \qquad\ \ | \\ \qquad\ R_c \end{array}\right]$$

wherein R, R_a, R_b and R_c are independently selected from the group consisting of hydrogen and alkyl having from 1 to 5 inclusive carbon atoms, R_d is selected from the group consisting of hydrogen and $-(C_3H_6O)_n(C_2H_4O)_mM$, wherein n is an integer of from 0 to about 5, m is an integer of from 3 to about 20 and M is selected from the group consisting of hydrogen, sodium, potassium and ammonium and R_e is selected from hydrogen and $-(C_3H_6O)_r(C_2H_4O)_sM$ wherein r is an integer of 0 to about 5, and s is an integer of from 3 to about 20 and with the proviso that when R_d is hydrogen then R_e is $-(C_3H_6O)_r(C_2H_4O)_sM$ and when R_e is hydrogen, then R_d is $-(C_3H_6O)_n(C_2H_4O)_mM$.

Preferably, the acidic polymer solution is one comprising an aqueous solution of about 2 to 30% by wt of a mineral acid selected from the group consisting of hydrochloric or sulfuric acid which may or may not include brine, and which contains dissolved therein between about 0.1 and 10% by wt based on the total solution weight of a water-soluble, oxyalkylated acrylamido alkanesulfonic acid polymer or copolymer.

The oxyalkylated acrylamido alkanesulfonic acid compounds utilized in preparing the polymers and copolymers employed in the treating compositions of this process can be prepared by methods well known in the art. For example, the alkylene oxide can be reacted with the acrylamido alkanesulfonic acid dissolved in a suitable solvent throughout which an alkaline catalyst, such as potassium hydroxide or sodium hydroxide, is uniformly dispersed. The quantity of the catalyst utilized generally will be from about 0.15 to 1.0% by wt of the reactants. Preferably, the reaction temperature will range from about 80° to 180°C while the reaction time will be from about 1 to 20 hours or more depending on the particular reaction conditions employed. This process is more completely described in U.S. Patent 2,425,845.

Oxyalkylated acrylamido alkanesulfonic acid compounds containing block polypropylene and polyethylene groups can be prepared by well-known methods such as taught, for example, in U.S. Patents 3,062,747; 2,174,761 or in 2,425,755. In general, the acrylamido alkanesulfonic acid initiator procedure consists in condensing with propylene oxide in the presence of an oxyalkylation catalyst until the required amount of the oxide has reacted, then continuing the oxyalkylation reaction with the ethylene oxide until the desired block oxyalkylated polymer is formed.

The oxyalkylated acrylamido alkanesulfonic acid monomers can be homopolymerized, for example, in distilled water at 30° to 95°C in 2 to 5 days or more and the reaction rate and extent of polymerization can be considerably increased by the addition of catalysts such as ferrous sulfate heptahydrate, hydrogen peroxide, etc.

In carrying out the method a solution of from about 3 to 30% by wt of the nonoxidizing mineral acid dissolved in water is first prepared. An inhibitor to prevent corrosion of acid on the metal equipment associated with the well is usually added with mixing in the next step. The polymer in an amount within the stated concentration range is then admixed with the aqueous acid solution employing a blender. The polymer dissolves rather rapidly in the acid solution and the thus-prepared composition is forced, usually via a suitable pumping system, down the well bore and into contact with the formation to be treated.

Example: A well drilled in a tight limestone formation is treated with an aqueous acidic polymer composition of this process in order to stimulate oil production. In preparing to treat the produeing formation of the well a packer is set above perforations located in the interval 6,725 to 6,740 feet. A solution of 1% by wt of a polymer having a molecular weight of about 72,000 and consisting essentially of recurring units of the formula:

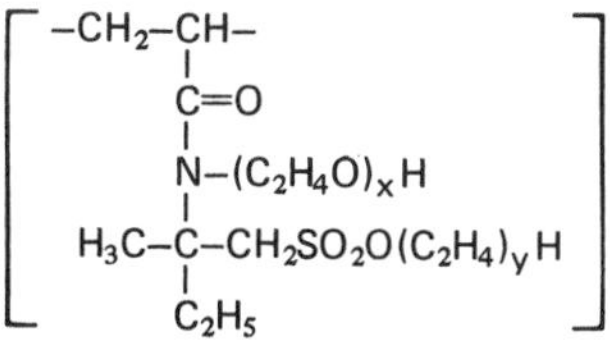

$$\left[\begin{array}{l} -CH_2-CH- \\ \qquad\quad | \\ \qquad\quad C{=}O \\ \qquad\quad | \\ \qquad\quad N-(C_2H_4O)_xH \\ \qquad\quad | \\ H_3C-C-CH_2SO_2O(C_2H_4)_yH \\ \qquad\quad | \\ \qquad\quad C_2H_5 \end{array}\right]$$

wherein the sum of x + y is about 7, is prepared by dissolving completely 500 pounds of the polymer in 6,000 gal of 15% by wt hydrochloric acid using cyclic turbulent circulation. A conventional corrosion inhibitor and nonemulsifying agent are present in the acid.

In the first part of the stimulation operation, a pad of 3,000 gal of lease water containing 25 gal of a scale inhibitor initially to prevent postprecipitation of carbonates is dissolved in the subsequent acidizing process and is pumped into the formation. In the next step, 1,500 gal of conventional 15% HCl is pumped into the formation to remove scale in the vicinity of the well bore. In the third step, 4,500 gal of the acidizing mixture previously described is pumped into the formation. Finally, the aqueous acidic polymer solution was displaced into the formation by pumping an additional 15,000 gal of lease water into it. The well is then shut in 10 hours after treatment and at the end of that time the production is measured and found to be substantially in excess of production prior to the acidization treatment.

J.F. Tate; U.S. Patents 4,200,154; April 29, 1980; and 4,163,476; August 7, 1979; both assigned to Texaco Inc. describes a similar process using an aqueous solution of an acid selected from the group consisting of hydrochloric and sulfuric acid and a fluorine-containing acid or salt and having dissolved therein an oxyalkylated acrylamido alkanesulfonic acid polymer or copolymer.

The fluorine-containing acid salt is added in an amount of 2.5 to 10% by wt and is selected from the group consisting of (a) fluoride salts such as ammonium fluoride and lithium fluoride as well as (b) acid fluorides as exemplified by ammonium acid fluoride (NH_4HF_2) and lithium acid fluoride ($LiHF_2$). The fluorine-containing salt, when added to the solution of the mineral acid, reacts to form hydrogen fluoride and the corresponding ammonium, or lithium chloride or sulfate. It has been found that a highly beneficial effect is achieved when the acidic aqueous polymer solution contains, in addition to the mineral acid, hydrofluoric acid, and the oxyalkylated acrylamido alkanesulfonic acid polymer, the ammonium or lithium ions derived from one or more of the fluorine-containing salts.

Polymers of Diallyldimethylammonium Chloride as Thickeners

K.W. Dixon; U.S. Patent 4,225,445; September 30, 1980; assigned to Calgon Corporation has found that branched emulsion or suspension polymers of diallyldimethylammonium chloride are useful as acid thickeners in oil well drilling and fracturing operations because of their acid stability, heat stability and salt stability. Suitable branching agents which may be used include, but are not limited to, triallylmethylammonium chloride, tetraallylammonium chloride and bisdiallyl ammonium salts, such as tetraallylpiperazinium chloride and N,N,N',N'-tetraallyl-N,N'-dimethyl hexamethylenediammonium chloride.

These polymers may be prepared by emulsion or suspension polymerization techniques such as that described in U.S. Patent 3,968,037 and may contain from about 95 to 99.99 mol % diallyldimethylammonium chloride and from about 0.01 to 5 mol % of one the aforementioned branching agents.

The thickening agents of the process are useful over a wide range of molecular weights, from as little as about 5,000 to as much as several hundred thousand to one million or more.

The thickening agents may be employed satisfactorily in concentration amounts as low as 0.01 % by wt of the acid-based liquid being thickened. Higher concentrations of 10% by wt or more may be employed, but the range of concentration amounts will ordinarily be from about 0.1 to 5% by wt. As with molecular weight of the polymeric thickening agents, the concentration of the thickening agent which is desired will depend on a number of factors, but especially upon the viscosity of the final thickened acid composition which is desired.

The acid solution fracturing fluids with which the thickening agents comprising the branched emulsion or suspension polymers of diallyldimethylammonium chloride are used can also contain fluid loss control additives, surfactants, propping agents and clay control chemicals which are compatible with the thickening agents.

Self-Breaking Acid Emulsion

G.A. Scherubel; U.S. Patent 4,140,640; February 20, 1979; assigned to The Dow Chemical Company describes an acidizing method using an emulsion of the type containing an effective amount of at least one C_{8-18} primary amine as a cationic surfactant to increase the normal reaction time of the acid with an acid-soluble formation. The improvement in both the composition and method is the inclusion in the emulsion of an effective amount of a nonionic surfactant comprising at least one diethanolamide of at least one C_{8-18} fatty acid, to cause the emulsion to break as the acidizing capacity of the emulsion becomes substantially depleted, i.e., spent, on the formation.

Specific cationic surfactants which can be employed include, for example, cocoamine, dodecylamine, tetradecylamine, decylamine, octylamine, and mixtures thereof. Preferably, the surfactant consists primarily of dodecylamine.

Suitable results can be obtained when as little as about 0.01, preferably about 0.05% by wt of the surfactant is employed. Economic considerations generally dictate a maximum amount of about 10% by wt of surfactant in the acidizing emulsion, preferably about 0.5%.

Specific amides which may be employed include the diethanolamides of any one of or a mixture of any two or more of acids such as octanoic, pelargonic, capric, undecylic, lauric, tridecylic, myristic, pentadecylic, palmitic, margaric, and stearic. Unsaturated fatty acid diethanolamides are also suitable, such as the diethanolamides of such acids as obtusilic, caproleic, 10-undecylenic, lauroleic, physeteric, myristoleic, palmitoleic, petroselinic, petroselaidic, oleic, elaidic, vaccenic, and linoleic. Preferably, the amide is derived from a C_{18} acid, most preferably oleic.

The amount by weight of amide employed depends upon the amount of the primary amine employed, and generally is from about 0.85 to 2 times the weight of the amine. Preferably, about 1 to 1.6 pbw of the diethanolamide is employed per part of primary amine.

Acidizing acids which can be employed include: HCl, HF, formic acid, acetic acid, sulfamic acid, various mixtures thereof and other acids which are compatible with the specific surfactants employed. The acid solutions can contain up to about 40% by wt or more of the acid. The surfactants of the process are particularly useful in acidizing emulsions wherein the aqueous phase contains an acidizing acid, e.g., HCl, in a concentration of more than about 15% by wt of the aqueous phase, especially where the concentration of the acid is greater than 25%, e.g., about 28% HCl, and the temperature of the formation to be acidized is about 150° to 250°F.

Any liquid hydrocarbon generally employed in the art to prepare acid-in-oil emulsions can be employed. Liquids which can be employed include, for example, crude oil, various grades of diesel oil, fuel oil, kerosene, gasoline, aromatic oils, petroleum fractions, mineral oils and varous mixtures thereof. Kerosene is preferred because corrosion rates using kerosene are generally somewhat less than with certain other hydrocarbons.

The liquid hydrocarbon phase can comprise, as percent by volume, from about 5 to 95%, preferably from about 10 to 50% of the emulsion, and most preferably, about 25 to 35%.

The method is practiced using standard acidizing equipment and procedures well-known in the art. The composition and method can be employed in matrix and fracturing acidizing techniques.

Gel-Forming Nonionic Surfactant

C.G. Inks; U.S. Patent 4,163,727; August 7, 1979; assigned to BASF Wyandotte Corporation describes a composition which consists essentially of three components: (1) hydrochloric acid or its equivalents, (2) water, and (3) a gel-forming nonionic surfactant, in a proportion capable of causing the composition to form a gel at a temperature of use, i.e., the temperature prevailing within the oil-bearing stratum or strata of rock being treated, in order to increase the rate of production of the well. The composition also usually contains a corrosion inhibitor, to the extent necessary.

More specifically, the composition consists of (A) a proportion effective to attack rock strata of an acid selected from the group consisting of hydrochloric acid, hydrofluoric acid, formic acid, acetic acid, and mixtures thereof; and (B) a proportion, effective to cause the composition to gel at the temperature of the rock strata but sufficiently low to cause the composition to remain liquid at ambient temperature, of a nonionic surfactant of molecular weight between 4,000 and 30,000, the surfactant being of a formula selected from the group consisting of $HO(C_2H_4O)_a(C_3H_6O)_b(C_2H_4O)_c$ and

$$[H(C_2H_4O)_y(C_3H_6O)_x]_2N{-}CH_2CH_2{-}N[(C_3H_6O)_x(C_2H_4O)_yH]_2$$

a, b and c are integers such that the poly(oxyethylene) hydrophilic portion of the molecule accounts for at least 25% of its molecular weight and the molecular weight of the poly(oxypropylene) hydrophobic portion of the molecular weight has a molecular weight greater than 2,150.

x and y are integers such that the poly(oxyethylene) hydrophilic portion of the molecule accounts for at least 40% of its molecular weight and the molecular weight of the poly(oxypropylene) hydrophobic portion of the molecule has a molecular weight of at least 3,250, and if the proportion of the molecular weight of the molecule accounted for by the polyoxyethylene units is less than 55%, greater than 5,250.

One example is the use of a composition consisting essentially of 12% hydrochloric acid, 3% hydrofluoric acid, 20% of Tetronic 1508 surfactant, and 65% water. Tetronic 1508 surfactant is a graft polymer based upon ethylene diamine which has first been oxypropylated to a typical molecular weight for the poly(oxypropylene) hydrophobe of 5,501 to 7,000 and then oxyethylated to such an extent that about 80% of the molecular weight of a typical molecule is provided by poly(oxyethylene) hydrophilic units. Such a material forms a gel upon heating to a temperature of 70°C.

Another composition is one consisting of 15% by wt of hydrochloric acid, 5% formic acid, 30% Tetronic 1304 surfactant and 50% water. Tetronic 1304 surfactant is a block polymer based upon ethylenediamine which has first been oxypropylated to a typical molecular weight by the poly(propylene) hydrophobe of 5,501 to 6,000 and then oxyethylated to such an extent that about 40% of the molecular weight of a typical molecule is provided by poly(oxyethylene) hydrophilic units. Such a composition forms a gel upon heating to 55°C.

Hydrohalic Acid Precursor

D.J. Watanabe; U.S. Patent 4,148,360; April 10, 1979; assigned to Union Oil Company of California describes a method for acidizing subterranean formations having temperatures between about 250° and 700°F, wherein a substantially anhydrous acid precursor is introduced through a well and into contact with the formation. The acid precursor is displaced from the borehole into the formation wherein it hydrolyzes in situ to generate a hydrohalic acid. The acid precursor is a normally liquid, halogenated hydrocarbon having a generalized formula: $C_xH_yX_z$ wherein X = Cl, Br, I, or F; x = 1 or 2; y = 0, 1 or 2, but $y \leq x$; and $z = 2x - y + 2$; and which is thermally stable under the high temperature and pressure conditions encountered prior to hydrolysis.

The method allows acidization of subterranean formations in which the prior art acidization methods are rendered impractical due to the high formation temperatures. The process provides an acidization method for high-temperature formations in which corrosion of well equipment is substantially eliminated and the undesirable consumption of acid by the formation immediately adjacent the borehole is avoided. The method can be employed in high-temperature formations having a large connate water concentration, such as a formation containing an aqueous geothermal fluid, or in high-temperature formations having little or no connate water. The method has the advantage of being operable with conventional well equipment and does not require the use of exotic alloys or other materials to avoid corrosion of the well equipment.

For the acidization of carbonate materials and other acid-soluble formation materials having high concentrations of calcium, magnesium or other multivalent cations, acid precursors which hydrolyze to generate hydrochloric, hydrobromic or hydriodic acids are preferred, particularly the hydrochloric acid precursors. However, for the acidization of siliceous materials, such as clay, hydrofluoric acid precursors are preferred and acid precursors which hydrolyze to generate a mixture of hydrofluoric and hydrochloric acids are particularly preferred.

In general, the halogenated hydrocarbons having one carbon atom are preferred over the halogenated hydrocarbons having two carbons, especially at formation temperatures above about 500°F, because various side reaction products of the hydrolysis of the halogenated hydrocarbons having two carbon atoms, such as acetic acid, can be pyrolyzed to form plugging solid residues at these very high temperatures. Of the halogenated hydrocarbons having one carbon atom, the acid precursors of the formula CX_4 are preferred, and tetrachloromethane (i.e., carbon tetrachloride) is particularly preferred due to its ability to hydrolyze readily over the temperature range 250° to 700°F, as well as its low cost and availability.

The preferred hydrochloric acid precursors are tetrachloromethane, trichloromethane, pentachloroethane and tetrachloroethane, with tetrachloromethane being particularly preferred. Preferred acid precursors which hydrolyze to form a mixture of hydrochloric and hydrobromic acids are bromotrichloromethane chlorodibromomethane, bromodichloromethane, trichlorodibromoethane, 1,1-dichloro-1,2-dibromoethane, 1,2-dichloro-1,2-dibromoethane, and 1,1-dichloro-2,2-dibromoethane. Preferred acid precursors which hydrolyze to form a mixture of hydrochloric and hydrofluoric acids are 1,1,2-trifluorotrichloroethane, fluorotetrachloroethane and fluorotrichloroethane, with 1,1,2-trifluorotrichloroethane being particularly preferred.

Example 1: A natural gas-bearing dolomite formation having a temperature of about 350°F and located at a depth of about 21,800 feet is acidized in accordance with this process. A production well penetrating the formation has a production tubing disposed therein. Nine discrete slugs of liquid are sequentially injected at a rate of about 20 barrels per minute through the production tubing into a mixing zone of the well adjacent the formation. The first, third, fifth, seventh and ninth slugs are each a 200-barrel slug of fresh water containing about 1,000 scf of nitrogen per barrel, and the second, fourth, sixth and eighth slugs are each a 25-barrel slug of tetrachloromethane containing about 1,000 scf of nitrogen per barrel. The injected fluids are at least partially mixed in the mixing zone to form a reaction mixture. The reaction mixture is displaced into the formation by injecting through the production tubing about 1 million scf of nitrogen.

The well is shut in for a period of 68 hours to allow substantially complete hydrolysis of the tetrachloromethane and complete reaction of the in situ-produced acid. At the end of this time period, the well is opened and the well effluent is contacted at the well site with a dilute ammonium hydroxide solution prior to returning the well to natural gas production.

Example 2: A natural gas- and condensate-bearing sandstone formation having a temperature of about 320°F and located at a depth of about 19,500 feet is fracture-acidized in accordance with this process. An injection tubing is positioned in a production well penetrating the formation and 20 barrels of a 5 wt % solution of ammonium chloride in fresh water is injected into the formation to stabilize the clay in the formation. Then, 350 barrels of water containing 1,000 scf of nitrogen per barrel of water are injected through the well annulus at a rate of about 10 barrels per minute, and simultaneously, 35 barrels of tetrachloromethane followed by 20 barrels of water are injected through the injection tubing at a rate of about 1 barrel per minute.

The injected fluids mix in a mixing zone of the borehole adjacent the formation to form a reaction mixture. The reaction mixture is displaced into the formation by simultaneously injecting through the tubing and the well annulus a total of 1 million scf of nitrogen. The well is shut in for a period of about 20 hours to allow some hydrolysis of the tetrachloromethane and reaction of the in situ-produced acid.

At the end of the 20-hour period, 2,500 barrels of water containing 1,000 ppm of polyacrylamide polymer, Pusher 1000 polymer (The Dow Chemical Company), and 1,000 scf of nitrogen per barrel of water, are injected through the well annulus at a rate of 50 barrels per minute, and, simultaneously, 250 barrels of a 2:1 mixture of tetrachloromethane and 1,1,2-trifluorotrichloroethane followed by 20 barrels of water are injected through the injection tubing at a rate of 5 barrels per minute, thereby hydraulically fracturing the formation.

During the injection operation, the injected fluids mix in the borehole adjacent the formation to form a reaction mixture which is displaced into the formation and newly opened fractures. The reaction mixture is overdisplaced from the formation immediately adjacent the borehole by simultaneously injecting through the tubing and well annulus a total of 1 million scf of nitrogen.

The well is shut in for a period of 68 hours to allow substantially complete hydrolysis of the acid precursor and complete reaction of the in situ-produced acid. Thereafter, the well is opened and the well effluent is contacted at the well site with a dilute ammonium hydroxide solution prior to returning the well to natural gas production.

D.J. Watanabe; U.S. Patent 4,203,492; May 20, 1980; assigned to Union Oil Company of California describes a related process for acidizing siliceous materials contained in subterranean formations having temperatures between about 250° and 700°F, wherein (1) a substantially anhydrous treating fluid consisting essentially of an acid precursor and (2) an aqueous fluoride salt solution are introduced through a well and into contact with the formation. The acid precursor is the same as described in the previous process.

The aqueous fluoride salt solutions suitable for use in the method are noncorrosive aqueous solutions of water-soluble alkali metal and/or ammonium fluoride salts. The fluoride salt must be capable of dissociating in situ to provide fluoride ions for the in situ generation of hydrofluoric acid. Suitable fluoride salts include the water-soluble alkali metal and/or ammonium salts of hydrofluoric acid, fluoroboric acid, hexafluorophosphoric acid, difluorophosphoric acid and fluorosulfonic acid. Preferred fluoride salts include ammonium fluoride, ammonium hexafluorophosphate, ammonium difluorophosphate, ammonium fluorosulfonate, cesium fluoride, cesium bifluoride, cesium hexafluorophosphate, cesium difluorophosphate and cesium fluorosulfonate.

These ammonium and cesium salts are preferred because the ammonium and cesium fluorosilicate salts formed in situ upon acidization of siliceous materials are relatively water-soluble as compared to the corresponding sodium, potassium and rubidium salts. Where sodium, potassium or rubidium fluoride salts are to be used, suitable precautions known in the art must be taken to avoid excessive precipitation of the corresponding fluorosilicate salts. For example, an overflush fluid, such as water, may be injected to displace these salts away from the vicinity of the well before they are precipitated.

Aqueous solutions of ammonium fluoride are particularly preferred due to their low cost and noncorrosivity, and aqueous solutions of ammonium fluoroborate are particularly preferred under circumstances in which the fluoroborate anion will serve to fuse movable formation fines and/or to desensitize clay particles. However, at very high temperatures, depending upon the pressure, ammonia gas may evolve from these aqueous solutions, which evolution has the effect of causing the solutions to become corrosive. In most cases the ammonia evolution can be controlled by pressurizing the fluoride salt solution. Where the ammonia evolution cannot be controlled or where the ammonium salts are otherwise deemed unsuitable, the use of the corresponding cesium salt is preferred.

The concentration of the fluoride salt in the aqueous solution may vary widely depending, inter alia, upon the desired hydrofluoric acid concentration in the acid solution produced in situ and whether the injected fluids will be diluted with connate water and/or water otherwise injected into the formation. Where no dilution of the fluoride salt solution is expected, the fluoride salt concentration is preferably sufficient to provide a fluoride ion concentration between about 0.1 and 25 wt %, more preferably between about 1 and 10 wt %. Conversely where substantial dilution of the injected solution is expected a proportionately higher concentration is needed in the injected solution to yield the desired fluoride ion concentration in the acid solution produced in situ.

Example: A subterranean formation contains a geothermal fluid at a temperature of about 450°F. A steam-producing zone of the formation is penetrated by a production well at a depth of about 5,500 feet, and is fracture-acidized in accordance with this method.

First, a 20-barrel slug of a 5 wt % ammonium chloride solution followed by an 80-barrel slug of fresh water are injected through the well into the formation in order to stabilize any clay contained thereon. A volume of the formation around the well is then preflushed to remove nonsiliceous acid-soluble materials by alternately injecting through the well and into the formation ten 3.5-barrel slugs of tetrachloromethane and ten 35-barrel slugs of fresh water. The injected fluids mix in a mixing zone of the well to form a preflush mixture which is then displaced into the formation by the later-injected fluids. A 20-barrel slug of fresh water is injected to displace the preflush mixture away from the well and a 20-barrel slug of a 15 wt % ammonium chloride solution is positioned in the well adjacent the producing zone to protect the well from the hydrochloric acid produced upon hydrolysis of the tetrachloromethane.

The preflush tetrachloromethane hydrolyzes in situ to generate hydrochloric acid which solubilizes nonsiliceous acid-soluble materials in the formation. The later-injected fluids will displace these solubilized materials away from the volume of the formation surrounding the well.

After an appropriate period of time to allow about 80% hydrolysis of the preflush acid precursor, such as about one hour or less, fifty 6.5-barrel slugs of tetrachloromethane and fifty 50-barrel slugs of an aqueous solution containing 5.5 wt % of ammonium fluoride and 1,000 ppm of the polyacrylamide polymer Pusher 1000 (The Dow Chemical Company) are injected through the well into the formation at a rate sufficient to hydraulically fracture the steam-producing zone of the formation. The injected fluids mix in the mixing zone of the well to form a reaction mixture which is then displaced into the formation. A 20-barrel slug of fresh water is injected to displace the last of the reaction mixture away from the well and a 20-barrel slug of a 30 wt % solution of ammonium chloride is positioned in the well adjacent the formation to protect the well.

The well is shut in for about two hours to allow substantially complete hydrolysis of the acid precursor and substantially complete reaction of the hydrofluoric acid produced in situ. Thereafter, the well is brought back on production.

In Situ Generation of Hydrofluoric Acid

W.M. Salathiel and C.M. Shaughnessy; U.S. Patent 4,136,739; January 30, 1979; assigned to Exxon Production Research Company describe a technique for generating hydrofluoric acid in a formation which overcomes the problems associated with rapid spending of the acid solution within a short radial distance from the well bore.

In accordance with the process, the hydrofluoric acid is formed by combining an aqueous solution of a fluoride salt and an aqueous solution of an acid in the pore spaces of the formation. This is accomplished by immobilizing one of the aqueous solutions in the pore spaces of the formation by displacing solution with a liquid phase that is substantially immiscible with the aqueous phase, thereby driving the aqueous solution to a saturation at or below its residual saturation. The second aqueous solution is then injected into the formation.

In a preferred embodiment, water containing an ammonium salt of hydrofluoric acid is injected into the formation to be treated. The salt solution is then followed by a hydrocarbon liquid such as diesel oil to reduce the saturation of the salt solution to residual saturation. After the hydrocarbon liquid has been injected, hydrochloric acid solution is injected into the formation. The acid solution contacts the ammonium fluoride salts to form hydrofluoric acid which is capable of dissolving siliceous material in the formation.

Other embodiments of this process include adding various additives such as viscosifiers, surface active compounds and corrosion inhibitors to the aqueous fluoride-containing solution and/or the injected hydrocarbon liquid and/or the aqueous acid-containing solution. It is particularly preferred to inject into the formation an aqueous fluoride-containing solution which also contains a viscosifier such as a salt of poly-2-acrylamido-2-methyl propyl sulfonate and to inject into the formation a hydrocarbon which also contains one or more preferentially oil-soluble surface-active compounds such as ethylene glycol monobutyl ether and/or sorbitan monooleate.

It is also particularly preferred to follow the aqueous fluoride-containing solution with two sequential volumes of diesel oil, the first volume containing surface active compounds and the second volume being substantially free of surface-active compounds.

By this process, hydrofluoric acid is continuously generated at and behind the advancing hydrochloric acid front as it displaces the hydrocarbon liquid which immobilized the fluoride salt solution from the pore spaces. In this way hydrofluoric acid is generated at deeper radial depths into the formation.

Example: This example is provided to show how one embodiment of this process may be practiced in a field application. The sandstone formation treated in this example is penetrated by a well. The formation has a porosity of 35%. The treatment for this well is designed to penetrate a 4-foot radius surrounding the well over any desired formation interval. The formation has a pore volume of 130 gal/ft of the formation interval. If the formation contains carbonate minerals or it is desired to break down the perforations in the casing, a mixture containing 12 wt % HCl and 3 wt % HF acids may be injected as a preflush liquid. All fluid injection should be at a rate which maintains the injection pressure below the formation fracture pressure.

After the preflush, if carried out, 50 gal/ft of formation interval of an aqueous solution containing 4 wt % NH_4Cl is injected into the formation by means of the well. This amount of ammonium chloride is injected into the formation because generally about ⅓ pore volume is required to displace resident aqueous fluids from the zone to be treated. About 40 gal of an aqueous solution containing 10.5 molar ammonium fluoride which contains 3,000 ppm potassium salt of poly-2-acrylamido-2-methyl propyl sulfonate (PAMPS) is injected into the formation per foot of formation interval. The amount of ammonium fluoride solution injected is slightly more than the actual residual volume for the formation being treated. The PAMPS is included in the ammonium fluoride solution to increase the solution viscosity to about 5 cp. After injection of the ammonium fluoride solution, 50 gal of diesel oil with 10% by volume of ethylene glycol monobutyl ether (EGMBE) is injected into the formation per foot of interval. Thereafter, 50 gal/ft of diesel oil without EGMBE is injected into the formation. After the diesel oil is injected, 100 gal/ft of an aqueous solution containing 28 wt % of HCl is injected into the formation. The well is then returned to production.

Levulinic Acid-Citric Acid Sequestering Additive

When the formation being acidized contains deposits of metal compounds such as iron compounds or clays containing aluminum compounds, the acid solution dissolves such deposits as well as other reactive substances contained in the formation, but upon becoming spent, iron and/or aluminum contained in the solution precipitate as hydroxides which can reduce the permeability of the treated formation.

W.R. Dill and J.A. Knox; U.S. Patent 4,151,098; April 24, 1979; assigned to Halliburton Company describe sequestering additives, acidizing compositions containing the additives and methods of acidizing subterranean well formations which are more effective in preventing the precipitation of metal compounds in formations than the heretofore used additives, compositions and methods.

The sequestering additives are comprised of a mixture of a first ingredient selected from the group consisting of levulinic acid, a salt of levulinic acid and mixtures thereof, and a second ingredient selected from the group consisting of citric acid, a salt of citric acid and mixtures thereof. The relative proportions of the ingredients can vary, but generally fall within a levulinic acid and/or salt thereof to citric acid and/or salt thereof weight ratio in the range of from about 5:1 to 1:5. The specific amount of each ingredient used in the additive depends on the metal compound content of the formation to be acidized and other factors, but the relative proportions of the ingredients are adjusted so that they are sufficient to prevent precipitation of metal compounds from a spent aqueous acid solution when added to the live solution in a given amount for a longer period of time than obtainable with like amounts of either ingredient alone.

In most applications an additive concentration in the range of from about 10 to 400 lb per 1,000 gal of aqueous acid solution is sufficient. Preferably, the citric acid or salt thereof ingredient is combined with the acid solution in a solid state in an amount in the range of from about 10 to 300 lb per 1,000 gal of the solution. The levulinic acid ingredient is preferably combined with the acid solution in an amount in the range of from about 0.2 to 6% by volume of the solution. An equivalent amount of a salt of levulinic acid, preferably alkali metal or ammonium, can be utilized.

The most preferred acidizing composition is an aqueous solution of hydrochloric acid and a levulinic acid-citric acid metal sequestering additive, the hydrochloric acid being present in the composition in an amount of about 15% by wt of the composition, the citric acid being present in the composition in an amount of about 50 lb per 1,000 gal of the composition and the levulinic acid being present in the composition in an amount of about 1% by volume of the composition.

In preparing the acidizing compositions, a corrosion inhibitor, if utilized, is first added to water in a mixing tank. The acid or mixture of acids used is next combined with the water in an amount sufficient to obtain a solution of desired acid concentration and mixed thoroughly. The levulinic acid or salt thereof is next combined with the acid solution followed by combining the citric acid or salt thereof with the solution. The resulting composition is mixed or agitated for a period of time sufficient to completely dissolve the sequestering additive ingredients followed by the addition of other conventional formation-treating additives, if used.

Once the acidizing composition has been prepared as described above, it is introduced into a subterranean well formation to be acidized. The acidizing composition dissolves deposits of metal compounds as well as other reactive substances in the formation and maintains the metals in solution after becoming spent for a period of time sufficient to recover the spent composition, i.e., the spent composition is recovered from the subterranean formation by producing the formation, by driving the spent composition through the formation to a recovery well, or by driving the spent composition over such a wide area that any precipitate that forms cannot have a detrimental effect.

Chelating Iron with Sulfosalicylic Acid

W.A. McLaughlin and D.C. Berkshire; U.S. Patent 4,137,972; February 6, 1979; assigned to Shell Oil Company describe a process for acidizing a subterranean reservoir which contains an asphaltenic crude by injecting an aqueous strong acid solution into the reservoir. The improvement comprises dissolving enough 5-sulfosalicylic acid in the acid solution to chelate with ferric ions and avoid the formation of permeability-impairing deposits of iron-asphaltenic solids.

In general, the aqueous strong acid can be substantially any which is capable of dissolving solid materials encountered within a subterranean asphaltenic oil reservoir. Such acids generally comprise solutions and/or homogeneous dispersions or emulsions of an aqueous hydrochloric acid, or a mixture of hydrochloric acid with hydrofluoric acid and/or thickeners, corrosion inhibitors, wetting agents, or the like. The hydrochloric acid content of such solutions can range from about 1 to 30% by wt. Particularly suitable acids comprise aqueous hydrochloric acids containing from about 5 to 15% hydrochloric acid, and aqueous mud acids containing from about 5 to 15% hydrochloric acid mixed with from about 1 to 3% hydrofluoric acid.

In including the sulfosalicylic acid in a formation-treating aqueous acid in accordance with the process, various forms of starting materials and various procedures can be used to form a suitable solution and/or homogeneous dispersion. Sulfosalicylic acid and/or at least one alkali metal or ammonium sulfosalicylate in the form of a solid or solution (preferably aqueous) can be simply mingled with the formation-treating acid and agitated to an extent sufficient to provide a solution and/or homogeneous dispersion.

Alternatively, such a form of the acid or salt can be premixed with the formation-treating acid along with a solubilizing agent, e.g., a completely water-miscible monohydric or polyhydric alcohol. Alternatively, such an acid or salt can be mixed with the formation-treating acid along with an oil solvent liquid and agitated to provide an oil and water emulsion or dispersion. Alternatively, such a solution or dispersion of sulfosalicylic acid or its salts can be injected into the reservoir immediately ahead of the formation-treating acid (which may or may not be mixed with solubilizing agents or oil solvents) so that the formation-treating acid is mixed with the sulfosalicylic acid within the reservoir formation.

Particularly advantageous procedures comprise (a) premixing the reservoir formation-treating acid (and/or a mixture of it and an oil solvent) with the solid sulfosalicylic acid or an aqueous solution of its salt or (b) where the well or subterranean earth formation to be treated may contain aqueous solutions of ferric iron within the zone to be treated, injecting a slug of an aqueous solution of the acid or its salt before injecting the sulfosalicylic acid-containing portion of the formation-treating acid.

In general, the concentration of sulfosalicylic acid within the first injected portion of the formation-treating acid should be from about 0.2 to 1.0 mol/ℓ. The concentration of sulfosalicylate ions in an aqueous sulfosalicylate-containing pretreatment solution should be from about 0.01 to 1.0 mol/ℓ. Where the aqueous formation-treating acid is being mixed with a solution of sulfosalicylic acid (or at least one of its salts) the salicylate-containing solution is preferably relatively concentrated, e.g., a substantially saturated solution of the acid in a hot aqueous solution or a substantially saturated aqueous solution of a salt of the acid.

E.H. Street, Jr.; U.S. Patent 4,167,214; September 11, 1979; assigned to Shell Oil Company describes a related process which comprises dissolving in the acid to be injected (a) an amount of 5-sulfosalicylic acid which is at least sufficient to sequester significant proportions of ferric ions when the pH of the acid is from about 0.5 to 3 but is less than enough to cause a significant salting-out of solid materials, and (b) an amount of citric acid which is at least sufficient to sequester significant proportions of ferric ions when the pH of the acid is from about 3 to 6 but is less than enough to precipitate a significant amount of calcium citrate.

In general, it is preferable that an aqueous acid solution injected in accordance with the process contain from about 0.001 to 0.009 mol/ℓ of citric acid and from about 0.01 to 0.05 mol/ℓ of 5-sulfosalicylic acid.

In a particularly suitable procedure for conducting the process, a volume of liquid solvent for asphaltenic oil (e.g., toluene) sufficient to dissolve most of the oil within the first few feet around the well is injected along with the citric acid and salicylic acid-containing acidizing acid in the form of an oil-in-water emulsion. Alternatively, such a volume of such a solvent can be injected in the form of a slug preceding the injection of the acid.

Cleaning of Propped Fractures with Fluoboric Acid

R.L. Thomas and F.A. Suhy; U.S. Patent 4,160,483; July 10, 1979; assigned to The Dow Chemical Company describe a method for cleaning a proppant pack in a fracture in a subterranean formation. The improvement is based on the use of fluoboric acid as the, or one of the, treating fluids. The fluoboric acid is injected into the borehole, and thereafter the fluoboric acid is permitted to at least partially hydrolyze so that the propping agent pack is contacted with the hydrolysis products from the fluoboric acid, which hydrolysis products include, principally, hydrofluoric acid and hydroxyfluoboric acid. In one embodiment, e.g., where the propping agent is already in place in the fracture, the fluoboric acid is injected into the borehole and then directly into the prop pack where the hydrolysis is permitted to occur. Where the cleaning is carried out in conjunction with the fracturing operation, the fluoboric acid may be injected into the proppant-containing fracture as described in the preceding sentence, or, the fluoboric acid may be injected deep into the formation prior to placement of the propping agent and then at least a portion thereof back-flowed into the proppant pack after the pack has been deposited in the fracture.

The fluoboric acid solution may be prepared in any convenient manner. U.S. Patent 2,300,393, for example, teaches preparation of fluoboric acid by mixing boric and hydrofluoric acids. Alternatively, boric acid may be added to ammonium fluoride or ammonium bifluoride in the presence of an approximately stoichiometric amount of HCl. For example, an approximately 8 wt % solution of fluoboric acid may be prepared by admixing the following:

	U.S.	Metric
Water	340 gal	1.36 m^3
Ammonium bifluoride	500 lb	240 kg
35 wt % HCl	97 gal	0.388 m^3
Boric acid	250 lb	120 kg
Total	~500 gal	~2 m^3

Other variations will be readily apparent to those skilled in the art.

Generally, solutions of from about 1 wt % or less up to about 48 wt % HBF_4 may be employed. More preferably, the fluoboric acid solution consists substantially of fluoboric acid, i.e., optionally includes functional additives such as a corrosion inhibitor, diverting agent, or the like, but containing (when injected) less than about 2% HCl and less than about 1% HF.

In the treatment of a fracture containing a proppant prior to injection of the fluoboric acid, the fluoboric acid is injected into the wellbore and then into the formation. Injection into the formation may be carried out at either a matrix or fracturing rate and pressure. However, the acid is preferably injected at a matrix rate, most preferably at a rate of about ¼ barrel per 4 feet of perforations (about 33 ℓ/m of perforations) which assures that most of the acid will enter the propped fracture and also that migratory fines are not disturbed during the injection. The precise volume employed is not critical. Ideally, a sufficient volume is employed to penetrate the length of the fracture, although fracture volume calculations are at best close approximations. Moreover, since the greatest drawdown is near the wellbore, it is desirable but not essential to treat the entire fracture. Consequently, fracture volume may make complete treatment uneconomical. Therefore, from about 100 to 200 gal/ft of pay zone is typically employed.

The fluoboric acid may be displaced from the wellbore if desired with a suitable displacement fluid, e.g., an aqueous ammonium chloride or ammonium borate solution or a weak organic acid solution.

Among the advantages of using fluoboric acid as a prop cleaning treatment fluid is the slow rate of reaction, which permits injection to the extremities of a fracture. Also the slow rate of reaction makes fluoboric acid much less damaging to the principal propping agent. In addition to providing deep live acid (HF) generation, the treatment also apparently stabilizes any undissolved clays contacted. In contrast to conventional clay stabilizers which have been thought to act by ion exchange or adsorption, laboratory studies indicate fluoboric acid causes an actual chemical fusion of fines and clay platelets, so that they are mechanically much less likely to be disturbed by increased fluid flow. Also, the clays are desensitized and are not thereafter susceptible to swelling or dispersion by otherwise incompatible fluids. Finally, fluoboric acid does not significantly damage the larger propping agent particles.

FRACTURING COMPOSITIONS

Hydroxypropylcellulose and Poly(Maleic Anhydride/Alkene-1)

The productivity of oil and gas wells can be improved by increasing the area of communication within a selected producing zone. The drainage area can be increased by hydraulic fracturing of the producing zone to provide fractures and channels emanating from the well base area into the contiguous subterranean formations.

The hydraulic fracturing process is accomplished by rapid pumping of an aqueous fluid medium down a well which penetrates the subterranean formation where fracturing is desired. The rapid pumping of the aqueous fluid creates a hydrostatic pressure which energizes splitting forces in the confined zone. Pressures as high as 10,000 psi are employed to effect formation fracturing.

As cracks and channels are formed, a propping agent which is suspended in the high viscosity hydraulic fluid penetrates the newly created fissures and becomes lodged therein. The function of the propping agent is to support the fractures in an open position as a conduit for the flow of fluids such as oil, gas or water through the fractured zone. Various noncompressible materials are employed as proppants. These include sand, rounded walnut shells, glass beads, aluminum spheres, and the like.

After a fracturing operation has been completed and the propping agent has been deposited, the hydrostatic pressure is released, the flow of fluid is reversed, and the hydraulic fracturing fluid is withdrawn.

Hence, the hydraulic fracturing fluid composition functions to force fracturing under hydrostatic pressure, and it serves to transport the suspension of propping agent into the porous subterranean formations. The hydraulic well-treating fluid medium must exhibit advantageous viscosity and particulate solids transport properties.

R.N. DeMartino; U.S. Patent 4,172,055; October 23, 1979; assigned to Celanese Corporation describes well-treating hydraulic fluid compositions consisting essentially of (1) an aqueous medium, (2) hydroxypropylcellulose/poly(maleic anhydride/alkene-1) gelling agent, (3) a breaker additive, and (4) a propping agent, which are eminently suitable for application as well-fracturing fluid media.

Hydroxypropylcellulose is commercially produced by reacting alkali cellulose with propylene oxide at elevated temperatures and pressures. Methods of producing hydroxyalkylcellulose are described in U.S. Patents 2,572,039; 3,131,196; and 3,485,915.

The term "maleic anhydride" as employed herein is meant to include α,β-olefinically unsaturated dicarboxylic acid anhydride comonomers represented by the structural formula:

$$\begin{array}{l} \qquad\quad\;\; O \\ \qquad\quad\;\; \| \\ R_1-C-C \\ \qquad \| \quad\;\; \diagdown \\ \qquad \| \qquad\;\; O \\ \qquad \| \quad\;\; \diagup \\ R_2-C-C \\ \qquad\quad\;\; \| \\ \qquad\quad\;\; O \end{array}$$

wherein R_1 and R_2 are independently selected from the group consisting of hydrogen, halogen, cyano, and aliphatic and aromatic substituents such as alkyl, aryl, alkaryl, aralkyl, cycloaliphatic, and the like, containing between one and about ten carbon atoms.

Compounds corresponding to the above formula include maleic anhydride, chloromaleic anhydride, 2,3-dichloromaleic anhydride, cyanomaleic anhydride, 2,3-dicyanomaleic anhydride, methylmaleic anhydride, 2,3-dimethylmaleic anhydride, ethylmaleic anhydride, propylmaleic anhydride, butylmaleic anhydride, 2,3-di-n-butylmaleic anhydride, phenylmaleic anhydride, benzylmaleic anhydride, cyclohexylmaleic anhydride, and the like.

The quantity of gelling agent incorporated in the well-treating hydraulic composition can vary in the range between about 0.05 and 5 wt % based on the weight of the water component. A preferred range is between about 0.1 and 2 wt % of gelling agent, based on the weight of water.

The breaker additive can be employed in a quantity between about 0.01 and 25 wt %, based on the weight of gelling agent in the hydraulic fluid composition.

One type of breaker additive compound which can be employed is one which provides an acidic pH to the well-treating hydraulic fluid composition. Such breaker additives include inorganic and organic acids, and compounds such as esters which convert to acidic derivatives under well-treating conditions. Illustrative of suitable breaker additive of this type are sulfuric acid, hydrochloric acid, p-toluenesulfonic acid, acetic acid, triethyl phosphate, methyl formate, ethyl propionate, butyl lactate, and the like. This type of breaker additive can be employed in a quantity between about 0.5 and 20 wt %, based on the weight of hydroxypropylcellulose/poly(maleic anhydride/alkene-1) gelling agent in a hydraulic fluid composition.

Another type of breaker additive compounds which can be employed is oxidizing agents. Illustrative of suitable breaker additives of this type are ammonium persulfate, potassium dichromate, potassium permanganate, peracetic acid, tertiary-butyl hydroperoxide, and the like. This class of breaker additive can be employed in a quantity between about 0.5 and 20 wt %, based on the combined weight of hydroxypropylcellulose and poly(maleic anhydride/alkene-1) in a hydraulic fluid composition.

The hydraulic fluid compositions of the process exhibit excellent solution stability and heat stability in comparison with the corresponding hydraulic fluid compositions containing as a gelling agent any of the conventional industrial gums or cellulosic derivatives. The hydraulic fluid compositions have superior ability to hydrate and develop high viscosity in the presence of salts. Further, a breaker additive can degrade the hydroxypropylcellulose/poly(maleic anhydride/alkene-1) gelling agent at a convenient rate and with a resultant low yield of residue, e.g., a yield of less than about 2 wt % residue, based on the original weight of gelling agent.

Example: This example illustrates the large viscosity increase in an aqueous medium provided by a synergistic mixture of hydroxypropylcellulose and maleic anhydride/alkene-1 copolymer as a gelling agent.

A 1% aqueous solution of each of hydroxypropylcellulose, Klucel H (Hercules), and maleic anhydride/isobutylene copolymer, Isobam HH (Kuraray Co. Ltd.), is prepared. The viscosity of each solution is determined in centipoises with a Brookfield viscometer Model RVF, spindle No. 4 at 20 rpm.

In the preparation of the solutions, dissolution of the maleic anhydride/isobutylene copolymer is aided by adjusting the pH of the aqueous medium into the range of about 6 to 11 with sodium hydroxide.

Equal volumes of each solution are blended together and the viscosity of the blend solution is measured.

	Centipoises
Initial viscosities	
Hydroxypropylcellulose (1%)	2,000
Maleic anhydride/isobutylene copolymer (1%)	9,100
Blend viscosity (1:1)	
Calculated	4,200
Observed	9,200

Hydroxypropylcellulose and Poly(Maleic Anhydride/Alkyl Vinyl Ether)

In a related process, *R.N. DeMartino; U.S. Patent 4,169,818; October 2, 1979; assigned to Celanese Corporation* describes well-treating hydraulic fluid compositions consisting essentially of (1) an aqueous medium, (2) hydroxypropylcellulose/poly(maleic anhydride/alkyl vinyl ether) gelling agent, (3) a breaker additive, and (4) a propping agent, which are eminently suitable for application as well-fracturing fluid media.

The term "alkyl vinyl ether" as employed herein is meant to include comonomers represented by the formula: $CH_2{=}CH{-}O{-}R$ wherein R is preferably limited to alkyl groups such as methyl, ethyl, isobutyl, pentyl, octyl, decyl, and the like, which contain between one and about 10 carbon atoms.

The ratio of the comonomers in the maleic anhydride/alkyl vinyl ether copolymer hydrocolloid usually is in a 1:1 ratio, and in some cases the ratio will vary between 5:4 and 4:5. The molecular weight of the copolymer may vary over a wide range between several thousand and several million. In terms of viscosity as disclosed in U.S. Patent 3,781,203, the specific viscosity values can vary between about 0.1 and 10.

Illustrative of a preferred copolymer hydrocolloid is one corresponding to the structural formula:

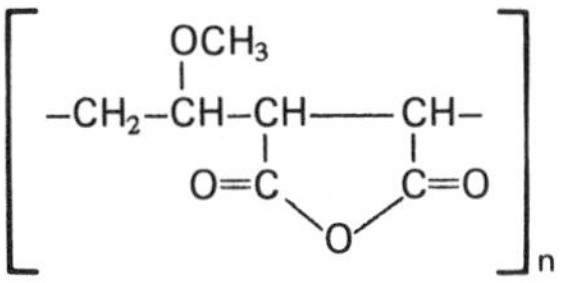

$$\left[-CH_2-\underset{}{\overset{OCH_3}{\overset{|}{C}}}H-CH-\!\!-\!\!-CH- \right]_n \quad \text{(with } O{=}C\text{ and } C{=}O \text{ of the two CH groups joined through O to form the anhydride ring)}$$

where n is an integer between about 11 and 700. A particularly preferred poly(maleic anhydride/alkyl vinyl ether) hydrocolloid is a high molecular weight maleic anhydride/methyl vinyl ether copolymer having a specific viscosity of at least 2.5 as determined in a solution of 1 g of copolymer in 100 ml of methyl ethyl ketone (MEK) at 25°C.

Example: This example illustrates the large viscosity increase in an aqueous medium provided by a synergistic mixture of hydroxypropylcellulose and maleic anhydride/alkyl vinyl ether copolymer as a gelling agent.

A 3% aqueous solution of maleic anhydride/methyl vinyl ether copolymer, Gantrez AN-179 (GAF), and a 1% aqueous solution of hydroxypropylcellulose, Klucel H (Hercules), are prepared. Dissolution of the methyl vinyl ether copolymer is aided by adjusting the pH of the aqueous medium into the range of 6 to 11 with sodium hydroxide.

The viscosity of each solution is determined in centipoises with a Brookfield viscometer Model RVF, spindle No. 4 at 20 rpm.

Equal volumes of each solution are blended together, and the viscosity of the blend solution is measured.

	Centipoises
Initial viscosities	
Hydroxypropylcellulose (1%)	1,800
Methyl vinyl ether copolymer (3%)	2,200
Blend viscosity (1:1)	
Calculated	1,989
Observed	5,900

Thickening Agent of Polygalactomannan Gum and Carboxylated Copolymer

In another similar process, *R.N. DeMartino; U.S. Patent 4,143,007; March 6, 1979; assigned to Celanese Corporation* describes a dry blend composition adapted for application as a thickening agent in aqueous solutions which comprises (1) a polygalactomannan gum and (2) a copolymer of an α,β-olefinically unsaturated dicarboxylic acid anhydride and a comonomer selected from α-olefinically unsaturated hydrocarbons and alkyl vinyl ethers.

Illustrative of a suitable polymer is Gantrez AN (GAF), which corresponds to the structural formula where n is an integer between about 11 and 700:

$$\left[-CH_2-\overset{OCH_3}{\overset{|}{C}}H-CH-\!\!-\!\!-CH- \right]_n \quad \text{(with } O{=}C\text{ and } C{=}O \text{ of the two CH groups joined through O to form the anhydride ring)}$$

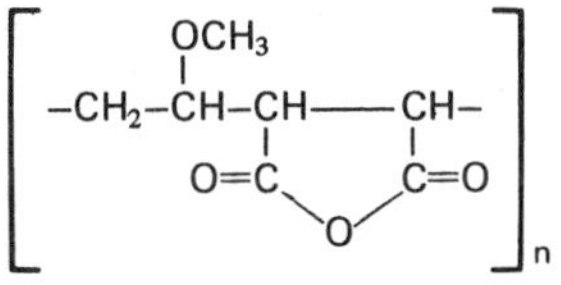

The polygalactomannan gum and copolymer components are employed in the thickener composition in a ratio between about 0.05 and 15 pbw of polygalactomannan gum per part by weight of copolymer.

The aqueous composition is prepared by adding the thickener composition to an aqueous medium in a quantity between about 0.2 and 5 wt %, based on the total aqueous composition weight. On the average the preferred quantity of added thickener composition will be in the range between about 0.3 and 2 wt %, based on the total aqueous composition weight.

A particularly important application of the thickener composition is as a gelling agent in a hydraulic well-treating fluid medium, e.g., a hydraulic fracturing fluid composition.

Methyl Ether of Polygalactomannan Gum

R.N. DeMartino; U.S. Patent 4,169,798; October 2, 1979; assigned to Celanese Corporation describes an hydraulic fluid composition adapted for fracturing of subterranean formations, which comprises (1) an aqueous medium, (2) methyl ether of polygalactomannan gum as a gelling agent, and (3) a breaker additive for subsequent reduction of fluid viscosity.

Polygalactomannan gums swell readily in cold water and can be dissolved in hot water to yield solutions which characteristically have a high viscosity even at a concentration of 1 to 1.5%. Guar gum and locust bean gum as supplied commercially usually have a viscosity (at 1% concentration) of around 1,000 to 4,000 centipoises at 25°C using a Brookfield viscometer Model LVF, spindle No. 2 at 6 rpm.

The breaker additive is preferably an enzyme which under formation-fracturing conditions autonomously degrades the polygalactomannan methyl ether gum gelling agent so as to reduce the viscosity of hydraulic fluid which is under hydrostatic pressure. Although the effect of the enzyme breaker additive commences immediately upon intimate admixture of the polygalactomannan methyl ether gum and the breaker additive, the time required to reduce the solution viscosity by 50% can range over a period between about one-half hour and two hours. The rate of polygalactomannan methyl ether gum degradation is affected by pH, temperature, and salt content of the hydraulic fluid system.

The enzyme breaker additive can be employed in a quantity between about 0.01 and 5 wt %, based on the weight of polygalactomannan methyl ether gum in a hydraulic fluid composition. Hemicellulase enzyme is illustrative of a suitable breaker additive for hydraulic fluid compositions containing methylated guar gum.

The polygalactomannan methyl ether gum component of the hydraulic fluid composition is produced by contacting the gum with a methylating reagent in the presence of a basic compound. The reaction proceeds readily at a temperature between about 0° and 100°C.

Suitable methylating reagents include methyl chloride, methyl bromide, methyl iodide, dimethyl sulfate, and the like. The methylating reagent is employed in a quantity sufficient to provide the desired degree of substitution of hydroxyl groups with methoxyl groups, e.g., a degree of substitution (DS) between about 0.1 and 1.5, and preferably between about 0.2 and 0.5.

By the term "degree of substitution" as employed herein is meant the average substitution of ether groups per anhydro sugar unit in the polygalactomannan gums. On the average, each of the anhydro sugar units contains three available hydroxyl sites.

The basic compound in the reaction process is employed in a quantity which is at least stoichiometrically equivalent to the strong acid which is generated in situ during the course of the reaction. Suitable basic compounds include inorganic and organic derivatives such as alkali metal and alkaline earth metal hydroxides, quaternary ammonium hydroxides, alkoxides, organic acid salts, and the like. Illustrative of basic compounds are sodium hydroxide, potassium hydroxide, calcium hydroxide, barium hydroxide, sodium carbonate, potassium acetate, sodium methoxide, tetramethylammonium hydroxide, and the like.

In conducting the methylation reaction, it is advantageous to form a slurry of the polygalactomannan gum in an aqueous alkanol medium and effect the reaction without completely hydrating the gum reactant. This facilitates recovery of the methylated polygalactomannan gum product. After neutralization of any excess basic compound in the reaction medium, the solid gum product is readily recovered by filtration. Methods of preparing saccharidic ether derivatives are disclosed in U.S. Patents 2,140,346; 2,609,367; and 3,170,915.

The quantity of polygalactomannan methyl ether gum incorporated in the hydraulic composition can vary in the range between about 0.05 and 5 wt % based on the weight of the water component. A preferred range is between about 0.1 and 2 wt % of gum, based on the weight of water.

A.A. DeGuia, R.W. Stackman and A.B. Conciatori; U.S. Patent 4,169,945; October 2, 1979; assigned to Celanese Corporation describe a related process for preparing alkyl ethers of polygalactomannan gum which comprises reacting polygalactomannan gum with an alkyl halide in the presence of (1) at least a stoichiometric quantity of a strongly basic metal hydroxide, based on the weight of alkyl halide, and (2) between about 0.05 and 5 wt % of a water-soluble quaternary ammonium halide phase-transfer agent, based on the weight of polygalactomannan gum; in an aqueous reaction medium at a temperature between about 10° and 100°C for a reaction period sufficient to achieve a degree of substitution by alkyl ether groups between about 0.01 and 3.0.

The term "alkyl halide" is meant to include alkenyl and aralkyl halides such as allyl halide and benzyl halide. The preferred alkyl halides are alkyl chlorides and alkyl bromides containing between 1 and about 20 carbon atoms, e.g., methyl chloride and bromide, pentyl chloride and bromide, decyl chloride and bromide, eicosyl chloride and bromide, and the like.

The quantity of alkyl halide employed is determined by the degree of substitution which it is desirable to achieve. For example, the etherification of 5 pbw of guar gum with 1 pbw of alkyl chloride nominally yields guar gum ether having a 0.3 degree of substitution. A higher relative weight ratio of alkyl halide reactant to galactomannan gum yields a higher degree of substitution. Generally, the preferred degree of substitution is in the range between about 0.05 and 2.5.

The hydroxide component which functions as a reactant/catalyst can vary in quantity between about 1 and 3 mols of hydroxides per mol of alkyl halide present in the reaction system. Any stoichiometric excess of hydroxide corresponds to the quantity not consumed in the Williamson etherification reaction. Illustrative of strongly basic metal hydroxide catalysts are alkali metal hydroxides such as sodium hydroxide, potassium hydroxide and lithium hydroxide.

In addition to the metal hydroxide catalyst, the process is conducted in the presence of a water-soluble quaternary ammonium halide phase-transfer agent. The quaternary ammonium halide transfer agent is employed preferably in a quantity between about 0.1 and 3 wt %, based on the weight of polygalactomannan gum in the reaction system.

Preferred water-soluble quaternary ammonium halide phase-transfer agents are those in which the anion is either chloride or bromide, and the total number of carbon atoms in the phase-transfer agents is in the range between about 4 and 40, respectively. Illustrative of suitable water-soluble phase-transfer agents are tetramethylammonium chloride and bromide, benzyltrimethylammonium chloride and bromide, tetraethylammonium chloride and bromide, tetrabutylammonium chloride and bromide, methylpyridinium chloride and bromide, benzylpyridinium chloride and bromide, trimethyl-p-chlorobenzylammonium chloride and bromide, and the like.

The process is conveniently conducted as a two-phase reaction system comprising an aqueous solution, or an aqueous solution of a water-miscible solvent, in contact with solid polygalactomannan gum. In the case of the water-miscible solvent medium, the water content of the medium preferably is maintained in the range between about 10 and 60 wt %, depending on the particular solvent of choice. If more than an optimum quantity of water is present in this type of reaction medium, then the polygalactomannan gum may swell or enter into solution, thereby complicating product recovery and purification.

Suitable water-miscible solvents for suspension of polygalactomannan gum in the reaction medium include alkanols, glycols, cyclic and acyclic alkyl ethers, alkanones, dialkylformamide, and the like, and mixtures thereof. Illustrative of suitable water-miscible solvents are methanol, ethanol, isopropanol, secondary butanol, secondary pentanol, ethylene glycol, acetone, methyl ethyl ketone, diethyl ketone, tetrahydrofuran, dioxane and dimethylformamide.

Dialkylacrylamide Ether Adduct of Polygalactomannan Gum

R.N. DeMartino and A.B. Conciatori; U.S. Patent 4,137,400; January 30, 1979; assigned to Celanese Corporation describe an N,N-dialkylacrylamide ether adduct of polygalactomannan gum. As employed herein, the term "N,N-dialkylacrylamide" is meant to include N,N-dialkylmethacrylamide.

Illustrative of a preferred class of polygalactomannan ether adduct gum is one represented by the formula:

$$\text{guar}-O-CH_2-CH_2-\overset{\overset{\displaystyle O}{\|}}{C}-N\begin{matrix} \diagup R \\ \diagdown R \end{matrix}$$

wherein R is an alkyl group containing between 1 and about 4 carbon atoms.

The N,N-dialkylacrylamide ether adduct of polygalactomannan gum is produced by contacting natural polygalactomannan gum with N,N-dialkylacrylamide (or N,N-dialkylmethacrylamide) in the presence of a basic compound. The etherification reaction proceeds readily at a temperature between about 0° and 100°C.

The N,N-dialkylacrylamide reactant is employed in a quantity sufficient to provide the desired degree of substitution of hydroxyl groups with oxypropionamido groups, e.g., a degree of substitution between about 0.01 and 3.0, and preferably between about 0.1 and 1.5.

The basic compound is employed in a quantity sufficient to provide an alkaline reaction medium having a pH between about 7.5 and 12. Suitable basic compounds include inorganic and organic derivatives such as alkali metal and alkaline earth metal hydroxides, quaternary ammonium hydroxides, alkoxides, organic acid salts, and the like. Illustrative of basic compounds are sodium hydroxide, potassium hydroxide, calcium hydroxide, barium hydroxide, sodium carbonate, potassium acetate, sodium methoxide, tetramethylammonium hydroxide, and the like.

In conducting the etherification reaction, it is advantageous to form a slurry of the polygalactomannan gum in an aqueous alkaline medium and effect the reaction without completely hydrating the gum reactant. This facilitates recovery of the etherified polygalactomannan gum product. After neutralization of any excess basic compound in the reaction medium, the solid gum product is readily recovered by filtration.

Hydraulic fluid compositions consisting essentially of (1) an aqueous medium, (2) N,N-dialkylacrylamide ether adduct of polygalactomannan gum, (3) an enzyme breaker additive, and (4) a propping agent, are eminently suitable for application as well-fracturing fluid media.

Reaction Product of Hydroxy Ether, Pentavalent Phosphorus and Alcohol

T.J. Griffin, Jr.; U.S. Patents 4,153,649; May 8, 1979; 4,153,066; May 8, 1979; 4,152,289; May 1, 1979; and 4,174,283; November 13, 1979; all assigned to The Dow Chemical Company describes a compound formed by reacting an essentially anhydrous hydroxy ether of the formula ROR_1OH wherein R is a C_{1-6} alkyl group, R_1 is a C_2 or C_3 alkylene group and the total carbon atoms of R_1 and R range from 3 to about 8 with a pentavalent phosphorus compound which is substantially free from acid groups such as Cl, F and the like. When the total carbon atoms in the hydroxy ether is 3 or 4, there is also reacted with the hydroxy ether and phosphorus compounds a long chain aliphatic monohydric alcohol containing at least 5 carbon atoms. A short chain aliphatic monohydric alcohol (C_{1-4}) can also be reacted therewith if desired. When the total carbon atoms in the hydroxy ether is 5 or more, there is reacted with the hydroxy ether and phosphorus compound either a long chain aliphatic alcohol (at least 5 carbons) or a short chain aliphatic monohydric alcohol (C_{1-4}) or a mixture thereof.

The above-defined compounds are reacted with a pentavalent phosphorus compound for a period of time ranging from about 1.5 to 6 hours at a temperature ranging from about 70° to 90°C to form the complex reaction product.

To gel an organic liquid, or to reduce the friction pressure generated by an organic liquid flowing through a confining conduit, the reaction product is dispersed into an organic liquid along with a compound containing a multivalent cation capable of having a coordination number of 6, the reaction product and metal cation compound being employed in amounts and a specific ratio to each other to impart to the organic liquid a desired viscosity, or reduction in friction pressure, under substantially neutral conditions.

The gelled organic liquid can be employed as a fracturing fluid, as a carrying liquid for solids, and other utilities where organic liquids having a viscosity which is greater than the normal viscosity of the organic liquid is useful.

The reactants should be reacted together in certain molar ratios to provide reaction products having the most favorable gelling characteristics. The molar ratios which are operable are set forth in the following table wherein P_2O_5 is the pentavalent phosphorus compound. Where a mixture of P_2O_5 and polyphosphoric acid is employed, the ratios set forth below are based on the total of the mols of P_2O_5 provided by the P_2O_5 component plus the equivalent mols of P_2O_5 provided by the polyphosphoric acid component.

Mol Ratio of Reactants to Total P_2O_5*

	Short Chain Alcohol	ROR_1OH	Long Chain Alcohol
Operable	0-5.0	0.4-4.5	0-4.0
Preferred	0.9-2.0	0.8-1.8	0-1.4

*Or P_2O_5 equivalent.

The mol ratio of the total of the short chain and/or long chain alcohol and the hydroxy ether to total phosphorus pentoxide ranges from about 2.8:1 to 7.0:1 with the most preferred ratio being about 3.64:1.

Suitable compounds containing a multivalent metal cation which can be employed with a separate source of base include ferric nitrate, aluminum nitrate, and rare earth metal salts of the elements of atomic numbers 57 to 71. Preferred are basic aluminum salts such as aluminum isopropoxide (also known as aluminum isopropylate) or an alkali metal aluminate.

Mixture of Polymer, Phenol and Aldehyde

B.L. Swanson; U.S. Patent 4,246,124; January 20, 1981; assigned to Phillips Petroleum Company describes a composition identical to that described in U.S. Patent 4,244,826, page 112 without the acid component. The gelled compositions comprise water; a water-dispersible polymer selected from cellulose ethers, polyacrylamides, biopolysaccharides, and polyalkylene oxides; one or more phenolic components such as resorcinol, catechol, and the like; as well as selected oxidized phenolic components such as 1,4-benzoquinone of natural or synthetic origin and natural and modified tannins.

Any suitable water-dispersible aldehyde meeting the compatibility requirements can be used. Thus, under proper conditions of use, both aliphatic and aromatic monoaldehydes, and also dialdehydes, can be used. The aliphatic monoaldehydes containing from 1 to about 10 carbon atoms per molecule are presently preferred. Representative examples of such aldehydes include formaldehyde, paraformaldehyde, acetaldehyde, propionaldehyde, butyraldehyde, isobutyraldehyde, valeraldehyde, heptaldehyde, decanal, and the like. Representative examples of dialdehydes include glyoxal, glutaraldehyde, terephthaldehyde, and the like. Various mixtures of the aldehydes can also be used. Formaldehyde is the preferred aldehyde compound for use in the process.

Any suitable water-dispersible phenol or naphthol meeting the compatibility requirements can be used. Suitable phenols include monohydroxy and polyhydroxy naphthols. Phenolic compounds suitable for use include phenol, catechol, resorcinol, phloroglucinol, pyrogallol, 4,4'-diphenol, 1,3-dihydroxynaphthalene, and the like. Other phenolic components that can be used include at least one member of selected oxidized phenolic materials of natural or synthetic origin such as 1,4-benzoquinone; hydroquinone or quinhydrone; as well as a natural or modified tannin such as quebracho or sulfomethylated quebracho (SMQ) possessing a degree of sulfomethylation (DSM) up to about 50. (See U.S. Patent 3,344,063.)

Any suitable amount of aldehydes and phenolic compounds can be used. In all instances the amounts of aldehyde and phenolic compound used will be a small, but effective amount which is sufficient to cause gelation of an aqueous dispersion of the polymer, the aldehyde, and the phenolic compound. As a general guide, the amount of aldehyde used in preparing the gelled compositions will be in the range of from about 0.02 to 2, preferably 0.1 to about 0.8 wt %, based on the total weight of the composition. The weight ratio of sulfomethylated quebracho to polymer is in the range of 0.1:1 to 5:1, preferably 0.5:1 to 2:1. The polymer concentration is in the broad range of 1,000 to 50,000 ppm, preferably 3,000 to 20,000 ppm. The concentration of phenolic material (other than SMQ) will be in the range of 0.005 to 2, preferably 0.04 to 1 wt %.

Propping agents can be included in the gelled compositions. Generally speaking, it is desirable to use propping agents having particle sizes in the range of 8 to 40 mesh (U.S. Sieve Series). However, particle sizes outside this range can be employed.

Polymer and Chromium(III), Carbonate and Oxalate Ions

H.S. Golinkin; U.S. Patent 4,137,182; January 30, 1979; assigned to Standard Oil Company (Indiana) describes a method and composition for fracturing of subterranean formations with an aqueous liquid containing a polymeric composition of acrylamide-sodium methacrylate copolymer crosslinked with chromium(III) ion in the presence of carbonate ion as an activating agent and oxalate ion as a gel stabilizer. The polymer concentration (content) is within the range of 0.4 to 1.0 wt % of the total gel weight including aqueous component. The chrome alum concentration (content) is within the range of 0.12 to 1.1 wt % of the total gel weight and the weight ratio of chrome alum concentration to polymer concentration is within the range of 0.26 to 1.1. Preferred range of chrome alum to polymer is 0.50 to 0.55 by weight for 65:35 acrylamide-sodium methacrylate copolymer.

The concentration (content) of carbonate ion as sodium carbonate is in the range of from 0.003 to 0.2% of total gel weight or 0.3 to 50 wt % of polymer weight. The concentration (content) of oxalate ion as sodium oxalate is within the range of 0.01 to 0.03% of total gel weight or 1.0 to 7.5 wt % of polymer weight. If persulfate ion is used as a breaker, the concentration (content) of persulfate ion expressed as ammonium persulfate is within the range of from 0.0005 to 0.2 wt % of total gel weight.

The method of fracturing a subterranean formation comprises the steps of (1) contacting the subterranean formation with an aqueous liquid of the above composition, (2) applying sufficient pressure to fracture the formation, (3) maintaining the pressure while injecting the aqueous liquid into the fracture, (4) and the formed gel breaking within 24 hours of gel formation. Approximately the same method can be used for fluid control such as to control water in subterranean formations, to consolidate sand in incompetent formations and other typical uses by formulating the aqueous gel without the breaker component.

The term "acrylamide-methacrylate copolymer" is defined as a high molecular weight water-soluble polymeric salt of ammonia, sodium or potassium, of a molecular weight range of greater than 1,000,000 consisting essentially of from about 50 to 90 wt %, preferably from about 60 to 70 wt %, acrylamide and from about 10 to 50 wt %, preferably from about 30 to 40 wt %, methacrylate as the sodium salt. The most preferred copolymer consists essentially of about 65 wt % acrylamide and about 35 wt % sodium methacrylate. The defined acrylamide-sodium methacrylate copolymers are further characterized as having a molecular weight sufficient to provide a brine viscosity of at least 1.8 centipoises (measured in Cannon viscometer) when dissolved in a concentration of about 500 ppm in an aqueous brine containing from about 3,650 to 3,750 ppm sodium chloride and from about 365 to 375 ppm calcium chloride made with deionized water. The copolymers can be produced by any known method of conducting polymerization reactions provided substantially no crosslinking or formation of water-insoluble polymer occurs. Solution, suspension or emulsion techniques can be used. The physical form of the copolymer is not critical but it is used in accordance with conventional practices known in the art.

A preferred water-soluble trivalent chromium salt is chrome alum, chromium potassium sulfate, $CrK(SO_4)_2 \cdot 12H_2O$. Substantially any water-soluble chromium salt can be used including chromium acetate, chromium chloride and chromium sulfate.

Carbonate ion can be incorporated into the aqueous gel as an aqueous solution of any carbonate compound in which the carbonate ion appears as $-CO_3$ or $-HCO_3$, and in which the cation of the carbonate compound does not exhibit a detrimental effect upon the aqueous gel. Suitable carbonate salts include sodium carbonate, sodium bicarbonate, potassium carbonate, potassium bicarbonate, ammonium carbonate, and ammonium bicarbonate. The preferred carbonate salts are sodium carbonate and sodium bicarbonate. The carbonate ion is present in an amount sufficient to actuate rapid formation of the gel upon addition of the chromium salt to the copolymer.

Oxalate ion can be incorporated into the aqueous gel composition as an aqueous solution of any water-soluble oxalate salt in which the cation of the oxalate compound does not exhibit a detrimental effect upon the aqueous gel. Suitable oxalate salts include sodium oxalate, $Na_2C_2O_4$, potassium oxalate, $K_2C_2O_4$, and ammonium oxalate, $(NH_4)_2C_2O_4$. The oxalate ion is present in an amount sufficient to stabilize the gel against syneresis upon standing and heating.

Any water-soluble perborate or persulfate can be employed in which the cation of the perborate or persulfate compound does not exhibit a detrimental effect upon the aqueous gel. A preferred source of persulfate ion is ammonium persulfate $(NH_4)_2S_2O_8$, but suitable water-soluble persulfate salts include sodium persulfate, $Na_2S_2O_8$, and potassium persulfate, $K_2S_2O_8$.

Propping agents which can be used include any of those known in the art, e.g., sand grains, walnut shell fragments, tempered glass beads, aluminum pellets, nylon pellets or any mixture of two or more thereof, and similar materials. Such agents can be used in concentrations of about 0.1 to 10 pounds per U.S. gallon of fracturing fluid. In general, propping agents with particle sizes of 6 mesh to about 400 mesh, most preferably 20 to 60 mesh, are employed.

Example 1: The following example shows the slow gelling rate when sodium carbonate or sodium bicarbonate is not used.

A quantity of 1.5 g of 65:35 acrylamide-sodium methacrylate copolymer was stirred into 297 g (297 ml) of 1.0 N sodium chloride brine to make 0.5 wt % copolymer solution. This solution had a pH of 5.5 to 6.0. 4.0 ml of 20 wt % chrome alum solution containing 0.8 g of chrome alum were added. Weight ratio of chrome alum weight to total gel weight was 0.0026. Weight ratio of copolymer weight to total gel weight was 0.0050. Weight ratio of chrome alum weight to copolymer weight was 0.53. The Fann viscosity was determined.

Time (min)	Viscosity (poises)
1	0.67
2	0.59
3	0.56
4	0.58
5	1.4
6	3.6
7	3.5
8	3.9
9	4.2
10	4.4

Example 2: The procedure of Example 1 was repeated except that prior to addition of the chrome alum, 0.079 g Na_2CO_3, 0.026% of total gel weight, was added and the pH determined to be 8.5. After addition of the chrome alum solution, the Fan viscosity reached 8.04 poises in 50 seconds.

Example 3: The procedure of Example 2 was repeated using 0.061 g $NaHCO_3$, 0.020% of total gel weight, in place of the Na_2CO_3. The initial pH was 5.5 to 6.0. The Fann viscosity reached 8.04 poises in 51 seconds.

Example 4: A fracturing fluid consisting of 0.71 wt % copolymer aqueous solution with 0.017 wt % Na_2CO_3, 0.012 wt % $Na_2C_2O_4$ and 0.005 wt % ammonium persulfate is prepared using 1.5 lb of Na_2CO_3, 1.0 lb of $Na_2C_2O_4$ and 0.43 lb of ammonium persulfate per 1,000 gal of 1.0 N NaCl brine solution containing 62 lb of copolymer. The fracturing fluid is continuously mixed by a circulation pump. The fluid is pumped with a metering pump into an in-line mixer at the rate of 0.15 bbl/min simultaneously with 0.0013 bbl/min of a 20 wt % chrome alum solution, which is also pumped with a metering pump, and 9.5 lb/min of 40 to 60 mesh Ottawa washed sand. The sand acts as a proppant to maintain the open fracture after the gel breaks. The fluid is fed into the high-pressure fracturing pump. The resulting stabilized aqueous gel is injected into a 25-ft section sandstone formation in Crawford County, IL, through an injection well at 900 psig surface pressure to fracture the surrounding formation rock. The fracture treatment provides improved fluid injectivity. The aqueous gel breaks within 24 hours of gel formation and the fluid is pumped out from the borehole.

Aluminum Salt of a Phosphate Ester

J.W. Burnham; U.S. Patent 4,200,540; April 29, 1980; assigned to Halliburton Company describes a method for fracturing subterranean formations which are at relatively higher temperatures, i.e., above 150°F. The method is especially effective for fracturing formations in which the temperature at the location to be fractured exceeds about 200°F.

The method can be broadly described as initially forming an aged gel by blending an aluminum salt of a phosphate ester with a hydrocarbon base liquid in a sufficiently low concentration of salt to enable the viscosity of the gel to remain low enough that the gel can be relatively easily handled and pumped. To the aged gel is then added a gel-insoluble solid aluminum salt of a phosphate ester to form a dispersion or suspension of the solid particles of the salt in the gel. The preparation of the gel, and the addition of the solid salt to form the dispersion are generally carried out at above-ground ambient temperatures and, in any event, at a temperature which is less than about 150°F.

After formation of the dispersion, it is pumped downhole to the subterranean formation to be fractured, and is employed, while under elevated pressure, for creating or propagating a fracture in the formation. At the relatively higher temperature of the formation, and particularly at a temperature above about 150°F, the fracturing fluid undergoes an increase in viscosity, and concurrently the suspended solid aluminum salt commences to dissolve in the hydrocarbon base liquid so that a homogeneous gel having a relatively high viscosity is developed and functions effectively in the fracturing operation.

The aluminum salts used in initially making up the aged gel, as well as those then added in solid form to the aged gel to form a dispersion, are broadly aluminum aromatic phosphates, aluminum oxaalkyl phosphates, aluminum oxaalkyl alkyl phosphates and the aluminum aliphatic salts described in U.S. Patent 3,757,864. The salts employed in the preparation of the aged gel are selected from the group of salts having the structural formulas:

$$(1)\qquad \left[\begin{array}{c} \text{O} \\ \| \\ R_1\text{–O–P–O–} \\ | \\ R_2\text{–O} \end{array}\right]_b \text{–Al(OH)}_a$$

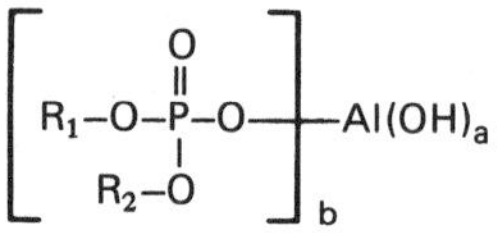

a is 0 to 2,

b is 1 to 3, and the sum of a + b is 3,

R_1 is an aryl or aliphatic-aryl group containing from 6 to 24 carbon atoms, and

R_2 is an aryl, aliphatic-aryl or aliphatic group containing from 1 to 24 carbon atoms, or H,

provided, however, that where R_1 is an aryl group, then R_2 must be either an aliphatic group containing at least 6 carbon atoms, or an aliphatic-aryl group containing at least 12 carbon atoms; and provided that if R_2 is an aryl group, then

R_1 is an aliphatic-aryl group containing at least 12 carbon atoms; and provided that if both R_1 and R_2 are aliphatic-aryl groups, one of R_1 and R_2 contains at least 12 carbon atoms; and provided that where R_1 is an aliphatic-aryl containing less than 12 carbon atoms and R_2 is aliphatic, R_2 must contain at least 6 carbon atoms; and

$$(2)\qquad \left[\begin{array}{c} O \\ \| \\ R_2O-(R_1O)_c-P-O- \\ | \\ R_4O-(R_3O)_d \end{array} \right]_b -Al(OH)_a$$

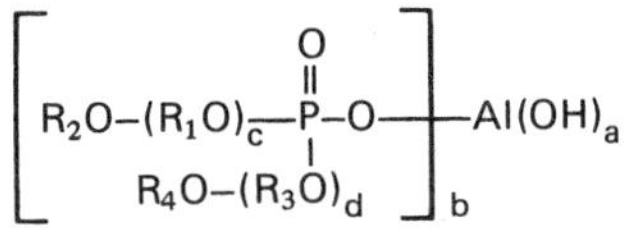

a is 0 to 2,

b is 1 to 3,

c is 1 to 5,

d is 1 to 5, and the sum of a + b is 3;

R_1O and R_3O are alkyloxy, alkenyloxy or alkynyloxy groups containing from 1 to 18 carbon atoms, or $CH_2CH(CH_3)O$, or CH_2CH_2O, and

R_2O and R_4O are alkyloxy, alkenyloxy or alkynyloxy groups containing from 1 to 18 carbon atoms.

R_1O and R_2O may differ from each other but shall together contain from 1 to 24 carbon atoms, and R_3O and R_4O may differ from each other but shall together contain from 1 to 20 carbon atoms, provided that at least one of R_1O and R_3O shall be either $CH_2CH(CH_3)O$ or CH_2CH_2O, and provided further that where either R_1O or R_3O is neither $CH_2CH(CH_3)O$ nor CH_2CH_2O, then the respective R_2O or R_4O group otherwise bonded thereto shall be deleted; and

$$(3)\qquad (HO)_nAl- \left[\begin{array}{c} O \\ \| \\ -O-P-OR \\ | \\ OR_1 \end{array} \right]_m$$

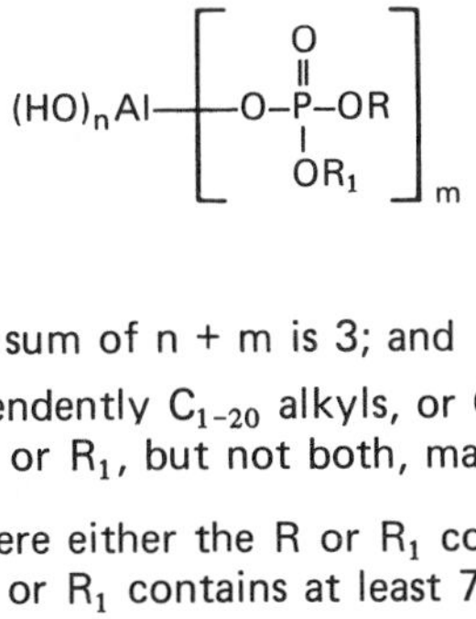

m is 1 to 3,

n is 0 to 2, and the sum of n + m is 3; and

R and R_1 are independently C_{1-20} alkyls, or C_{2-20} alkenyls, or C_{2-20} alkynyls, or R or R_1, but not both, may be H;

provided, however, that where either the R or R_1 contains fewer than 6 carbon atoms, then the other of R or R_1 contains at least 7 carbon atoms.

The aluminum aromatic phosphates having the structural formula (1) appearing above include aluminum aryl phosphates, aluminum aliphatic-aryl phosphates, aluminum aryl aliphatic phosphates, aluminum aliphatic-aryl aliphatic phosphates and aluminum aryl aliphatic-aryl phosphates. The most preferred salts within this group are aluminum aryl aliphatic phosphates in which the aliphatic hydrocarbon substituent contains from 6 to 18 carbon atoms.

The aluminum salts which conform to structural formula (2) above can be the aluminum salts of both mono- and diesters having either one or two oxaalkyl substituents, with the term "alkyl" included within the term "oxaalkyl" being used in the generic sense to include straight and branched chain, saturated and unsaturated aliphatic groups. The most preferred salts of this type are the aluminum salts of phosphate diesters in which the oxaalkyl substituent contains from about 16 to 24 carbon atoms and the second ester substituent in the salt contains from about 1 to 16 carbon atoms.

Examples of aluminum salts conforming to formula (3) above are the aluminum salts of octylethylorthophosphoric acid ester, propynyldecynylorthophosphoric acid ester, methyltetradecylorthophosphoric acid ester, methyldodecylorthophosphoric acid ester, ethyltetradecylorthophosphoric acid ester and octyldodecylorthophosphoric acid ester.

The amount of the aluminum salt placed in the oil-base liquid in making up the initial aged gel in the preliminary steps of the process can be varied in accordance with a number of factors. In general, however, the concentration of the aluminum salt in the oil-base liquid should not exceed about 6.0 wt %.

Example: An oil and gas producing well 15,000 ft in depth is fracture-treated using a slurry or dispersion (at surface ambient temperature) of solid aluminum aliphatic phosphate in a kerosene-base gel. The well has a static bottom hole temperature of 280°F.

The aged gel, as prepared on the surface, and prior to the addition of the solid aluminum salt thereto, contains an aluminum aliphatic phosphate salt prepared by adding 8 gal of a phosphate ester (prepared by reacting phosphorus pentoxide with a mixture of hexanol, octanol, decanol and ethanol) and 2 gal of sodium aluminate-sodium hydroxide solution to 72,000 gal of kerosene. The gel as thus prepared is aged for about 3 hours. 0.9 wt % (based on the total weight of the dispersion formed) of solid aluminum eicosyl octadecyl hexadecyl ethyl phosphate is then blended into the aged gel to form the slurry. The slurry is pumped downhole at a rate of 12 barrels per minute and a pumping pressure of approximately 12,000 psi. In the course of pumping the slurry into the well bore, a high-density proppant material is added to the slurry to enable a total of 47,400 lb of the proppant to be placed in the fracture zone.

J.W. Burnham and R.L. Tiner; U.S. Patent 4,200,539; April 29, 1980; assigned to Halliburton Company describe a related process whereby compounds corresponding to Formula 1 in the previous process are added to an oil-base liquid in an amount of 0.25 to 6.0% without the further addition of aluminum salt. This forms a fracturing composition exhibiting low frictional resistance.

Alkali Metal Silicate Gel

E.A. Elphingstone, M.D. Misak and J.E. Briscoe; U.S. Patent 4,215,001; July 29, 1980; assigned to Halliburton Company describe a method for treating a subterranean well formation which comprises combining an aqueous acid solution with an aqueous alkali metal silicate solution having a pH greater than about 11 in an amount sufficient to lower the pH of the resulting mixture to a level in the range of from about 7.5 to 8.5 thereby forming a polymerized alkali metal silicate gel, shearing the gel to obtain a highly viscous treating fluid having thixotropic properties and then introducing the treating fluid into the subterranean well formation.

A variety of alkali metal silicates can be utilized, e.g., sodium, potassium, lithium, rubidium and cesium silicate. Of these, sodium silicate is preferred, and of the many forms in which sodium silicate exists, those having an $Na_2O:SiO_2$ weight ratio in the range of from about 1:2 to 1:4 are most preferred. A specifically preferred material is a commercially available aqueous sodium silicate solution having a density of 11.67 lb/gal, an $Na_2O:SiO_2$ weight ratio of about 1:3.22 (Grade 40) and having the following approximate analysis:

Component	Percent by Weight
Na_2O	9.1
SiO_2	29.2
Water	61.7
Total	100.0

Of the acids and acid-forming materials which can be used, hydrochloric acid, sulfuric acid, phosphoric acid and mixtures of such acids are preferred with hydrochloric acid being the most preferred. As will be understood by those skilled in the art, hydrofluoric acid cannot be utilized in that its reaction with silicates has an adverse effect on the formation of polymerized silicate gel.

In preparing a highly viscous fluid having thixotropic properties and a high pH for treating a subterranean well formation, an aqueous alkali metal silicate solution having a pH of greater than about 11 is first prepared. Such a solution using Grade 40 sodium silicate solution starting material is prepared by mixing about 5 parts by volume Grade 40 sodium silicate solution with about 95 parts by volume water. The resulting solution has a pH in the range of from about 11 to 12, and a viscosity of about 1 cp. To this solution is added an aqueous acid solution, such as a 20°Bé aqueous hydrochloric acid solution, while agitating the mixture, to lower the pH of the mixture to a value in the range of from about 7.5 to 8.5 whereby the alkali metal silicate polymerizes to form a highly viscous rigid gel. While a polymerized silicate gel will form at pH levels other than from about 7.5 to 8.5, the rate of formation of the gel is greatest in such range.

Upon polymerization of the alkali metal silicate in the manner described above, a highly crosslinked rigid gel structure is formed which is not soluble in water, but which is gelatinous due to water being entrapped in the polymer structure. In order to impart thixotropic properties to the polymerized silicate gel, it is sheared by mixing or agitation, preferably while the polymerization reaction is taking place. It is believed the shearing of the gel divides it into fine particles carrying static charges which will not agglomerate into a mass and which exhibit thixotropic properties, i.e., a low viscosity in turbulent flow but a high viscosity when at rest or at low shear rates.

After being sheared, conventional well-treating additives such as surfactants, friction reducers, fluid loss additives, etc., can be added to the gel as can a propping agent such as sand, and the resulting high viscosity thixotropic fluid is introduced into a subterranean well formation to carry out a treatment therein.

The polymerized silicate treating fluids can be prepared in batch or they can be prepared continuously while being pumped or otherwise introduced into a subterranean well formation. After being introduced into the formation, the polymerized silicate gel dehydrates at a relatively rapid rate, and consequently it is not necessary to include a chemical for breaking the sodium silicate gel in the fluids. The time required for the gel to dehydrate depends on the rate of water loss to the formation and other factors, but generally is within the range of from about 4 to 24 hours. Upon dehydrating, some powdered silicate remains in the treated formation which can readily be removed by contacting the formation with hydrofluoric acid. Prior to the dehydration of the polymerized silicate gel, it has excellent stability, i.e., retains its high viscosity over a wide temperature range (up to about 500°F). The treating fluids are particularly suitable for treating subterranean well formations of low permeability in that they are relatively nondamaging as compared to conventional fluids to such formations, i.e., do not appreciably reduce the permeability thereof.

Fracturing Fluid for Low-Temperature Formations

J. Chatterji; U.S. Patent 4,144,179; March 13, 1979; assigned to Halliburton Company describes gelled aqueous compositions which are comprised of an aqueous liquid, a water-soluble organic gelling agent, a water-soluble free radical generating agent for generating free radicals to degrade the gelling agent, and a water-soluble reducing agent for accelerating the generation of free radicals at low temperatures.

Water-soluble organic gelling agents which readily form viscous gels with aqueous liquids and which are suitable for use in accordance with the process are water-soluble synthetic polymers, water-soluble derivatives of cellulose, water-soluble polysaccharides, water-soluble derivatives of polysaccharide, and mixtures of the foregoing compounds.

The preferred gelling agents which are polyacrylamide polymers which have been hydrolyzed and neutralized, hydroxyethylcellulose, carboxymethylhydroxyethylcellulose and guar gum are readily commerically available and form stable gels of desired viscosity when added to aqueous liquids at relatively low concentrations.

When using the preferred gelling agents mentioned above, they are preferably added to the aqueous liquid in an amount of about 0.25 to 1.5 pbw gelling agent per 100 pbw aqueous liquid.

The free radical generating agents useful in accordance with the process are water-soluble oxidizing agents having the property of generating free radicals for degrading the gelling agent. Such oxidizing agents include water-soluble peroxide compounds, persulfate compounds, or mixtures thereof. Examples of preferred and particularly suitable water-soluble peroxide compounds include, but are not limited to, hydrogen peroxide, tertiary-butyl hydroperoxide and ditertiary-butyl peroxide. Preferred water-soluble persulfates include ammonium persulfate and the alkali metal persulfates.

When utilizing the preferred free radical generating agents mentioned above, they are combined with the aqueous liquid and gelling agent in an amount of about 0.001 to 0.75 pbw free radical generator per 100 pbw aqueous liquid.

Reducing agents which are suitable for use in accordance with the process and which function to accelerate the generation of free radicals over an extended period of time at low temperatures are water-soluble metal salts wherein the oxidation number of the metal ion is less than the highest possible oxidation number for that ion. Examples of such salts are the cuprous, ferrous, stannous, cobaltous, chromous, nickelous, titanous, manganous and arsenous salts of the halides, sulfates and nitrates. Of these, cuprous chloride, ferrous chloride and cobaltous chloride are preferred in that they have a limited solubility in aqueous fluids, and as a result the aqueous treating compositions including these salts remain gelled for periods of time long enough to place and utilize the compositions in subterranean well formations. That is, the reducing agent dissolves slowly so that the reactivity of the free radical generating agent is not accelerated too quickly. The metal salt used is preferably included in the treating composition in an amount at least equal to the amount required to stoichiometrically react with the free radical generating agent present in the composition. This concentration is generally about 0.1 to 7.5 pbw metal salt per 100 pbw aqueous liquid.

While the gelled aqueous compositions of the process have a variety of uses in treating low-temperature subterranean well formations, the compositions are particularly suitable as fracturing fluids. That is, the gelled aqueous compositions can be formed with a viscosity sufficient to bring about the fracturing of a well formation using conventional techniques. Further, the gelled aqueous compositions are able to maintain solid particulated propping agents in suspension so that the propping agent can be placed in the fractures thereby preventing them from closing.

In fracturing a low-temperature subterranean formation, a gelled aqueous composition of this process can be prepared with a quantity of solid particulated propping agent suspended therein. In preparing the composition, the propping agent is preferably added to a quantity of aqueous liquid prior to or simultaneously with the gelling agent and reducing agent and the mixture vigorously agitated to form the gel with propping agent uniformly distributed therein. The free radical generating agent is added to the composition while being agitated, and the composition is then introduced into the formation and forced into fractures created therein so that the propping agent is placed in the fractures.

After the propping agent has been placed, the composition is allowed to revert to a thin fluid and recovered leaving the propping agent in the fractures so that they are held open. As will be understood by those skilled in the art, the recovery of the treating composition can be accomplished in a variety of ways, but generally simply involves placing the treated well formation on production so that the fluids are produced back along with natural fluids from the formation.

Example: A gelled aqueous composition is prepared in the laboratory by dissolving a mixture of 4.8 g (the equivalent of 80 lb per 1,000 gal) of guar gum and 0.3 g of ferrous sulfate in 500 ml of tap water. A highly viscous aqueous composition is formed at a temperature of 70°F. 0.3 g of potassium persulfate is then added to the aqueous composition and the mixture is vigorously agitated with the result that the composition reverts to a thin fluid having a viscosity of less than 2 cp.

High-Temperature Crosslinking Agent

J.W. Ely and J.M. Tinsley; U.S. Patent 4,210,206; July 1, 1980; assigned to Halliburton Company describe a method for treating high-temperature well formations which comprises preparing a composition which forms a highly viscous crosslinked gel at formation temperatures in the range of from about 120° to

350°F, and then introducing the composition into the well formation whereby it is heated and the gel formed.

The compositions are aqueous solutions comprised of water, a gelling agent and a crosslinking agent which is activated at temperatures above about 120°F, namely, hexamethoxymethylmelamine. At temperatures of about 120°F and above, the composition forms a highly viscous crosslinked gel which remains highly viscous for long periods of time at temperatures up to and including about 350°F.

Gelling agents useful herein are selected from the group consisting of water-soluble hydratable polysaccharides having a molecular weight of at least about 100,000, preferably from about 200,000 to 3,000,000 and derivatives thereof; water-soluble synthetic polymers such as high molecular weight polyacrylamides; water-soluble hydratable polysaccharides which have been crosslinked with a compound selected from the group consisting of dialdehydes having the general formula: $OHC(CH_2)_nCHO$, wherein n is an integer within the range of 0 to about 3; 2-hydroxyadipaldehyde; dimethylolurea; water-soluble urea formaldehyde resins; water-soluble melamine formaldehyde resins; and mixtures thereof.

In preparing the compositions, one or more of the abovedescribed gelling agents are added to water in an amount in the range of from about 0.1 to 13 pbw gelling agent per 100 pbw of the water used. The temperature-activated hexamethoxymethylmelamine crosslinking agent is preferably combined with the aqueous gelling agent solution in an amount in the range of from about 0.05 to 5 pbw crosslinking agent per 100 pbw of water used. When the resulting composition reaches a temperature of about 120°F, the hexamethoxymethylmelamine reacts with the gelling agent to form a crosslinked highly viscous semisolid gel.

The crosslinking reaction takes place at a high rate when the aqueous composition is maintained at a pH in the range of from about 2 to 6. A pH of about 4 to 5 is preferred. In order to insure that the desired pH is retained for a period of time sufficient to permit the composition to be introduced into a well formation, a buffer can be incorporated into the composition. Examples of suitable buffers are potassium biphthalate, sodium biphthalate, sodium hydrogen fumarate, and sodium dihydrogen citrate. Of these, sodium biphthalate is preferred and is preferably combined with the composition in an amount in the range of from about 0.05 to 2 pbw buffer per 100 pbw of water utilized.

A particularly suitable composition for use as a treating fluid in high-temperature well formations is comprised of water; a cationic mixture of copolymers of acrylamide and a quaternary amine acrylate having $\overline{M}w$ 1,000,000 in an amount from about 0.05 to 3 pbw/100 pbw of water used; hydroxyethylcellulose having an ethylene oxide substitution of 1.5 mols per anhydroglucose unit in an amount from about 0.05 to 5 pbw/100 pbw of water; hydroxyethylcellulose having an ethylene oxide substitution of about 1.5 mols per anhydroglucose unit crosslinked with about 0.8 pbw glyoxal per 100 pbw hydroxyethylcellulose in an amount from about 0.05 to 5 pbw/100 pbw of water; hexamethoxymethylmelamine in an amount from about 0.05 to 4 pbw/100 pbw of water; and sodium biphthalate in an amount from about 0.05 to 0.5 pbw/100 pbw of water.

While this composition has some viscosity after being prepared at ambient temperatures ordinarily encountered at the surface, the glyoxal crosslinked hydroxyethylcellulose component is not appreciably hydrated until reaching a temperature of about 100°F. Further, as stated above, the crosslinking agent, hexamethoxymethylmelamine, does not begin to crosslink the hydrated gels until reaching a temperature of at least about 120°F.

In carrying out the method for propping a fracture in a high-temperature well formation, after the treating composition having propping agent suspended therein is prepared, it is introduced, such as by pumping into the formation and into one or more fractures therein so that the composition is heated, the gelling agents contained in the composition are hydrated and the resulting hydrated gel crosslinked. The highly viscous semisolid gel formed in the fracture or fractures maintains the propping agent in uniform suspension and as a result, when the fracture or fractures are caused to close on the propping agent they are uniformly held open. The crosslinked gel is next caused to break or revert to a less viscous fluid so that it is removed from the formation without disturbing the propping agent distributed within the fracture or fractures.

Friction-Reducing Polymers

Friction-reducing polymers are added to fracturing fluids to reduce turbulence and consequent energy loss in the flow of the fluid from the surface to the formation.

K.G. Phillips and W.E. Hunter; U.S. Patent 4,152,274; May 1, 1979; assigned to Nalco Chemical Company have found that water-soluble copolymers of acrylamide and either dimethylaminoethyl methacrylate (DMAEM) quaternary salts or dimethylaminopropyl methacrylamide (DMAPM) quaternary salts are useful as friction reducers in fresh water, in brines containing polyvalent cations, as well as in acidic environments.

The DMAEM copolymers should contain from about 40 to 95% by weight acrylamide and the remainder DMAEM quaternary salt; preferably, the copolymers comprise 70 to 85% acrylamide. The DMAPM copolymers should contain 40 to 95% by weight acrylamide; 70 to 90% by weight acrylamide is preferred.

The quaternary salts of DMAEM AND DMAPM are formed by reacting these molecules with C_{1-4} alkyl salts. Preferred alkyl salts include methyl chloride and methyl sulfate. Other useful quaternary agents would include ethyl bromide and propyl chloride.

The molecular weight of the copolymers described above should be at least 100,000. Preferably, the molecular weight will lie in the range 1,000,000 to 3,000,000. Treatment dosages will be at least 50 ppm by weight. Preferably, the treatment dosage will lie in the range 100 to 300 ppm.

Emulsion of Low Oil Content

W.M. Salathiel, T.W. Muecke, C.E. Cooke, Jr., and N.N. Li; U.S. Patent 4,233,165; November 11, 1980; assigned to Exxon Production Research Company describe a well treatment method employing a dispersion of a water-in-oil emulsion in an aqueous medium. The emulsion comprises an internal aqueous phase and an external hydrocarbon phase containing a liquid hydrocarbon and a surfactant soluble in the hydrocarbon. For most well treatments, the water-in-oil emulsion dispersed in the aqueous medium should comprise from about 30 to 95 vol % of the well treatment composition, with a more preferable range being between about 60 and 80 vol %.

The water-in-oil emulsion is prepared by vigorously mixing a liquid hydrocarbon-surfactant blend with an aqueous fluid to form a stable, fine-grained emulsion. The hydrocarbon-surfactant blend contains soluble surfactant, the surfactant concentration generally being in the range of 0.5 to 40 wt % and preferably 3.0 to 25 wt %. The mixing operation for the emulsion should be designed to form an emulsion having internal aqueous droplets with an average diameter of from about 0.01 to 100 μ and preferably from about 0.1 to 10 μ. The external oil phase comprising the hydrocarbon-surfactant blend should amount to from about 3 to 50% of the total volume of the emulsion and preferably from about 5 to 25%. Once a stabilized emulsion is prepared, it is dispersed in an aqueous suspending medium to form the well treatment fluid. Since the overall oil phase concentration in the dispersion will generally be less than about 25%, the well treatment compositions used in the process are much lower in cost than conventional oil-in-water emulsions having significantly higher oil contents.

The well treatment compositions can be used in hydraulic fracturing and acidizing operations as well as numerous other well treatment techniques. In a hydraulic fracturing operation, the water-in-oil emulsion is uniformly blended into an external aqueous medium to form a fracturing fluid. The fracturing fluid is then injected into a subterranean formation at sufficient pressure to fracture the formation.

Viscosifiers and other chemicals may be added to the external aqueous phase to increase the viscosity of the fracturing fluid to delay the onset of turbulence and to stabilize the dispersion of the emulsion in the aqueous phase. For acidizing operations, acid may be added to the external aqueous phase or to the internal aqueous droplets of the emulsion. If added to the internal aqueous droplets, the acid will not be released until the droplets break apart upon entering the smaller pore spaces found deep in the formation. Well treatment fluids such as contaminant scavengers, can also be dissolved in the aqueous droplets. Well contaminants soluble in the external aqueous medium can permeate through the external oil phase of the emulsion and react with the contaminant scavenger entrained in the aqueous droplets of the emulsion. The aqueous droplets can also serve to encapsulate particulate matter such as plugging particles used for well diversion treatments.

The liquid hydrocarbon selected can be a crude oil or a refined petroleum fraction such as diesel oil, gas condensate, gas oil, kerosene, gasoline, and the like. Particular hydrocarbons, such as benzene, toluene, ethylbenzene, cyclohexane, hexane, decane, hexadecane and the like can also be used. Crude oil, however, is normally preferred because it is usually readily available at a well site and compatible with hydrocarbon-bearing formations. However, if crude oil is unavailable, then hydrocarbon liquids which have a viscosity less than about 10 cp at formation temperatures (e.g., diesel oil) are preferred.

The liquid hydrocarbon can be blended with a wide variety of different oil-soluble surfactants. Surfactants useful in forming a compatible mixture with liquid hydrocarbons include anionic, cationic and nonionic surfactants. Suitable anionic surfactants include fatty acid soaps which are the salts of long chain fatty acids derived from naturally occurring fats and oils and salts of alkylbenzene sulfonic acids. A preferred anionic surfactant is the morpholinium salt of tetracosanylbenzene sulfonic acid. The ammonium and alkali metal salts are also suitable. Cationic surfactants include amine salts such as polyoxyethylene amine as well as quaternary ammonium compounds. Particularly useful cationic surfactants include high molecular weight alkyl imides and amides of polybasic amines. A highly preferred cationic surfactant of this type is ECA 4360 (Exxon Chemical Company). Another useful surfactant is an amine surfactant ENJ 3029 (Exxon Chemical Company). Suitable nonionic surfactants include derivatives of glycerides, glucosides, polyoxyethylene and polyoxypropylene. Typical nonionic surfactants include ethoxylated linear alcohols and ethoxylated alkyl phenols. A preferred nonionic surfactant is a sorbitan fatty acid, Span 80 (Atlas ICI). Mixtures of surfactants can also be used. For example, mixtures of Span 80 and ECA 4360 are particularly suited to forming stable, strong water-in-oil emulsions.

Example: A well treatment composition was prepared by finely dispersing a water-in-oil emulsion in an aqueous solution. The water-in-oil emulsion was prepared by first blending together No. 2 diesel oil and ENJ 3029 in a ratio of 3:1 to form a hydrocarbon-surfactant mixture. An aqueous saline solution having 2 wt % NaCl was then slowly and continuously added to the hydrocarbon-surfactant mixture while virorous shearing was applied until the volumetric water-to-oil ratio was about 9:1. Vigorous mixing in a Waring blender was maintained at 3,000 rpm for 30 minutes to form a stable, fine-grained water-in-oil emulsion. The average size of the internal phase water droplets was about 2.25 μ while the largest droplets were almost 14 μ.

After the emulsion was prepared, an additional 4 parts by volume of the 2% saline solution was gently mixed into the emulsion. Some of the aqueous solution initially added to the emulsion continued emulsifying and increased the droplet size of the aqueous phase in the emulsion. However, most of the excess saline water added did not emulsify and instead remained externally separate from the original emulsion.

Shear was then applied to the emulsion which caused the larger water droplets to interconnect and form a continuous external aqueous phase in which globules of water-in-oil emulsion are dispersed. The globules of emulsion generally contain the smaller droplets of water which are in the 10 to 15 μ range. The individual globules are quite coarse and generally vary in size from about ½ to 4 inches. The final composition of the dispersion contained 93.8 wt % water, of which about 60 wt % was present in the dispersed water-in-oil emulsion. The diesel oil-surfactant blend amounted to only 6.2 wt % of the total composition.

Mixture of Water, Chloroform and Surfactant

J.J. Meister; U.S. Patent 4,148,736; April 10, 1979; assigned to Phillips Petroleum Company describes a thixotropic fluid obtained by suitably emulsifying a mixture of water and chloroform in the presence of minor amounts of an anionic surfactant, a cationic surfactant, an acid, and a salt, to form a viscosified surfactant fluid which has thixotropic properties.

This thixotropic fluid decreases in viscosity after a period of time, depending upon the pH of the aqueous phase. It has been found that this viscosified fluid loses viscosity rapidly, e.g., within about 8 hours, if this pH is 0; within about 1 to 2 days if this pH is about 1; and within about 1 to 2 months if this pH approaches the upper limit of 4.5. Thus the fluid can be modified to have a low viscosity after a time when the high viscosity originally achieved is no longer desired by merely adjusting the pH of the fluid.

The major component of the composition consists essentially of an emulsified mixture of water and chloroform containing 10 to 90%, preferably 50 to 85%, water by weight, based on the total combined weight of the water and chloroform. The water can be fresh water or a brine which is compatible with the system.

The cationic surfactant present in the composition is selected from the group of compounds consisting of the quaternary ammonium salts of amines, including aliphatic primary, secondary, and tertiary monoamines, aliphatic polyamines, rosin-derived amines, alkyl amino-substituted 2-alkylimidazolines, polyoxyethylene alkyl amines, polyoxyethylene alicyclic amines, N,N,N',N'-tetrakis-substituted ethylenediamines, and amines with amide linkages. The aliphatic polyamines preferably are alkylene diamines or dialkylene triamines. The number of carbon atoms in the cationic surfactant preferably is in the range of 10 to 30. A preferred example for a cationic surfactant is diisobutylphenoxyethoxyethyldimethylbenzylammonium chloride monohydrate. This compound is commercially available as Hyamine 1622 (Rohm and Haas) in 98.8 wt % purity.

The anionic surfactants in the preferred embodiment comprise those organic carboxylates, sulfonates, sulfates, or phosphates which are known to have surfactant properties. The preferred anionic surfactants are the ammonium and sodium salts of petroleum sulfonic acids having an average molecular weight in the range of from about 200 to 550.

Both the cationic surfactant and the anionic surfactant are each present in the composition in a concentration of 1 to 5,000 ppm, preferably 50 to 2,000 ppm, by weight, based on the combined weight of the water and chloroform. The molar ratio of anionic surfactant to cationic surfactant is in the range of 4:1 to 0.1:1, preferably about 2:1 to 0.5:1.

The acids which are applicable for use in the process can be any convenient acids which are sufficiently ionizable to reduce the pH of the resulting composition to the desired level in the range of pH 0 to 4.5. Thus, depending on the desired pH, the acids can be inorganic acids such as hydrochloric, phosphoric, sulfuric, as well as suitable organic acids such as formic, acetic, chloroacetic acid, and the like. Mineral acids are preferred.

The amount of acid incorporated into the viscosified compositions will depend upon the nature of the acid, the nature of the other ingredients, and upon the specific pH level desired. Ordinarily the amount of acid will be in the range of 50 to 20,000 ppm, more usually 100 to 5,000 ppm, by weight based on the combined weight of the water and chloroform. As an additional requirement, the amount of acid incorporated will be sufficient to adjust the pH of the mixture to a value in the range of from about 0 to 4.5, preferably about 0.5 to 4.5.

The composition must contain a water-soluble alkali metal buffering salt. This salt should be water-soluble and capable of acting to some extent as a buffer. Preferably the salt's anion is the same as that of the acid. The cation of the salt preferably matches that of the anionic surfactant. The amount of salt incorporated into the viscosified composition will vary depending upon the amount and nature of the acid present, as well as upon the amounts and nature of the other ingredients. Ordinarily, the amount of salt will be in the range of 50 to 20,000 ppm, more usually 1,000 to 12,000 ppm, by weight based on the combined weight of the water and chloroform. For best effectiveness, the molar ratio of the salt to the acid should be in the range of about 0.1:1 to 5:1. Molar ratios of about 2:1 are particularly effective.

Examples of salts useful as buffer salts in the composition are lithium sulfate, lithium chloride, sodium chloride, sodium formate, sodium sulfate, sodium nitrate, sodium acetate, potassium chloride, potassium sulfate, and potassium nitrate. The preferred buffer salts are sodium chloride and sodium sulfate.

Other additives of organic or inorganic nature can be incorporated into the composition in small amounts if they have no detrimental effect upon the viscosity or thixotropy.

The composition is prepared by mixing the components in any suitable manner. In order to obtain a viscous composition, the components have to be vigorously agitated. The agitation preferably is carried out for at least one minute and should be sufficient to emulsify the composition. The temperature used for the mixing procedure is not critical, provided the components of the mixture are maintained in a substantially liquid state. Temperatures around the ambient temperature, such as 24°C, are preferred.

The composition can have an initial viscosity in the range of about 5 to 40,000 cp at 24°C. The composition can form a virtually solid gel after agitation. The viscosity is measured on a Brookfield viscosimeter using the procedure disclosed in ASTM D 1824-66.

These compositions are valuable drilling fluids, fracturing fluids, and surfactant fluids for chemical flooding in tertiary oil recovery.

Delayed Release Breaker Additive

After a high-viscosity aqueous gel or emulsion has been pumped into a subterranean formation and one or more fractures formed therein, it is desirable to convert the gel or emulsion into a low-viscosity fluid, referred to herein as "breaking" the gel or emulsion, so that it can be recovered from the formation through the well bore. While a variety of chemicals which are added to high-viscosity fracturing fluids have been utilized for breaking the fluids, hereinafter referred to as "breakers," problems caused by insufficient breaking or the breaking of the fluid too quickly are often experienced.

Ideally, a high-viscosity fracturing fluid or other well treating fluid has an initial high viscosity, stability during the well treatment and controlled breaking after the treatment. Heretofore, the control of the breaking of gelled aqueous fluids, and particularly water-hydrocarbon emulsions, has been inadequate resulting in less than desirable treatment results and/or clean-up problems after the treatment has been carried out. For example, attempts to use demulsifiers for breaking water-hydrocarbon emulsions have generally been unsuccessful in that rapid or instantaneous breaking of the emulsion takes place when the demulsifier is added thereto decreasing the viscosity of the emulsion.

Described by *J.W. Burnham, J.E. Briscoe and E.A. Elphingstone; U.S. Patent 4,202,795; May 13, 1980; assigned to Halliburton Company* are methods and additives for delaying the release of chemicals such as demulsifiers, gel breakers and the like in aqueous fluids whereby such release is accurately controlled and the desired results achieved. In accordance with the process, the chemical to be released in an aqueous fluid is combined with a solid hydratable gelling agent and a breaker for the gel formed by the gelling agent when hydrated. The mixture is then formed into prills or pellets, preferably having a size in the range of from about 20 to 40 mesh (U.S. Sieve Series). Upon combining the pellets with an aqueous fluid into which the chemical is to be released, the gelling agent in the pellets hydrates and forms a protective gel around each of the pellets which prevents the release of the chemical into the aqueous fluid for the time period required for the protective gel to be broken by the gel breaker in the pellets. Once the gel breaker has broken the protective gel, the chemical in the pellets is released into the aqueous fluid. The time required for the protective gel to be broken can be varied by varying the quantities of hydratable gelling agent and gel breaker utilized in the pellets and by using different gelling agents and gel breakers.

A variety of solid hydratable gelling agents can be utilized, such as hydratable polysaccharides, polyacrylamides, polyacrylamide copolymers and polyvinyl alcohol.

Breakers for the gels produced by the gelling agents mentioned above when hydrated which are preferred for use in accordance with the process are mild oxidizing agents, enzymes, acids and mixtures of such compounds. Examples of particularly suitable oxidizing agents are sodium persulfate and ammonium persulfate. Examples of suitable enzymes which can be utilized are α- and β-amylases, amyloglucosidase, oligoglucosidase, invertase, maltase, cellulase and hemicellulase. An example of a suitable acid is fumaric acid. Of these, a mixture of sodium persulfate, hemicellulase and cellulase is most preferred.

In using the pelletized additives for delayed demulsification of water-hydrocarbon emulsions, the demulsifying surface active agent utilized in the additives is preferably of opposite charge from the charge of the emulsifying agent used to form the emulsion or nonionic. That is, if an anionic emulsifying agent is utilized in forming the emulsion, the demulsifying agent utilized in the additive of the process is preferably either cationic or nonionic. Conversely, if the emulsifying agent is cationic, the demulsifying agent is preferably anionic or nonionic.

The most preferred additive for demulsifying gelled water-hydrocarbon emulsions formed with cationic emulsifying agents is comprised of hydroxypropylguar present in the additive in an amount in the range of from about 40 to 60% by wt of the additive, a breaker comprised of a mixture of hemicellulase, cellulase and sodium persulfate, the enzymes being present in the additive in an amount in the range of from about 0.25 to 5.0% by wt, and the sodium persulfate being present in an amount in the range of from about 1 to 4% by wt and sodium lauryl sulfate present in the additive in an amount of about 40 to 60% by wt of the additive.

In applications for hydraulic fracturing and/or fracture-acidizing fluids, the type of breaker used and the quantity of breaker included in the pellets is preferably such that complete breaking of the fluids takes place when pellets are combined with the fluids in an amount of about 0.005 to 0.5% by wt of the fluid.

Low Fluid Loss Foam

G.E. King; U.S. Patent 4,217,231; August 12, 1980; assigned to Standard Oil Company (Indiana) describes a process which employs certain alcohols and/or organic acids (those unsubstituted alcohols and unsubstituted aliphatic monocarboxylic acids having between 5 and 10 carbon atoms), malonic acid, lower n-alkyl diesters of malonic acid, and their mixtures in connection with the use of aqueous foams in operations such as enhanced recovery, hydraulic fracturing, well cementing, acidizing, and drilling.

Generally, the foam is to be used in about 55 to 95% quality (55 to 95% of the total volume at working pressure is gas) and about 65 to 90% quality is preferred. Tests comparing 70%, 80%, and 90% quality foams showed that fluid loss did not vary significantly with foam quality.

Normally, the fluids to be introduced into a well will contain several ingredients (possibly including suspended solids) as appropriate for the particular operation being performed, as known in the art.

It is generally more convenient to mix the additives with the water and other liquid ingredients first, then add any solids which are to be suspended in the fluid. The liquid is then pressurized and gas added to produce a pressurized, foamed fluid. Generally, it is convenient to add the gas as the fluid is being introduced into the well.

The amount of fluid loss additive is generally from about 0.0005% by volume of the aqueous liquid (before the gas is added) up to the maximum soluble amount.

By way of example, the following formulation is appropriate for a 50,000 gal foam hydraulic fracturing job on a formation at 3,950 ft down, 5½ inch casing. In this particular design a 5,000-gal pad is introduced first, followed by 45,000 gal of sand-carrying foam. An initial solution is prepared by mixing 8.75 gal of octyl alcohol and 87.5 gal of surfactant in 17,500 gal of water. 10% of this mixture is used for the pad with approximately 120,000 scf of nitrogen gas being added to produce the foam. 67,500 lb of sand are added to the remaining 90% of the original water-octyl alcohol mixture and then approximately 1,080,000 scf of nitrogen is added to produce approximately 1.5 lb of sand/gal in 45,000 gal of sand-carrying foam.

As an additional example, the following is a design of a 25,000 gal foam acidizing treatment at 3,100 ft. It uses 8,750 gal of 15% concentration HCl, 44 gal of foaming agent (an alkyl aryl sulfonate), 0.5 gal of hexanoic acid, and 297,896 scf of nitrogen. The hexanoic acid and foaming agent are mixed in the hydrochloric acid. This acid mix is then pumped at 5.25 barrels per minute to the wellhead, where nitrogen is injected at a rate of 7,507 scf per minute. This results in a 65 quality foam being introduced into the well at a rate of 15 barrels per minute.

Similarly, of course, these fluid loss additives can be used in aqueous forms for well cementing purposes, for enhanced recovery, or for a drilling fluid for drilling wells.

ENHANCED OIL RECOVERY METHODS

SURFACTANT WATERFLOODING SYSTEMS

Sulfated Polyoxyethylene Alcohol Thickener

The recovery of oil from a subterranean oil formation may be divided into stages. In the first stage, known as primary oil recovery, the oil is forced out by natural flow, gas lifting, gas repressurization or by pumping methods. When the ability of these methods to recover the oil is exhausted there still remains a considerable quantity of oil in the formation and in order to recover some of this oil alternative recovery techniques may be brought into play in the secondary oil recovery stage.

The secondary oil recovery techniques may vary in detail but broadly most involve waterflooding the formation in order to force the oil towards the producing well or wells. The main difficulty with this technique arises from the location of the oil which is usually in small pores and capillaries of the formation where it tends to adhere to the rock and from which by reason of the high interfacial tension between the water and the oil it is difficult to displace. One method of enhancing the waterflooding process is to contact the oil with a surfactant composition in front of the waterflood as the interfacial tension between the surfactant solution and the oil is much less than between water and the oil so that the latter is now more readily displaced. Sulfated polyoxyethylene alcohols are included among the surfactants which have been suggested for this purpose.

T. Cox and R.I. Hancock; U.S. Patent 4,199,027; April 22, 1980; assigned to Imperial Chemical Industries, Ltd., England describe a process for the recovery of oil from a subterranean formation which comprises the step of treating the formation with a salt of a sulfated polyoxyethylene alcohol in which at least 40% by weight of the alcohol portion of the sulfated polyoxyethylene alcohol is a C_8 to C_{20}, preferably C_{10} to C_{16}, primary alcohol having a C_1 to C_3 alkyl substituent in the 2-position of an otherwise linear chain.

The 2-(C_{1-3} alkyl) branched alcohol may comprise 40 to 90% by weight of the alcohol portion of the sulfated polyoxyethylene alcohol, particularly 45 to 80%

by weight. The balance of the alcohol portion suitably comprises a linear primary alcohol which is preferably 5 to 55 wt % of the alcohol portion and generally of the same carbon number as the 2-(C_{1-3} alkyl) branched chain alcohol. Mixtures of alcohols which may be ethoxylated and sulfated to make the sulfates used in the process may be made by carbonylation of a linear α-olefin or mixture of linear α-olefins so that a product is obtained which contains, for example, 50% by weight linear alcohol and 45% by weight 2-(C_{1-3} alkyl) branched primary alcohol, the remaining 5 wt % being 2-($>C_3$ alkyl) branched alcohol. In particular the linear α-olefins may be a mixture of C_{12} and C_{14} α-olefins which give a mixture of linear and 2-(C_{1-3} alkyl) branched C_{13} and C_{15} alcohols [50% by weight linear, 45% by weight 2-(C_{1-3} alkyl) branched; 67% by weight C_{13} and 33% by weight C_{15}].

If desired, the mixture obtained by the carbonylation process may be further refined by distilling out some of the linear alcohol to increase the proportion of 2-(C_{1-3} alkyl) branched chain alcohol in the mixture, e.g., so that it contains up to 50 to 60% by weight branched chain product, the balance including 37 to 27 wt % linear alcohol. The C_{1-3} alkyl substituent in the 2-(C_{1-3} alkyl) branched alcohol generally comprises a mixture of methyl, ethyl and propyl groups, the methyl group predominating.

The sulfated polyoxyethylene alcohol has the general formula

$$[RO(CH_2CH_2O)_nSO_3]_mX$$

where n indicates an average value and is an integer lying in the range 1 to 50 preferably 1 to 10, X is an alkali metal cation preferably sodium, an alkaline earth metal cation, e.g., calcium magnesium, a higher valent metal cation, e.g., aluminum, or an ammonium or amine cation, R is an alkyl radical derived from the alcohol portion of the molecule and m is the valency of X. An amine cation may be derived from an aliphatic or aromatic mono-, di- or polyamine, preferably from a mono-, di- or polyamine containing up to 20 carbon atoms, e.g., stearylamine, ethylenediamine, hexamethylenediamine, diethylenetriamine or triethylenepentamine.

The salt of the sulfated polyoxyethylene alcohol which is used in the process may be used in an aqueous solution or dispersion in a concentration of up to 5% by weight. Clearly, it is desirable for cost reasons to use as little alcohol sulfate as possible and in general concentrations in the range 0.1 to 1.5% by weight are satisfactory.

It is one of the advantages of the sulfates pf polyoxyethylene alcohols in general that they are effective in water of high ionic strength, i.e., greater than 1%, e.g., 5 to 10% by weight salinity and in the presence of calcium, magnesium and iron (ferric) ions. In addition to this advantage the polyoxyethylene alcohol sulfates used in the process confer, as compared with their analogues derived from linear and other branched chain alcohols, an enhanced thickening of an aqueous solution and in comparison with the linear alcohol derivatives a greater resistance to shear stress, i.e., the thickening effect has less tendency to diminish under the mechanical forces experienced as the composition is forced into the formation.

Adjustment of Surfactant System Using Distribution Coefficients

It is known in the art to form a surfactant system using a petroleum sulfonate, brine, and a cosurfactant, such as an alcohol. It is known that the cosurfactant should be a material having moderate solubility in water and excellent surfactant systems can be made using a cosurfactant with a solubility in water within the range of 0.5 to 20 g/100 g of water. However, a cosurfactant having what would otherwise appear to be ideal solubility characteristics, may in fact not be optimum for a given type of oil.

J.E. Vinatieri; U.S. Patent 4,239,628; December 16, 1980; assigned to Phillips Petroleum Co. describes a process to tailor a surfactant system to a specific oil based on cosurfactant distribution coefficients.

By distribution coefficient as used herein, is meant the numerical ratio of the concentration of cosurfactant in the upper oil phase divided by the concentration of cosurfactant in the lower aqueous phase, after mixing oil and surfactant and allowing time for separation into three phases, the predominantly oil upper phase, the predominantly aqueous lower phase and a microemulsion middle phase.

Systems having a distribution coefficient within the range of 0.3 to 3 and even systems outside of this range are highly effective in tertiary oil recovery operations. However, in accordance with this process any given system can be improved by varying the cosurfactant so that the distribution coefficient is nearer to 1. The preferred systems are those in which the distribution coefficient is within the range of 0.6 to 1.5, preferably 0.8 to 1.2, most preferably about 1.

The process for preparing an aqueous surfactant system comprises:

(A) Preparing a surfactant system comprising brine, a petroleum sulfonate, and a cosurfactant having a solubility in water within the range of 0.5 to 20 g/100 g of water;

(B) Mixing the resulting surfactant system with oil corresponding to that to be produced, allowing the system to equilibrate and form three phases, an upper predominantly oil phase, a middle microemulsion phase, and a lower predominantly aqueous phase;

(C) Analyzing the upper and lower phases for cosurfactant content; and

(D) Preparing a new surfactant system utilizing a cosurfactant of differing solubility from that of (A) so as to give a system which on equilibration with oil will result in the distribution of the cosurfactant more nearly equally into the upper and lower phases.

Example: A 49.2 g portion of an aqueous saline solution containing 1.4 wt % sodium chloride (14,000 ppm) was mixed with 30.8 g of a stock solution prepared by mixing 3 g 3-pentanol, 3 g of petroleum sulfonates having an equivalent weight of about 420 (Witco TRS 10-410), and 51.75 g of dodecane. After equilibration, the three-phase system consisted of a top oil phase, a middle microemulsion phase, and a bottom water phase.

Analysis of the top phase and the bottom phase for 3-pentanol gave a distribution coefficient of about 1.06 which indicated that the concentration of 3-pentanol was about the same in the top oil phase as in the bottom water phase. The middle microemulsion phase was found to be dilutable with either 8 vol % of additional dodecane or 8 vol % of additional 1.4 wt % sodium chloride solution before phase separation. In contrast, a similar microemulsion using the same sulfonate formulated with isobutyl alcohol cosurfactant (distribution coefficient of 0.22) was only dilutable with either 4 vol % of additional dodecane or 4 vol % of additional 1.6 wt % sodium chloride solution before phase separation.

The foregoing results indicate that alcohols having distribution coefficients nearer to 1 than that of an initial trial run yield microemulsions with greater tolerance to dilution by either additional brine or additional oil. In both instances the salinity was selected so as to be a value near the center of the salt range which yields three-phase partitioning, which for 3-pentanol was 1.4 wt % and for isobutyl alcohol was 1.6 wt %. It is believed this is necessary to give a fair comparison of results.

Alkane Sulfonate Anionic Surfactant

S.A. Williams; U.S. Patent 4,138,345; February 6, 1979; assigned to Mobil Oil Corp. describes a surfactant waterflooding process employing a water-soluble anionic surfactant which is designed for use in oil reservoirs in which the connate waters exhibit high salinities and/or divalent metal ion concentrations or in instances in which the available injection waters exhibit high salinities and/or divalent metal ion concentrations.

In carrying out the process, at least a portion of the fluid introduced into the reservoir via a suitable injection system is an aqueous liquid which contains a water-soluble anionic surfactant comprising a hydrocarbylamino ether-linked alkane sulfonate wherein the hydrocarbyl group provides a lipophilic base. The ether linkage in the surfactant is provided by an alkoxy linkage having a ratio of carbon atoms to oxygen atoms within the range of 2 to 3. One form of surfactant for use in carrying out the process is characterized by the formula

$$R_1-\underset{\displaystyle R_2}{\overset{|}{N}}-(C_nH_{2n}O)_x-\underset{\displaystyle R_4}{\overset{|}{R_3}}-SO_3M \quad (1)$$

wherein

R_1 is a lipophilic base provided by a C_{8-20} aliphatic group or an aryl group substituted with a C_{6-18} aliphatic group;
R_2 is a C_{1-4} aliphatic group;
n is 2 or 3;
x is a number within the range of 1 to 10;
R_3 is a C_{1-4} alkane group;
R_4 is hydrogen, a hydroxy group or a methyl group; and
M is an alkali metal, ammonium, or substituted ammonium ion.

Another form of surfactant for use in carrying out the process is a disulfonated derivative characterized by the formula

$$R_a-N\begin{cases}(C_nH_{2n}O)_{x_1}-R_d(R_e)-SO_3M \\ (C_nH_{2n}O)_{x_2}-R_b(R_c)-SO_3M\end{cases} \quad (2)$$

wherein

R_a is a lipophilic base provided by a C_{10-25} aliphatic group or an aryl group substituted by a C_{8-23} aliphatic group;
n is 2 or 3;
x_1 and x_2 are each independently a number within the range of 1 to 10;
R_b and R_d are each independently a C_{1-4} alkane group;
R_c and R_e are each independently a hydrogen, a hydroxy group, or a methyl group; and
M is an alkali metal, ammonium, or substituted ammonium ion.

The surfactants characterized by formula (1) may be prepared by alkoxylation of a secondary amine. The resulting ethoxylated or propoxylated adduct may then be reacted with a sultone and base to produce the sulfonate derivative.

By way of example, sodium tetradecylmethylaminopolyethoxypropane sulfonate may be prepared by first reacting ethylene oxide with methyltetradecylamine to produce the ethoxylated adduct. The mol equivalent of ethylene oxide relative to the mol equivalent of the secondary amine is varied, as will be understood by those skilled in the art, in order to arrive at the desired number of ethylene oxide units in the ether linkage. Thereafter, the ethoxylated product is reacted with propane sultone and base (e.g., sodium hydroxide) to produce the sulfonate. The hydroxyalkane sufonates and the methyl alkane sulfonates can be prepared by a similar reaction procedure. Thus, the ethoxylated secondary amine may be reacted with 3-methylpropane sultone or 3-hydroxypropane sultone to form the methyl- or hydroxy-substituted alkane sulfonate, respectively.

The disulfonates characterized by formula (2) can be prepared by reactions similar to those employed for the monosulfonates except that a primary amine is employed as the starting material.

When a primary amine is employed as the starting material, both hydrogens on the primary amine are replaced by formation of the dialkoxylated adduct, i.e., two ethoxy chains are attached to the nitrogen. This adduct may then be reacted with a sultone and base, e.g., propane sultone and sodium hydroxide, to produce the disulfonate derivative.

By way of example, sodium hexadecylaminodi(polyethoxypropane sulfonate) may be prepared by first reacting ethylene oxide with hexadecylamine to produce the di-polyethoxylated adduct. The number of ethylene oxide units in the polyethoxy chain is varied by proper control of the mol ratio of ethylene oxide to primary amine, as will be understood by those skilled in the art. Thereafter, the di-polyethoxylated product is reacted with propane sultone and base (e.g., sodium hydroxide) to produce the di(polyethoxypropane sulfonate) derivative. The hydroxyalkane sulfonates and the methyl alkane sulfonates can be prepared by a similar reaction procedure. Thus, the dipolyethoxylated primary amine may be reacted with 3-methylpropane sultone or 3-hydroxypropane sultone to form the methyl- or hydroxy-substituted alkane sulfonates, respectively.

While the aqueous solution of hydrocarbylamino ether-linked alkane sulfonate, either alone or as a cosurfactant or with a cosolvent, may be employed as the sole displacing fluid, it will usually be injected as a discrete slug and then followed by a driving fluid. Preferably, the aqueous surfactant solution is injected in an amount of at least 0.1 pore volume. Typically the size of the surfactant slug will be within the range of 0.2 to 0.5 pore volume. Where a relatively viscous mobility control fluid is employed the mobility control fluid normally will be injected in an amount within a range of 0.2 to 0.5 pore volume. Thereafter, a driving fluid is injected in order to displace the previously injected fluid through the formation. The driving fluid typically may be any water which is locally available and is not incompatible with the formation. The driving fluid is injected in such amount as necessary to carry the recovery process to its conclusion.

Quaternary Ammonium Amphoteric Surfactants

S. Stournas; U.S. Patent 4,216,097; August 5, 1980; assigned to Mobil Oil Corp. describes a surfactant waterflooding process employing an amphoteric surfactant which is effective in reducing oil-water interfacial tensions in relatively saline aqueous media which include the presence of significant quantities of divalent metal ions. The process is carried out in a subterranean oil reservoir penetrated by spaced injection and production system. In carrying out the process, at least a portion of the injected fluid comprises an aqueous liquid containing an amphoteric surfactant characterized by the formula:

$$R_1-\overset{\overset{\large R_2}{|}}{\underset{\underset{\large R_3}{|}}{N^+}}-R_4A^-$$

wherein

R_1 is a hydrocarbyl group containing from 8 to 26 carbon atoms;
R_2 and R_3 are each independently a hydrocarbyl group containing from 1 to 8 carbon atoms or an alkoxy group containing from 2 to 10 carbon atoms and having a ratio of carbon atoms to oxygen atoms within the range of 2 to 3;
R_4 is an aliphatic group containing from 1 to 6 carbon atoms; and
A is a sulfonate group or a carboxylate group.

In a preferred embodiment of the process, the injected aqueous liquid contains the amphoteric surfactant in a relatively low concentration within the range of 0.001 to 0.1 wt % and is injected in a relatively large pore volume amount of at least 0.5 pore volume.

In one aspect of the process, the amphoteric surfactant employed is a hydrocarbyldimethylammoniumpropane sulfonate characterized by the formula:

$$R-\overset{\overset{\large CH_3}{|}}{\underset{\underset{\large CH_3}{|}}{N^+}}-CH_2CH_2CH_2-SO_3^-$$

wherein R is a hydrocarbyl group containing from 10 to 26 carbon atoms.

In yet a further embodiment of the process, the amphoteric surfactant is a hydrocarbyldihydroxyethylammoniumpropane sulfonate characterized by the formula:

$$R-\overset{\overset{\large CH_2CH_2OH}{|}}{\underset{\underset{\large CH_2CH_2OH}{|}}{N^+}}-CH_2CH_2CH_2-SO_3^-$$

wherein R is a hydrocarbyl group containing from 10 to 26 carbon atoms.

Amphoteric surfactants of the general type employed in carrying out the process are known in the detergent art and have been proposed for various uses such as liquid or solid soap additives, shampoo additives, lime soap dispersants, scale inhibitors, and bactericides. For descriptions of such compounds and their methods of preparation, reference is made to Parris, N. et al, "Soap Based Detergent Formulations. V. Amphoteric Lime Soap Dispersing Agents," *Journal of the American Oil Chemists' Society,* Vol. 50, pp 509-512 (1973), U.S. Patent 3,280,179 and U.S. Patent 3,660,470.

Quaternary Ammonium Sulfonates and Alcohol

P.M. Wilson and J. Pao; U.S. Patent 4,193,452; March 18, 1980; assigned to Mobil Oil Corp. describe a process for the recovery of oil employing a mixture of a C_{5-8} aliphatic alcohol and quaternary ammonium sulfonate. The amphoteric sulfonates and aliphatic alcohols interact synergistically in aqueous solution to produce a pronounced thickening effect even in the presence of high salt concentrations.

Amphoteric quaternary ammonium sulfonates suitable for use in carrying out this process are disclosed in the previous process in U.S. Patent 4,216,097. The aliphatic alcohol is present in a concentration to provide a ratio of the volume amount of alcohol to the weight amount of surfactant within the range of 0.1 to 0.6. Preferably, the alcohol is a C_{6-8} aliphatic alcohol and is employed in a concentration to provide an alcohol-surfactant ratio within the range of 0.1 to 0.4.

The thickened aqueous solution of alcohol and amphoteric sulfonate may be injected in any suitable amount as necessary for effective mobility control. Normally, the alcohol-surfactant solution will be injected in an amount within the range of 0.1 to 0.5 pore volume. The surfactant slug or slugs injected before or after the alcohol-surfactant mobility control slug normally will be injected in amounts of 0.1 pore volume or more but may range down to 0.02 pore volume. Where an amphoteric surfactant such as disclosed in the previous process is employed, it usually will be desirable to inject a relatively large amount, e.g., 0.5 pore volume or more, and employ the surfactant at a relatively low concentration in order to arrive at an optimum low oil-water interfacial tension.

Subsequent to the injection of the mobility control slug and surfactant slug or slugs, a driving fluid is injected in order to displace the previously injected fluids through the formation. The driving fluid may be any water which is locally available and is not incompatible with the formation and is injected in such amount as necessary to carry the process to its conclusion.

Amphoteric Sulfonium Sulfonates

In another process employing an amphoteric surfactant which is effective in reducing oil-water interfacial tensions in relatively saline aqueous media which include the presence of significant quantities of divalent metal ions, *S. Stournas; U.S. Patent 4,166,038; August 28, 1979; assigned to Mobil Oil Corp.* describes a water-soluble amphoteric surfactant characterized by the formula:

$$\begin{array}{c} R_1-\overset{+}{S}-R_3\ SO_3^- \\ | \\ R_2 \end{array}$$

wherein

R_1 is a hydrocarbon group containing from 8 to 24 carbon atoms;
R_2 is a hydrocarbon group containing from 1 to 4 carbon atoms or an alkoxy group containing from 2 to 10 carbon atoms and having a ratio of carbon atoms to oxygen atoms within the range of 2 to 3; and
R_3 is an aliphatic group containing from 1 to 4 carbon atoms.

In a preferred embodiment of the process, the amphoteric surfactant is a methylsulfonium or ethoxysulfoniumpropane sulfonate characterized by the formula:

$$R_1-\underset{\underset{R_2}{|}}{S^+}-CH_2CH_2CH_2-SO_3^-$$

wherein

R_1 is a hydrocarbon group containing from 10 to 18 carbon atoms; and
R_2 is $-CH_3$ or $-(CH_2CH_2O)_nH$ and n is a number within the range of 1 to 5.

Amphoteric surfactants of the general type employed in carrying out the process are disclosed in U.S. Patent 2,813,898. They are intended for use in biological toxicants, herbicides, and, in the case of aromatic sulfonium sulfonates substituted with a long chain alkyl group, as surface-active agents.

Alkarylsulfonic Acid Salts

P.D. Marin, M. Prillieux and R. Tirtiaux; U.S. Patent 4,171,323; October 16, 1979; assigned to Exxon Research & Engineering Co. describe a surface-active agent which is quite compatible with sodium chloride and makes it possible to lower the interfacial tension between water and hydrocarbons to 1 dyne/cm or even less.

The surface-active agent comprises a salt of an alkarylsulfonic acid, the mean molecular weight of which is between 250 and 1,000, and an organic base the molecules of which include at least one $-(C_2H_4O)_nH$ group, n denoting a number larger than 1 and smaller than 40.

It is possible to sulfonate alkylaromatic hydrocarbons by any appropriate means. A good process for sulfonating alkylbenzenes comprises using sulfur trioxide in solution in liquid sulfur dioxide at a temperature of less than -5°C.

The organic base from which the surface-active agent is derived is obtained by a reaction consisting of the polyaddition of ethylene oxide to ammonia or to an organic compound the molecule of which includes at least one basic function and at least one mobile hydrogen atom. The basic organic compounds that can be used include in particular primary or secondary amines, polyamines and alkanolamines and are preferably ethoxylated alkylamines or polyamines containing either a primary or secondary amine function.

Example of suitable basic compounds are triisopropanolamine and better still, triethanolamine. The average number of mols of ethylene oxide attached to each mol of the basic compound is selected as a function of the nature of the latter and as a function of the particular use for which the surface-active agent is intended. If the organic base contains alkyl groups they should be short chain (i.e., less than 6 carbon atoms) as long chains can lower compatibility with water.

Certain variants are especially suitable for extracting oil from a deposit through displacing with water. In this variant the mean molecular weight of the alkarylsulfonic acids is between 250 and 400, for preference, between 300 and 400. These acids can be obtained by sulfonation of an appropriate crude oil fraction. They are obtained preferably by the sulfonation of alkylbenzes derived from the alkylation of orthoxylene. The orthoxylene is alkylated with tetrapropylene. The base is the product that is obtained by condensing at least 1 mol of ethylene oxide on 1 mol of triethanolamine.

The higher the proportion of ethylene oxide, the better the compatibility of the surface-active agent with sodium chloride. The lowering of the interfacial tension remains excellent whatever the proportion of ethylene oxide contained in the base.

Sulfonation of Crude Oils

M.A. Plummer; U.S. Patent 4,147,638; April 3, 1979; assigned to Marathon Oil Company describes a process for the secondary recovery of oil which comprises forming a micellar dispersion containing hydrocarbon, water, and petroleum sulfonate produced by a process involving contacting sulfur trioxide with a hydrocarbon selected from the group consisting of whole crude oil, topped crude oil, and mixtures thereof in a reaction zone at a temperature of about 26.6° to about 121°C, at a pressure of about 0.01 to about 150 atm, and for a reaction time of about 0.001 to about 3,600 sec.

About 5 to about 30 lb of sulfur trioxide are contacted with each 100 lb of hydrocarbon and unreacted hydrocarbon after the contact with sulfur trioxide is removed by extracting the product mixture resulting from the contact of sulfur trioxide with the hydrocarbon with an extraction solvent selected from the group consisting of low molecular weight alcohol, ketone, ether, benzene, toluene, water or mixtures thereof. Thereafter the product is neutralized with sodium hydroxide, potassium hydroxide or ammonia. At least a portion of the micellar dispersion is injected into a petroleum-containing formation to displace petroleum.

Figure 4.1 schematically depicts the process and a number of optional procedures which can be used therewith. Basically, crude oil from well **11** and SO_3 vapor enter sulfonation reactor **12** through lines **13** and **14** respectively. Desired amounts of any solvent, unreacted hydrocarbons and sulfonic acids are removed in phase separator **15**. The sulfonic acids and a base are introduced into neutralization reactor **16** through lines **17** and **18** to form the desired sulfonation product which passes through line **19** to storage.

A number of optional steps (indicated by dashed lines) enhance the variety and characteristics of sulfonates which can be obtained, the purity of these products and the economics of the process. Thus, low molecular weight hydrocarbons can be removed in topping unit **21** and mixed, via line **22**, with the SO_3 for dilution purposes. The higher molecular weight hydrocarbon feedstocks are charged to reactor **12** through line **23**. Sulfonation additives can be introduced through line **24**.

Figure 4.1: Production of Petroleum Sulfonates

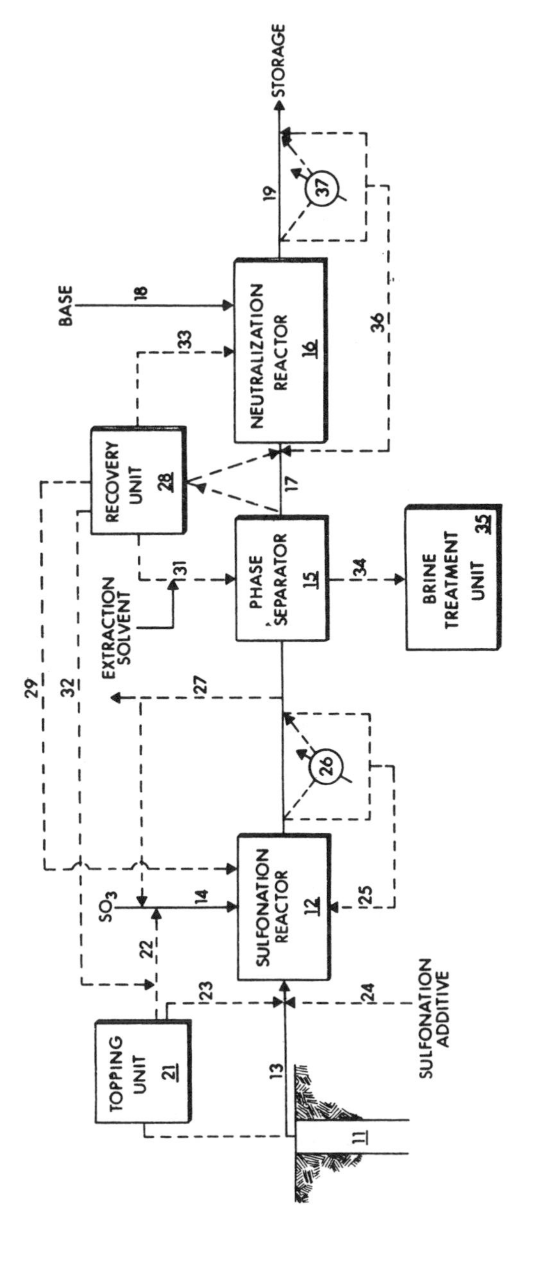

Source: U.S. Patent 4,147,638

Desired amounts of reaction products (sulfonic acids, SO_2, sulfuric acid, unreacted hydrocarbons and reaction solvent, if any) can be recycled to reactor **12** through line **25** either before or after passage through cooler **26**. Gaseous or vaporous products can be removed and, preferably, recycled as an SO_3 diluent through line **27**. One or more recovery units **28** can be used to separate sulfonic acids, reaction solvent, extraction solvent, diluents, unreacted hydrocarbons, etc.

Reaction solvent can be recycled to the process through line **29** to reactor **12** and extraction solvent to phase separator **15** through line **31**, diluent through line **32**, and sulfonic acids passed to neutralization reactor **16** through line **33**. Brine passing through conduit **32** can be cleaned up in brine treatment unit **35**. If necessary, a portion of the neutralized product can be recycled to neutralization reactor **16** through line **36**, either before or after passage through cooler **37**.

Those sulfonation reactors conventionally utilized for the hydrocarbon sulfonation including, for example, the falling film, scraped surface and stirred tank reactors may be used in the process. In those instances where an SO_3 diluent is used, it is preferred to use a back mixed tubular reactor. All materials should be in turbulent flow on entering the preferred tubular reactor.

Extraction of Wax from Crude Oil Sulfonates

In an improvement of the previous process, *M.A. Plummer and S.C. Jones; U.S. Patent 4,147,637; April 3, 1979; assigned to Marathon Oil Co.* describe a method of extracting wax from a micellar system containing crude oil sulfonates. Preferably the micellar system does not contain a cosurfactant.

The extraction is effected with an agent consisting essentially of paraffinic hydrocarbons or halocarbons having a molecular weight of about 50 to about 200. Aromatic extraction agents are not generally useful since they tend to be solubilized in the micellar system and have, under certain conditions, caused reduced oil recovery. Some aromatics can be tolerated, i.e., less than about 20% of the extraction agent. Extraction can be effected on the final micellar system, i.e., a micellar solution or microemulsion, before it is injected into the formation. The extraction also can be performed on the neutralized sulfonic acid stream from the sulfonation process simultaneously with removal of unsulfonated crude oil. The resulting micellar system can be subjected to slow cool-down rates or fast cool-down rates without adversely affecting injectivity of the system.

The micellar system is thoroughly contacted with the extraction agent at a temperature of about 5 to about 200°C and more preferably about 20 to about 150°C and most preferably at about 40° to about 100°C. Thereafter, the extraction agent and wax are separated, preferably by phase separation. Sufficient time should be allowed at the specified temperature to insure essentially complete separation of the wax. Typical separation times are about 0.4 to about 2.0 hr/ft of liquid (extraction agent and micellar system) height at 70°C. Preferably the micellar system is then filtered at temperatures within the range of about 10 to about 65 and preferably about 20° to about 40°C before it is injected into the reservoir.

Ether-Linked Sulfonates as Thickeners

J.G. Savins; U.S. Patent 4,181,178; January 1, 1980; assigned to Mobil Oil Corp. has found that certain ether-linked sulfonate surfactants act as viscosifiers for waterflood mobility control applications under certain conditions of salinity and reservoir temperature. The ether-linked sulfonate is employed in a brine having a salinity within the range of 5 to 18 wt %. Within this salinity range a significant thickening effect may be achieved and the salinity may be varied within this range to arrive at a value at which maximum thickening occurs. The thickening effect of the surfactant systems is also temperature dependent and the process is carried out at a reservoir temperature of at least 90°F.

The ether-linked sulfonates employed are sulfonated ethoxylated aliphatic alcohols characterized by an HLB (hydrophilic-lipophilic balance) within the range of 10.0 to 14.0 and wherein the aliphatic hydrocarbon group forming the lipophilic base of the surfactant contains from 16 to 20 carbon atoms.

The ether-linked sulfonates employed in this process can be prepared by ethoxylation of the appropriate aliphatic alcohol and subsequent sulfonation. The sulfonated polyethoxylated aliphatic alcohols used may be characterized by the formula:

$$R(OC_2H_4)_nOC_3H_6SO_3^-M^+$$

wherein R is an aliphatic hydrocarbon group containing from 16 to 20 carbon atoms; n is at least 2; and M is an alkali metal, ammonium, or substituted ammonium ion.

Where M is an alkali metal ion, it usually will take the form of sodium or potassium. Substituted ammonium ions which may be employed include mono-, di-, or tri-substituted alkylammonium or alkanolammonium ions.

The minimum surfactant concentration employed will depend upon the temperature and salinity conditions and of course the desired viscosity of the surfactant solution which normally should be equal to or greater than that of the reservoir oil for effective mobility control. As a practical matter the surfactant normally should be present in a concentration of at least 0.1 wt % and usually 0.2 wt %. For most applications the surfactant will be employed in a concentration within the range of 0.5 to 1.5 wt %.

The thickened aqueous surfactant solution may be injected in any suitable amount depending upon the conditions encountered in a particular reservoir and the concentration of the ether-linked surfactant. The thickened aqueous surfactant solution usually will be injected in an amount of at least 0.1 pore volume to provide for effective mobility control. Usually the aqueous solution of the ether-linked sulfonate will be employed in an amount within the range of 0.1 to 0.5 pore volume.

Organic Additive for Sulfonation Reactions

M.L. Nussbaum and E.A. Knaggs; U.S. Patents 4,148,821; April 10, 1979; and 4,177,207; December 4, 1979; both assigned to Stepan Chemical Co. describe a process whereby a select petroleum oil feed stock, such as a topped crude, a heavy vacuum gas oil or some other partially refined or whole crude is admixed with a small amount (i.e., about 0.5% to about 15% by weight of petroleum oil feed stock) of an additive, such as comprised of an unsulfonatable organic radical portion having an average molecular weight range extending from about 55 to 6,000, having a boiling point in the range of about 100° to 260°C, and a preponderance of such radicals each having attached thereto at least one portion replaceable by a sulfo group and at least one moiety selected from the group consisting of an aromatic nucleus, an olefinic carbon pair and an oxygen atom directly bonded to a carbon atom by at least one bond (i.e., a C_6 to C_{28} alcohol material, such as an Oxo alcohol still bottom).

Then the resultant additive feed stock mixture is sulfonated with SO_3 under sulfonation reaction conditions. A small amount (i.e., about 0.5 to 20% by weight of the resultant crude acidic reaction mixture) of water is then added to the resultant sulfonation reaction mixture and the sulfonation reaction mixture-water mixture is held at an elevated temperature (i.e., in the range of about 50° to 150°C) for a relatively brief period of time (i.e., ranging from about 2 to 60 minutes) and then neutralized with a base. The resultant petroleum sulfonate product, which may first be subjected to an extraction process, if desired, is then formulated into a micellar system, such as a microemulsion or the like system and injected into select petroleum reservoirs for enhanced oil recovery.

By practicing this process, one is able to attain enhanced oil recovery yields on the order of 60 to 90% or more, as compared to typically lower yields obtained with similar petroleum sulfonates which have been nonwater treated before neutralization by conventional prior art techniques.

The stabilized petroleum sulfonate products obtained by the process comprise, on a 100 organic wt % total weight basis from about 2 to 90 wt % of substantially nonsulfonated hydrocarbon material; from about 0 to 5 wt % of nonsulfonated but sulfonatable hydrocarbon material; from about 5 to 98 wt % of monosulfonated hydrocarbon material; from about 0 to 50 wt % of polysulfonated hydrocarbon material; and from about 0.5 to 15 wt % of an additive.

Generally, these additives comprise relatively high boiling organic compounds including unsaturated aliphatic hydrocarbons, substituted and unsubstituted aromatics, olefins, oxygen-containing compounds, esters (especially high boiling esters), ethers, ether esters (especially high boiling ether esters), certain catalytic phase oils, polymer distillation residues, mixtures of alkylated benzenes and naphthalenes, mixtures thereof, alkoxylated derivatives of such compounds, and the like.

These additives generally comprise organic compounds generally containing from 2 through 30 carbon atoms within their main hydrocarbon chain and may contain more carbon atoms, for example, in side chains or in alkoxylated additives condensed onto the main compound or radical. Such organic compounds are of a type which promote compatibility of unsulfonated (sulfonatable and nonsulfonatable) petroleum oil feed stocks with sulfonated components during SO_3 reaction conditions.

Preferred groups of organic additives useful in the process are selected from the classes consisting of alcohols, oxygen-containing compounds, hydroxy-containing compounds, substituted and unsubstituted hydrocarbons, high boiling esters, high boiling ethers, high boiling ester ethers, aromatic compounds, fatty acids and derivatives thereof, olefins, ketones, alkaryl compounds and mixtures thereof. A preferred class among this group is the oxygen or hydroxy-containing compounds.

Petroleum Sulfonate, Cosurfactant, and Secondary Alcohol

J.T. Carlin, J.A. Wells and T.N. Tyler; U.S. Patent 4,154,301; May 15, 1979; assigned to Texaco Inc. describe a method for the recovery of oil from a subterranean oil-bearing formation having a high salinity connate water wherein a slug of a mixture of a commercially available petroleum sulfonate, a cosurfactant and a monounsaturated secondary alcohol is injected into the formation prior to the undertaking of a waterflood. By this process, advantage is taken of the fact that the unreacted hydrocarbon portion of a commercially available petroleum sulfonate is used as a solvent for the monounsaturated secondary alcohol to enhance recovery.

By this process a monounsaturated secondary alcohol that is soluble in oil, is added to the surfactant slug composition whereby the alcohol is partitioned into the unreacted oil phase that is present in a commercially available petroleum sulfonate material. The oil phase thus acts as a carrier for the alcohol, and is thereby effectively utilized in the method of recovery. Furthermore, the presence of the monounsaturated secondary alcohol results in substantial improvement in the oil recovery.

It is recommended that the surfactant slug containing the commercially available petroleum sulfonate and the cosurfactant and the monounsaturated secondary alcohol be injected in the amount of from about 5% to about 50% of the reservoir pore volume.

The cosurfactant, which is present to enhance the compatibility of the surfactant slug with the formation fluids, is present in amounts of from about 0.3% to about 3.0% by weight. The cosurfactant may be selected from the group comprising ethoxylated phenols, ethoxylated alkyl phenols, ethoxylated alcohols, and sulfated and sulfonated derivatives of the abovementioned phenols and alcohols. Two examples of specific cosurfactants that have been used are an ethoxylated sulfonated nonylphenol and an ethoxylated sulfated tridecyl alcohol.

The concentration of the monounsaturated secondary alcohol will depend upon, among other things, the nature of the specific alcohol being employed and the quantity of unreacted oil in the commercially available petroleum sulfonate. Generally, the concentration should be in the range of from about 0.001% to about 1.0% by weight, with the preferred range being from about 0.002% to about 0.006% by weight. A wide variety of monounsaturated secondary alcohols may be used provided they are soluble in hydrocarbon or oil. For example, the monounsaturated secondary alcohol may be a cyclic secondary alcohol such as cyclohexenol and its alkyl derivatives, such as methylcyclohexenol, propylcyclohexenol and pentacyclohexenol. In addition, the alcohol may be a dialkyl derivative such as dimethylcyclohexenol or trialkyl derivatives of cyclohexenol.

A second group of alcohols suitable for practicing this process are the terpene alcohols which are alcohols derived from simple terpene hydrocarbons such as pulegol, isoborneol, menthol and piperitol. Other monounsaturated secondary alcohols that are suitable are those having a condensed ring structure such as a polyalicyclic alcohol, i.e., cholesterol, and derivatives thereof.

The method of operation is applicable to a formation being produced in pattern arrangement, as for example, a pattern arrangement wherein a central well may serve as the production well and the offset wells may serve as the injection wells. One of the more common pattern arrangements is the 5-spot pattern in which four offset wells form the corners of a square and the fifth well is centrally located within the square. The method is also applicable to a line drive where one line of wells serve as the injection wells and the two adjacent lines of wells serve as production wells.

Dialkylbenzene Polyethoxyalkyl Sulfonate Cosurfactant

J.W. Hughes, M.V. Kudchadker and N.G. Dunn; U.S. Patent 4,220,204; Sept. 2, 1980; assigned to Texaco Inc. describe an enhanced oil recovery process useful in reservoirs containing fluids having a salinity ranging from about 10,000 to 250,000 ppm total dissolved solids of which up to about 20,000 ppm are divalent ions. The process comprises mixing a water-soluble petroleum sulfonate with an effective amount of a solubilizing cosurfactant which comprises a dialkylbenzene polyethoxy alkyl sulfonate. This surfactant mixture is then injected into and forced through the subterranean petroleum reservoir in order to recover petroleum therefrom.

A preferred solubilizing cosurfactant is described by the following formula:

R, R' (on benzene ring) —$(OCH_2CH_2)_mC_2H_4SO_3X$

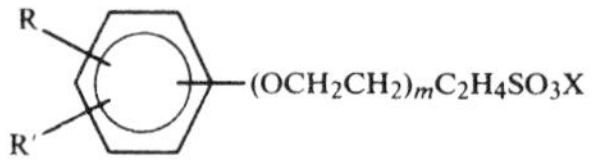

where R and R' are alkyl groups, m is a number from about 3 to about 15 and X is a cation.

The alkyl groups in the dialkyl portion of the above solubilizing cosurfactant may be linear or branched chain alkyl groups, preferably with from 6 to about 14 carbon atoms in each group. It is felt that it is this dialkyl formulation of the solubilizing cosurfactant that lends to it sufficient lipophilicity to effectively solubilize the companion petroleum sulfonate surfactant. Experiments with the monoalkyl form of the solubilizing cosurfactant, when combined with the water-soluble sulfonate have failed to show any of the oil recovery characteristics such as low interfacial tension, high viscosity or good capillary displacement.

Example: Three separate tests were conducted on a surfactant system consisting of brine, a water-soluble petroleum sulfonate, and a dialkylbenzene polyethoxy alkyl sulfonate. The brine contained 90,000 ppm total dissolved solids, which include 10,000 ppm divalent ion. The water-soluble petroleum sulfonate was TRS 40 (Witco Chemical Company) held at a concentration of 2.0% by weight. The dialkylbenzene polyethoxy alkyl sulfonate was added to the above components in varying amounts and is described by the following formula:

$H_{19}C_9$, $H_{19}C_9$ (on benzene ring) —$(OCH_2CH_2)_{4.8}C_2H_4SO_3Na$.

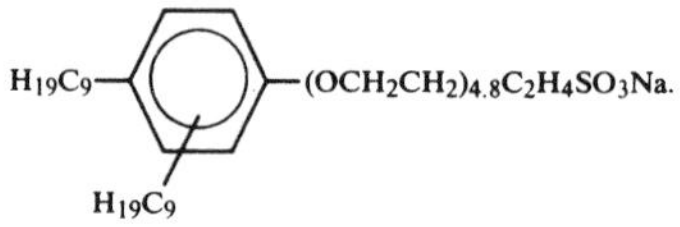

This mixture was then tested for capillary displacement viscosity and interfacial tension. The results are reported in the following table.

Solubilizing Cosurfactant (%)	Capillary Displacement (mm/20 min)	Viscosity (cp)	Interfacial Tension (md/cm)
0.35	*	–	–
0.43	*	–	–
0.52	*	–	–
0.60	18.2	4.56	19
0.69	20.0	3.88	20
0.78	11.8	4.74	20
0.86	5.6	5.66	15
1.04	6.4	6.02	15

*Precipitation.

Petroleum Sulfonate and Dipolyethoxylated Alkyl Catechol

G. Kalfoglou; U.S. Patent 4,194,565; March 25, 1980; assigned to Texaco Inc. has found that surfactant waterflooding may be accomplished effectively in high temperature, high salinity formations, e.g., formations whose temperature is in the range of from about 80° to about 180°F (27° to 82°C), the water salinity being from 40,000 to 140,000 ppm total dissolved solids which may include from 2,000 to 12,000 ppm divalent ions such as calcium and/or magnesium, by employing in the surfactant waterflood process, an aqueous fluid having a salinity and divalent ion concentration approximately equal to the salinity and divalent ion concentration of the formation water, and containing two surfactants.

The first surfactant is an organic sulfonate anionic primary surfactant such as petroleum sulfonate, preferably a water-soluble sodium, potassium, lithium or ammonium salt of petroleum sulfonate whose medium equivalent weight is in the range of from about 325 to about 480, or a synthetic sulfonate having the following formula: RSO_3X, wherein R is an alkyl group, linear or branched, containing from 5 to 25 carbon atoms and preferably from 8 to 18 carbon atoms, or an alkylaryl group such as benzene, toluene or xylene having attached thereto an alkyl chain, linear or branched, containing from 5 to 20 and preferably from 6 to 16 carbon atoms in the alkyl chain; and X is sodium, potassium, lithium or ammonium.

The second surfactant is dipolyethoxylated catechol nonionic surfactant having the following formula:

R— (benzene ring) with substituents $O(R'O)_xH$ and $O(R''O)_yH$

wherein R is an alkyl group, linear or branched, containing from 5 to 25 and preferably from 8 to 20 carbon atoms, R' and R'' are each ethylene or a mixture of ethylene and higher alkylene such as propylene with relatively more ethylene than higher alkylene, preferably at least 60% ethylene, and x and y are each 1 to 12 and the sum of x and y is from 2 to 24 and preferably from 4 to 14.

The balance between the alkyl group and the polyalkoxylated alkyl catechol and the total number of ethoxy groups in the ethoxy or alkoxy chain are both selected to ensure the nonionic surfactant is soluble and capable of reducing interfacial tension in an aqueous fluid having a salinity and hardness about equal to the salinity and hardness of the formation water, and further adjusted to ensure that the cloud point of the nonionic surfactant is from 2° to 40°F (1° to 22°C) greater than the temperature of the formation into which the fluid is injected.

While optimum results are obtained employing a fluid whose concentration of each component is determined by carefully measuring the variation in surfactant flooding effectiveness with surfactant concentration, the folloiwng general guidelines are helpful. The concentration of the primary anionic surfactant, e.g., the petroleum sulfonate or synthetic organic sulfonate will ordinarily be in the range from about 0.05 to about 10 and preferably from about 0.2 to 5% by weight. The concentration of the dipolyethoxylated alkyl catechol nonionic surfactant will ordinarily be from 0.01 to 10 and preferably from 0.2 to 6% by weight.

Once the optimum effective concentration of each of these components is established, it is sometimes necessary to include additional amounts of each material in the first 10 or 20% of the pore volume of surfactant fluid injected into the formation, where it is determined or known that a significant surfactant loss will occur as a consequence of adsorption of surfactant from solution by the formation rock or other factors. Ordinarily the concentration may be increased as much as two or three times the values determined to be optimum, with the ratio being held constant or varied if it is determined that adsorption is preferential for one or the other of the surfactant species.

Petroleum Sulfonate and Sulfated Polyethoxylated Alkylthiol

G. Kalfoglou; U.S. Patent 4,191,253; March 4, 1980; assigned to Texaco Inc. describes an aqueous, saline, surfactant-containing fluid, and an oil recovery method using the fluid. The fluid contains a sulfated or sulfonated, polyethoxylated alkyl or alkylarylthiol having the following formula: $R{-}S{-}(OR')_nSO_3^-M^+$ wherein R is an alkyl having from 6 to 26 and preferably 8 to 20 carbon atoms or an alkylaryl having from 5 to 24 and preferably from 6 to 16 carbon atoms in the alkyl chain, S is sulfur, O is oxygen, R' is ethylene or a mixture of ethylene and propylene with relatively more ethylene than propylene, and M is a monovalent cation preferably sodium, potassium, lithium and ammonium, and n is a number from 1 to 12 and preferably 3 to 8, or an alkylpolyalkoxyalkylene sulfonate or alkylarylpolyalkoxyalkylene sulfonate having the following formula: $R{-}S(OR')_nR''SO_3^-M^+$ wherein R'' is ethylene, propylene, hydroxypropylene or butylene, and R, S, O, R', M and n have the same meaning as above.

The abovedescribed surfactant may be used alone, but is preferably used in combination with a conventional primary anionic surfactant, generally an organic sulfonate, preferably a sodium, potassium, lithium or ammonium salt of petroleum sulfonate which is at least partially water-soluble and which has an average equivalent weight in the range of from about 300 to about 500 and preferably from 350 to 450. The sulfated or sulfonated polyethoxylated alkylthiol or alkylarylthiol exhibits oil recovery activity over a broader salinity range than an equivalent sulfated or sulfonated, polyethoxylated aliphatic alcohol or alkylphenol, and is more stable when used in certain subterranean oil-containing formations.

In applying the process to a conventional, petroleum-containing formation, a quantity of surfactant fluid equivalent to from 0.01 to 2 and preferably 0.05 to 0.5 pore volume based on the volume of formation to be contacted by the injected fluid, is injected into the formation. The surfactant fluid contains from

0.1 to 6.0 and preferably 0.2 to 3.0% by weight of the sulfated or sulfonated, polyethoxylated alkyl or alkylarylthiol. If the surfactant is utilized in combination with petroleum sulfonate or other organic sulfonate, the concentration of petroleum sulfonate should be from 0.1 to 10 and preferably from 0.5 to 5% by weight. The ratio of the sulfated or sulfonated polyethoxylated thiol to the organic sulfate should be from 0.1 to 1.0 and preferably 0.2 to 0.5, depending on formation water salinity, with relatively more sulfated or sulfonated ethoxylated thiol being required at greater formation water salinities.

Petroleum Sulfonates of Prescribed Equivalent Weight Distribution

J.T. Carlin, J.W. Ware, M.E. Mills, Jr. and T.N. Tyler; U.S. Patent 4,214,999; July 29, 1980; assigned to Texaco Inc. have found that when petroleum sulfonates are used in oil recovery processes, whether used as substantially the only surface active agent in a surfactant fluid injected into the formation, or when the petroleum sulfonate is combined with more complex synthetic surfactants which function as solubilizing cosurfactants in order to permit the use of surfactant flooding techniques in formations containing relatively high salinity and/or high hardness water, optimum results are obtained if the average equivalent weight of the petroleum sulfonate utilized is lower than that which would have been predicted by interfacial tension measurements, and generally should be in the range of from about 325 to about 425 and preferably from about 350 to about 400.

Moreover, the distribution of equivalent weight within the range of from about 250 to about 700 should be relatively uniform. For example, from about 40 to about 70% of the petroleum sulfonates should have equivalent weights of less than 400 and from about 60 to about 30% should have equivalent weights of 400 and above. Preferably, from about 15 to 35% should have equivalent weights of less than 350 with from 30 to 50% having equivalent weights from 350 to less than 500 and from 10 to 40% having equivalent weights of 500 or greater.

The especially preferred embodiment employs a mixture of petroleum sulfonates having equivalent weights so evenly distributed that from 5 to 15% of the molecular species have equivalent weights less than 300; from 25 to 50% have equivalent weights of from 300 to less than 400; from 15 to 25% have equivalent weights from 400 to less than 500; and from 5 to 20% having equivalent weights of 500 and greater.

These preferred ranges apply to processes in which petroleum sulfonate is utilized as substantially the only surface active agent in the aqueous surfactant fluid injected into a formation containing water having relatively low salinity and hardness values, e.g., salinities equal to or less than about 20,000 ppm total dissolved solids and divalent ion concentrations less than about 2,000 ppm, or in formations containing water having salinity and hardness somewhat higher than this but which can be adjusted downward to the operable range by preconditioning the reservoir such as by injecting a quantity of lower salinity and lower hardness water into the formations to displace the higher salinity, higher hardness waters therefrom prior to injecting the surfactant solution into the formation.

The foregoing equivalent weight ranges also apply to processes using complex combinations of petroleum sulfonate with synthetic surfactants which function as solubilizing cosurfactants for the purpose of rendering the petroleum sulfonate samples soluble in high salinity, high divalent ion concentration waters. Effective solubilizing cosurfactants include nonionic surfactants such as polyethoxylated alcohols or alkyl phenols; alkyl or alkylaryl polyethoxyl sulfates; or alkyl or alkylaryl polyethoxyethyl sulfonate.

Sulfonated, Oxyalkylated Quinoline Derivatives

J.H. Park and W.D. Hunter; U.S. Patent 4,187,185; February 5, 1980; assigned to Texaco Development Corp. describe an aqueous alkaline flooding medium which comprises an aqueous drive fluid which can be, for example, steam, hot water, a mixture of hot water and steam or cold water together with an alkaline agent and a sulfonated interfacial tension reducer. Useful alkaline agents include compounds selected from the group consisting of an alkali metal hydroxide, alkali metal hypochlorites, an alkaline earth metal hydroxide and a basic salt of the alkali metal or alkaline earth metal which is capable of hydrolyzing in an aqueous medium to give an alkaline solution, the concentration of the alkaline agent being about 0.001 to about 0.5 molar to give the required alkaline solution.

Examples of the especially useful alkaline agents include sodium hydroxide, potassium hydroxide, lithium hydroxide, ammonium hydroxide, sodium hypochlorite, potassium hypochlorite, sodium carbonate and potassium carbonate, etc.

Interfacial tension reducers which are highly useful in the process include sulfonated compounds of the formula:

(1) Quinoline ring substituted with $-N(R)-(OC_2H_4)_tC_2H_4SO_3M$

wherein t is an integer of from 3 to about 40, M is selected from the group consisting of hydrogen, sodium, potassium and ammonium and R is selected from the group consisting of hydrogen and $-(OC_2H_4)_sC_2H_4SO_3M$, where s is an integer of from 3 to about 40 and M has the same meaning as previously described.

Interfacial tension reducers of this type can be formed by first reacting an aminoquinoline such as 5-aminoquinoline, 6-aminoquinoline, etc. with ethylene oxide in the presence of a catalyst such as sodium hydroxide to form the corresponding ethoxylated aminoquinoline (2) which in turn is reacted with chlorosulfonic acid, for example, to yield the sulfated ethoxylated aminoquinoline (3). Reaction of Compound (3) with sodium hydroxyethane sulfonate in the presence of sodium hydroxide as described in U.S. Patent 2,535,678 gives the desired sulfonated ethoxylated quinoline (Compound 1).

The interfacial tension reducer should be present in the aqueous solution in sufficient concentration to effect the emulsification of the hydrocarbon material and maintain them in this state during passage through the formation. Concentrations of the aqueous solution of from about 0.05 to about 5.0% by weight of the interfacial tension reducer are usually sufficient, although smaller or larger amounts may be employed satisfactorily in some cases.

Example: An oil formation is nearing the end of waterflooding, the formation being located at a depth of 8,730 ft to 8,763 ft. Consideration is being given to application of an enhanced recovery process involving flooding with an aqueous solution of an interfacial tension reducer. The average oil saturation after waterflooding is 38%. The porosity of the formation is 39% and the permeability is 85 millidarcies. The formation water salinity is 9,000 ppm total dissolved solids including 1,300 ppm hardness (calcium and magnesium). The temperature of the formation is 168°F (75.6°C).

The field has been developed using an inverted five-spot pattern and only a single pattern is treated herein. Each pattern unit is square, 275 feet on a side, with an injection well in the center and production wells on each corner of the square. Since an inverted five spot achieves about 75% sweep efficiency, the pore volume of each pattern unit is: $(275)^2(0.75)(32)(0.38) = 689{,}700\ ft^3$. The interfacial tension reducer chosen for use in this application has the formula:

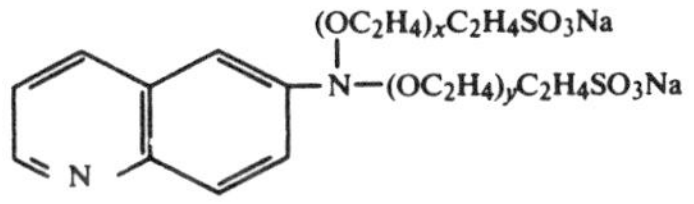

$(OC_2H_4)_xC_2H_4SO_3Na$
$N-(OC_2H_4)_yC_2H_4SO_3Na$
N

wherein the average sum of x and y is about 12.

A 20 pore volume % of oil recovery fluid containing the interfacial tension reducer (137,940 ft^3 or 1,030,930 gallons) is prepared. The first 10% or 103,193 gallons contains 2.5% by weight of the interfacial tension reducer, whereas the remaining volume contains 1.0% by weight. The higher concentration in the first fraction of fluid injected is employed to offset loss of the interfacial tension reducer from solution due mainly to adsorption. A total of 98,850 lb of interfacial tension reducer (on a 100% active material basis) is required. The interfacial tension reducer is dissolved in the abovedescribed formation brine.

The oil recovery fluid is injected into the formation at a pressure well below the known fracture pressure of 3,200 psi, which results in an average injection rate of 20 gpm or 28,800 gpd. The time required for injection of oil recovery fluid is 35 days.

The oil recovery fluid is followed by a one million gallon slug of viscous mobility buffer fluid comprising relatively fresh water containing a copolymer of acrylate and acrylamide, the polymer concentration being 900 ppm in the first half million gallons of the fluid and 500 ppm in the second half of the volume of fluid.

The abovedescribed polymer fluid is displaced through the formation with brine until the water cut rises to a value above 90, signifying the end of this pilot. The average residual oil saturation is reduced to 11% by this enhanced recovery process.

Mono- and Dialkylbenzene Polyethoxyalkyl Sulfonate Mixture

V.H. Schievelbein; U.S. Patent 4,217,957; August 19, 1980; assigned to Texaco Inc. describes a method of enhanced oil recovery utilizing a surfactant mixture comprising a mixture of mono- and dialkylbenzene polyethoxyalkyl sulfonates wherein each alkyl group has the same number of carbon atoms in both components of the mixture.

The method is practiced in a petroleum reservoir which also contains brine by injecting into the reservoir an aqueous surfactant composition comprising a mixture of two chemicals, A and B, A being characterized by the formula $R_A(OCH_2CH_2)_mR'SO_3X$ wherein m is a number from 1 to 6, X is a cation, R' is ethylene, propylene or hydroxypropylene, and R_A is an alkylbenzene radical containing from 8 to 15 carbon atoms in the alkyl group and B being characterized by the formula $R_B(OCH_2CH_2)_mR'SO_3X$ wherein m, R' and X have the same meaning as in A and R_B is a dialkylbenzene radical containing within each alkyl group a number of carbon atoms, that is, from 2 less than to 2 more than, and preferably from 1 less than to 1 more than, that found in the single alkyl chain of A. This aqueous surfactant solution is then driven through the reservoir, and petroleum is then recovered from production wells.

Example: In this example mixtures of the following two chemicals were employed:

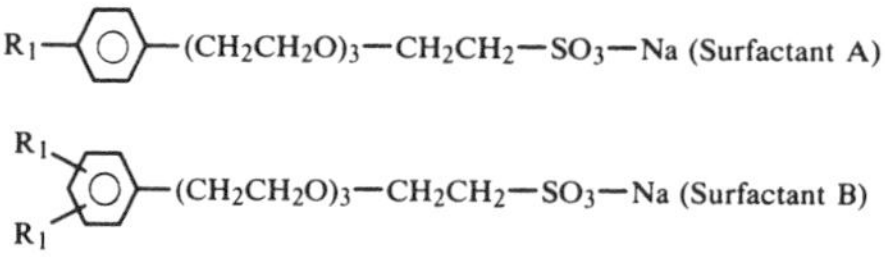

$$R_1-C_6H_4-(CH_2CH_2O)_3-CH_2CH_2-SO_3-Na \text{ (Surfactant A)}$$

$$(R_1)_2C_6H_3-(CH_2CH_2O)_3-CH_2CH_2-SO_3-Na \text{ (Surfactant B)}$$

where R_1 is a dodecyl group.

These two chemicals in varying mol ratios were then diluted with an oil field brine which contained 85 kg/m^3 total dissolved solids with 8 kg/m^3 hardness to form the aqueous surfactant mixture utilized in the following experiments.

Interfacial tension tests were run using the above surfactant mixture at a concentration of 10 kg/m^3. Crude oil from the same oil field was used for the hydrocarbon phase and measurements were made at 23°C. Results of these tests are reported in the table below.

These surfactant mixtures were also tested for enhanced oil recovery effectiveness. Berea sandstone cores, 5.08 cm in diameter and about 16 cm in length, were cleaned and dried. The cores were then saturated with oil field brine and crude oil thinned with 0.25 m^3/m^3 heptane to match reservoir viscosity. The cores were then waterflooded with brine to irreducible waterflood oil saturation. One pore volume of the various surfactant solutions at a concentration of from 9 to 10 kg/m^3 was then driven through each of the cores by a polymer flood comprising about 2.0 pore volumes of a 1.0 kg/m^3 solution of Xanflood polymer, a commercially available polysaccharide polymer, mixed with brine.

The core floods were run at a rate of 1.5 meters per day at a temperature of 43°C. The polymer solution was followed by further injection of the brine until a final irreducible oil saturation was reached. This value was then compared with the irreducible oil saturation at the end of the initial waterflood step and is reported in the table as the recovery efficiency, E_R.

It is clearly shown in the table that certain mixtures of the surfactants A and B are much more effective than a surfactant mixture containing surfactant A alone. One preferred mixture of surfactants A and B comprises a mol ratio of B to A of from about 0.05 to about 0.45. An especially preferred mixture of surfactants A and B comprises a mol ratio of B to A of from about 0.11 to about 0.25.

Mol Ratio B/A	Interfacial Tension (millidynes/cm)	E_R (m^3/m^3)
0.00	68	0.34
0.05	31	0.45
0.11	18	0.62
0.18	8	0.71
0.25	>1	0.56
0.33	3	–
0.45	326	0.49

Alkenyl Succinic-Tertiary Amine Reaction Product

J.J. Valcho and R.E. Karll; U.S. Patent 4,253,974; March 3, 1981; assigned to Standard Oil Company (Indiana) describe a method for recovering oil from an oil-bearing formation which comprises injecting into the formation an aqueous fluid containing an aqueous surfactant to displace the oil in the formation, the surfactant comprising an effective amount of the reaction product of an alkenyl succinic anhydride wherein the alkenyl group has an average molecular weight of about 100 to 600 and a tertiary amine.

The tertiary amine reactant has the formula: $N(R_1)_3$ wherein each R_1 substituent is independently selected from lower alkyls having less than about 7 carbon atoms per molecule, hydroxy-substituted lower alkyls and hydroxy-substituted ethoxylated lower alkyls, etc. In particular, lower alkyl materials which can comprise each R_1 substituent include materials such as methyl, ethyl, propyl, butyl, pentyl, isopropyl, isobutyl, etc., which themselves may contain hydroxy substituents such as hydroxyethyl and variously ethoxylated amyl alcohols, or mixtures thereof, but do not contain additional nitrogen atoms. R_1 preferably contains less than 6 carbon atoms to insure water-solubility. Preferably, the tertiary amine is triethanolamine for high surfactancy and low cost.

The alkenyl succinic anhydride can generally be prepared from the "ene" reaction of maleic anhydride with olefinic polymers such as polybutenes or polypropylenes of average molecular weights of from about 100 to 600 or greater. In particular, these olefinic polymers can be produced from the cationic polymerization of olefins having from 3 to 6 carbons such as propene, 1-butene, or 2-butene, pentene, isopentene, hexene, 3,3-dimethylbutene, etc. or mixtures thereof.

Depending upon the extent of polymerization and the catalyst and reaction conditions utilized, the alkenyl substituent will possess a varying range of molecular weights. It is necessary that the molecular weight range of the alkenyl and tertiary amine substituents be selected so as not to unduly interfere with the solubility of the produced succinate in both oil and water. A certain balance of solubility for water and oil is needed when this material is used either as a surface-active surfactant by itself or in conjunction with an anionic surfactant as a thickening agent possessing surfactant properties.

Especially useful alkenyl substituents include materials produced from viscous polybutene polymers having average molecular weights depending upon their source.

The resulting polybutenyl succinic anhydride is preferred for reason of low cost and ease of preparation. Materials specifically contemplated will have average molecular weights of around 280, 320, 340, and 420. These specific molecular weights are those from commercially available viscous polybutenes. However, other sources of such viscous polymers are not precluded, as are other molecular weight materials in variance from those described above. Polypropylene is also an excellent choice for the alkenyl substituent.

The alkenyl succinic anhydride is then reacted, preferably in a nonaqueous environment, with anywhere from about 0.7 to 1.3 equivalents of a tertiary amine to yield the desired reaction product. This material can form acid salts when placed in a suitable aqueous environment. In certain instances the above compound can form half-acids when contacted with water.

When using the reaction product as the primary surface-active agent in miscible flooding for recovery of crude oil from underground formations the surfactant can be mixed with the connate water recovered from the reservoir or from the brine available from other sources. The aqueous mixture of the tertiary amine-alkenyl succinic reaction product can include materials known in the art, e.g., water-soluble alcohols such as isopropyl alcohol, the oil-soluble alcohols containing no more than about 10 carbon atoms, and the 2 to 12 mol ethylene oxide adducts of primary alcohols and amines having from 4 to 16 carbon atoms, including such materials as n-butanol, 2-ethylhexanol, n-hexanol, n-octanol, n-decanol, and the like. In general, it is preferred to use the 6 to 8 mol ethylene oxide adducts of n-hexanol. These materials and others in this context are known as cosurfactant materials.

The cosurfactants can vary anywhere from a few tenths of a percent to 25 wt % or more of the succinamate material when it is the primary surface-active agent used in the miscible flooding process.

When the reaction product is used as an additive component in an aqueous mixture containing another anionic surfactant, its concentration can vary depending upon its molecular weight, reservoir conditions and type of other anionic surfactant used, from less than 1 to 200% by weight or more of the other anionic surfactant or surfactants. Preferably, for maximum performance the weight ratio of the reaction product to sulfonate surfactant is from 0.1 to 1.5.

In instances in which the alkenyl succinic-tertiary amine reaction product described above is itself used in an aqueous mixture as the primary surfactant for treating a reservoir, the aqueous mixture contains about 1 to 15% by weight of the reaction product. In instances in which the reaction product is added to an anionic surfactant, the aqueous mixture containing the surfactant is preferably followed by a mobility buffer slug.

The mobility buffer slug can be an aqueous solution containing one or more mobility-reducing agents including materials such as partially hydrolyzed high molecular weight polyacrylamides, high molecular weight polyalkylene oxide polymers, high molecular weight acrylamide polymers containing sulfo groups, copolymers of sodium acrylate or sodium methacrylate and acrylamide, biopolymers especially the polysaccharides, and other materials well known in the art. The conditions under which these mobility buffer slugs are used will vary depending upon the reservoir conditions.

Polybasic Carboxylic Acids as Sacrificial Agents

M.S. Doster and M.V. Kudchadker; U.S. Patent 4,217,958; August 19, 1980; assigned to Texaco Inc. describe a process whereby sacrificial material is injected through an injection means comprised of one or more injection wells into a subterranean petroleum-containing reservoir to preferably occupy or cover all potential adsorption sites within the reservoir thereby reducing the extent of adsorption of the more expensive chemical oil recovery agent. A sacrificial agent performs best when it exhibits adsorption on active sites of rock surfaces and thus diminishes surfactant and/or polymer adsorption.

In this process, the sacrificial agent is preferably injected in admixture with the surfactant slug into the petroleum formation. This surfactant/sacrificial agent mixture may be preceded by a slug of sacrificial material in the aqueous solution. It has been found that this technique is superior to the preflush method of injecting a slug or sacrificial material followed by a slug or surfactant solution without sacrificial material included. However, the preflush method is superior to using no sacrificial material at all.

The sacrificial materials useful in the process are the polybasic carboxylic acids and their water-soluble salts. Especially preferred are oxalic acid, malonic acid, succinic acid, maleic acid, malic acid, tartaric acid, citric acid and their water-soluble salts.

Generally, it has been found that the sacrificial agent in the surfactant slug will be effective in concentrations of from about 0.01 to about 10.0% by weight of the total surfactant solution while an effective volume of the above materials will range from about 0.01 to 1.0 pore volume of the aqueous solution containing the sacrificial agent or the surfactant-sacrificial agent solution.

Lignosulfonate Salts as Sacrificial Agents

G. Kalfoglou; U.S. Patent 4,157,115; June 5, 1979; assigned to Texaco Inc. describes a process of producing petroleum from subterranean formations having an injection well and a production well in communication therewith.

The process comprises injecting into the formation via the injection well an aqueous solution of lignosulfonate salts in admixture with surfactant solutions of alkylbenzene alkoxylated sulfonates and/or sulfates, sulfonated or sulfated alkoxylated sulfonates and/or sulfates, sulfonated or sulfated alkoxylated alkyl surfactants, mixtures of petroleum sulfonates and alkylbenzene alkoxylated sulfonates or sulfates, mixtures of petroleum sulfonates and alkoxylated alkyl sulfonates or sulfates or mixtures of petroleum sulfonates and alkoxylated organic alcohols or organic alcohols and petroleum sulfonates alone.

It is the usual practice to then inject a fluid such as water to sweep the chemical components through the reservoir to the production well, thereby displacing oil from the subterranean formation to the surface of the earth.

In one embodiment of the process the surfactant comprises an aqueous solution of alkylbenzene alkoxylated sulfonates or sulfates having the following general formulas:

(1) R_1, R_2-substituted benzene ring $-(OCH_2\underset{\displaystyle R_3}{\underset{|}{C}}H)_xSO_3^-M^+$

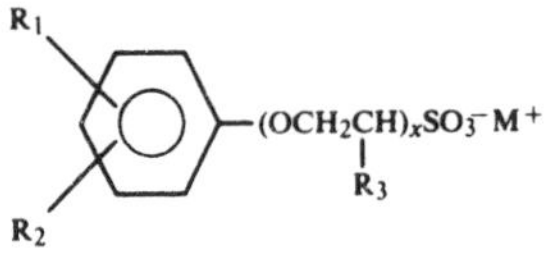

(2) R_1, R_2-substituted benzene ring $-(OCH_2\underset{\displaystyle R_3}{\underset{|}{C}}H)_xOSO_3^-M^+$

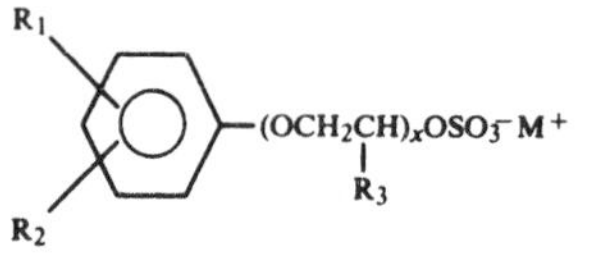

wherein

R_1 and R_2 are hydrogen or alkyl with at least one being an alkyl group of from 6 to 20 carbon atoms;
x is a number from 1 to 10;
M^+ is a cation selected from the group consisting of sodium, potassium, lithium, and ammonium; and
R_3 is either $-CH_3$ or hydrogen.

In another embodiment the surfactant comprises an aqueous solution of sulfonated or sulfated alkoxylated alkyl surfactants having one of the following general formulas:

(3) $R_1(OCH_2\underset{\displaystyle R_2}{\underset{|}{C}}H)_xSO_3^-M^+$ or (4) $R_1(OCH_2\underset{\displaystyle R_2}{\underset{|}{C}}H)_xOSO_3^-M^+$

wherein

R_1 is an alkyl group of from 8 to 22 carbon atoms;
R_2 is either CH_3 or hydrogen; and
x is a number from 1 to 10.

In another embodiment the surfactant comprises an aqueous solution of a mixture of petroleum sulfonates and solubilizers of formulas (1) to (4) above. In yet another embodiment the surfactant comprises petroleum sulfonate or a mixture of petroleum sulfonate and a water-soluble organic alcohol or a mixture of petroleum sulfonate and an organic alcohol which has been alkoxylated and which displays amphiphilic properties.

The amount of surfactant which must be employed in the practice of any chemical flood is generally known in the art and is to be found in published literature. However, the slug size of surfactant generally will range from about 0.01 to 1 pore volume of an aqueous surfactant solution having dissolved therein from about 0.01 to about 10.0% by weight of the surfactant itself.

In a specific embodiment of this process, a sacrificial material comprising lignosulfonate salts is injected via the suitable injection means, i.e., through one or more injection wells completed in the subterranean hydrocarbon formation, in admixture with a surfactant solution. By injecting the sacrificial material and surfactant together oil recovery is maximized.

The broad term lignosulfonates used herein refers to both sulfonated alkali lignins and sulfite lignosulfonates (sulfite lignins). Since the alkali lignins require sulfonation after extraction of the material from woody products it is proper to call them sulfonated alkali lignins. Likewise since sulfite lignins emerge from the extraction process already sulfonated it is proper to refer to this class of materials as sulfite lignins or sulfite lignosulfonates.

Modified sulfonated alkali lignins and sulfite lignosulfonates, such as those with ring sulfomethylation, oxidation, ethoxylation, formaldehyde condensation, phenolation, and/or carboxylation are also useful as sacrificial agents. Lignosulfonates having degrees of sulfonation from about 2.0 to saturation are acceptable. Cations which are acceptable include Na^+, K^+, NH_4^+, Ca^{++} and Mg^{++}. The degree of sulfonation is the weight percentage of sulfonate (SO_3^-) compared to the total molecular weight.

The quantity of sacrificial lignosulfonate materials to be injected into the subterranean hydrocarbon formation may be any amount up to and including an amount sufficient to occupy substantially all of the active sites of the formation matrix.

The amount of sacrificial lignosulfonate salts needed in the process depends on the particular formation, the area or pattern to be swept and other formation characteristics. Those skilled in the art can determine the exact quantity needed to afford the desired amount of protection.

In more specific applications of this process, *G. Kalfoglou; U.S. Patents 4,196,777; April 8, 1980; and 4,133,385; January 9, 1979; both assigned to Texaco Inc.* describes the use of oxidized lignosulfonates as sacrificial agents.

The terms oxidized or oxidation refer to those reactions which have the following effect on the unmodified lignosulfonate molecule. Oxidation of the unmodified lignosulfonates results in the formation of carboxyl groups from carbonyl groupings, alcoholic hydroxyl groups and at least a portion of terminal $-CH_3$ groupings on alkyl side chains and also by a demethylation of at least a portion of the methoxy groupings which appear as a portion of the ring structure. Oxidation also results in conversion of aromatic methoxy groups to form phenolic groups.

Air, air enriched with oxygen and/or oxygen is useful for oxidation of the unmodified lignosulfonate molecule as described above. Alternatively, where air, enriched air or oxygen will not produce the oxidative effects noted above it has been found necessary to react unmodified lignin type materials with an oxygen-containing gas which also contains ozone. U.S. Patent 3,726,850 describes such a reaction scheme on lignin type materials.

G. Kalfoglou; U.S. Patent 4,172,497; October 30, 1979; assigned to Texaco Inc. describes the use of lignosulfonates carboxylated with chloroacetic acid. Carboxylation of lignosulfonates by reaction with chloroacetic acid or its salts by methods known in the art yields products rich in carboxylated groups. The chloroacetic acid reacts with hydroxyl groups to yield acetates and with phenolic groups to yield phenoxy acetates and with sulfonates to yield ester acetates.

The use of sulfomethylated lignosulfonates is described by *G. Kalfoglou; U.S. Patent 4,172,498; October 30, 1979; assigned to Texaco Inc.* Sulfomethylation of lignosulfonates resulting in sulfomethylated lignosulfonates may take place by various routes known to those skilled in the art. Any route which adds the methylene sulfonate radical is acceptable. In one route, formaldehyde and a bisulfite salt is reacted with crude unmodified lignosulfonates under basic conditions to yield a product with additional sulfonate groups attached to the lignosulfonate through methylene groups. For example, sodium, potassium, calcium or magnesium bisulfites are useful. It is believed that these methylene sulfonate groups attach predominantly at the ortho sites of aryl groups in the lignosulfonate although some methylene sulfonate groups attach at the meta site.

Chrome Lignosulfonate Complexes

G. Kalfoglou; U.S. Patent 4,142,582; March 6, 1979; assigned to Texaco Inc. describes a related process where the sacrificial agents used are chrome, iron, aluminum and/or copper lignosulfonate complexes. The use of chrome complexes of sulfonated alkali lignins and sulfite lignosulfonates is preferred.

One way of obtaining chrome lignosulfonate complexes is to react dichromate with lignosulfonate. As a result of the reaction lignosulfonates are oxidized to structures containing functional groups such as phenolates, carboxylates and carbonyls. The presence of these groups on lignosulfonates improves their adsorptive effectiveness on rock surfaces. Dichromate is reduced to chromium(III) which is complexed with the lignosulfonate. Complexation also increases the macromolecular size of the lignosulfonate complexes by intermolecular crosslinking through chromium coordination. This is evidenced by the resulting higher solution viscosities of chrome lignosulfonate complexes.

Chrome complexation of lignosulfonates renders the product more tolerant to elevated temperatures. Chrome lignosulfonate complexes can also be formed by reacting chromium sulfate with sodium and/or ammonium lignosulfonate. These lignosulfonate chromium complexes will coordinate with hydroxy groups. The complexes will have improved adsorptive properties on rock surfaces.

Injection of Fatty Alcohol Followed by Soap

R.L. Cardenas and J.T. Carlin; U.S. Patent 4,213,500; July 22, 1980; assigned to Texaco Inc. describe a method for recovering crude oil from a subterranean reservoir containing brine by injecting fatty alcohol followed by soap to form an oil-in-water emulsion. This emulsion is recovered by continued water injection which forces the emulsion through the formation and out through the producing wells.

Two advantages are readily apparent in this method. First, the oil-in-water emulsion formed has significantly greater viscosity than does water alone, leading to a much more favorable mobility ratio between the displacing and the displaced fluids, a factor that indicates a more efficient sweep of the reservoir by the displacing waterflood. A second advantage of this method is that it will function in the presence of brine. This is especially important because it eliminates the need for the involved and expensive procedures that must be undertaken to condition a brine-containing oil reservoir for treatment by brine-intolerant enhanced oil recovery techniques.

The first step of this process comprises the injection of fatty alcohol into the petroleum reservoir. The alcohol combines with the oil in place to form a mixture that will readily emulsify when contacted with an aqueous surfactant solution containing soap.

Of the many fatty alcohols tested, those found effective include heptyl, octyl, nonyl, decyl, undecyl, dodecyl, hexadecyl and oleyl alcohol. Additionally, two branched alcohols, 2-octyl and 2-decyl, were found to be effective. Especially preferred are the straight chain fatty alcohols containing from 8 to 16 carbon atoms and oleyl alcohol. Effective concentration of the fatty alcohol can vary from about 0.1 to about 10 wt % with the preferred range being from about 0.5 to about 4 wt %.

Any one of a number of different soaps can be used. Sodium dodecyl sulfate was found to be especially effective in emulsifying the oil-alcohol mixtures. One important criterion that the soap should satisfy is that it should be brine tolerant. The sodium dodecyl sulfate systems tolerated sodium chloride concentrations up to and exceeding 3%. The effective concentration of the sodium dodecyl sulfate soap can range from 0.05 to 5% by weight. The preferred range is from about 0.2 to about 2.0% by weight.

Pretreatment of Surfactant Fluid by High Shear Rate

V.H. Schievelbein; U.S. Patent 4,187,073; February 5, 1980; assigned to Texaco Inc. describes a method of treating a surfactant-containing fluid prior to injecting it into a subterranean petroleum formation by subjecting the fluid at the surface to very high shear rate for at least a minimum period of time, prior to injecting it into the formation.

Specifically, it was found that a surfactant fluid should be subjected to a shear rate of at least 150 reciprocal seconds and preferably at least 250 reciprocal seconds for a period of time of at least 0.1 minute and preferably at least 0.5 minute, prior to injecting it into the formation. This is substantially greater shear than occurs during normal mixing and pumping during the injection process, and is substantially greater than the shear which results from flowing the fluid through formation flow channels.

The surfactant fluid which is especially benefited by this treatment is an aqueous fluid having a salinity of at least 20,000 ppm total dissolved solids, which may include substantial quantities of divalent ions such as calcium and magnesium, and containing as the primary surfactant, or essentially the only surfactant, a water-soluble sodium, potassium, lithium, or ammonium salt of an aliphatic polyalkoxyalkyl sulfonate or an alkylarylpolyalkoxyalkyl sulfonate having the following formula:

$$RO(R'O)_xR''SO_3^-M^+$$

wherein R is an aliphatic group such as an alkyl group, linear or branched having from 9 to 25 and preferably from 12 to 18 carbon atoms, or an alkylaryl group, such as a benzene, toluene or xylene having attached thereto at least one alkyl group having from 9 to 15 carbon atoms; R' is ethylene or a mixture of ethylene and propylene or other higher alkylene with relatively more ethylene than higher alkylene and R'' is ethyl, propyl, hydroxypropyl or butyl; x is a number from 2 to 10 and preferably from 3 to 7 including fractions; and M is a monovalent cation, preferably sodium, potassium, lithium or ammonium.

By shearing the abovedescribed surfactant-containing fluid for the prescribed period of time and at the stated shear rate, the turbidity of the fluid is reduced significantly, and the fluid may be injected into lower permeability formations over long periods of time with reduced injectivity or well plugging problems. Moreover, the fluid exhibits greater phase stability as a result of shearing, which increases the period of time that the fluid is effective in the formation while recovering petroleum. Finally, the interfacial tension of the fluid is reduced significantly, which increases its effectiveness as a low surface tension oil displacing fluid.

The surfactant fluid may be sheared in the field by pumping the fluid through one or more plates each having one or more relatively small orifices, e.g., in the range of 0.60 mm (0.0236") to 2.0 mm (0.0787") diameter, with high pressure differentials across the plates, e.g., in the range of 300 kPa (429.81 psi) to 7,000 kPa (1,002.9 psi). Very high speed rotary mixing devices may also be used.

In the case of using the abovedescribed preferred surfactant, an alkylarylpolyalkoxyalkyl sulfonate or alkylpolyalkoxyalkyl sulfonate, the concentration will ordinarily be from about 0.05 to 5.0 and preferably from about 0.5 to about 2.0% by weight.

Micellar Flooding Using Plugging Agents

Enhanced oil recovery processes for heterogeneous reservoirs are known in which aqueous solutions of plug-forming reagents, at least one of which contains a polyvalent cation, are injected to plug the more permeable strata. Subsequently, injected micellar solutions and drive fluids thus desirably penetrate the less permeable strata of the reservoir and increase the volumetric sweep efficiency of the enhanced oil recovery fluid. In some instances, such as due to inadequate mixing in the reservoir, incomplete reaction or use of an excess of one reagent, there remains in the reservoir, after the plugging material has formed, an appreciable amount of unreacted polyvalent cations.

When micellar solutions are injected into reservoirs containing unreacted polyvalent cations and contact the polyvalent cations, the surface-active component of the micellar solution could react with the polyvalent cations. This would decrease the efficiency of the micellar solution in reducing the interfacial tension between the reservoir oil and the aqueous drive fluid, which in turn would decrease the ability of the enhanced oil recovery fluids to mobilize oil in the reservoir.

To maintain the integrity of the micellar solution, *L.W. Holm; U.S. Patent 4,140,183; February 20, 1979; assigned to Union Oil Company of California* describes a method to produce the well treated with the plug-forming reagents after the plug has formed and before injection of additional micellar solution. The well is produced for a length of time and at a rate calculated to remove from the reservoir unreacted polyvalent cations which were injected in the plug-forming step. Production of the well is continued until the polyvalent cation content of the produced fluids falls below a predetermined amount. The remaining amount of the micellar solution and the drive fluid can then be injected with a reduced possibility that the micellar solution will contact and react with polyvalent cations in the reservoir.

The micellar solution injected to miscibly displace oil from the reservoir can be an anhydrous or substantially anhydrous soluble oil, a water-containing soluble oil, a water-external microemulsion, or an aqueous micellar solution. Such micellar solutions are well described in the literature, such as in U.S. Patents 4,011,908 and 4,037,659.

The substantially anhydrous soluble oil contains less than 10 vol % water, and preferably less than 5 vol % water. The water-containing soluble oil can contain water in an amount up to the inversion concentration, i.e., that concentration of water at which the oil-external microemulsion is inverted to a water-external microemulsion. Inversion typically occurs at water concentrations between about 40 and 70%, depending upon the particular composition and the ambient conditions.

The anhydrous soluble oil and water-containing soluble oil also contain about 30 to 90 vol % hydrocarbon, about 0.5 to 8 vol % stabilizing agent and about 4 to 30 vol % surface-active material. The water-external microemulsions typically contain more than about 40 vol % water, usually more than 60 vol % water and preferably 75 to 95 vol % water. The aqueous micellar solutions are aqueous systems, substantially free of oil, in which the surface-active agents are present in the form of macromolecular micelles. The water-external microemulsion and the aqueous micellar solution also contain about 0.5 to 8 vol % stabilizing agent and about 2 to 25 vol % surface-active material.

Soluble oils are compounded from a hydrocarbon component, a stabilizing agent and a surface-active material. Suitable liquid hydrocarbon components include crude petroleum, especially crude having an API gravity between about 27° and 50°; distillate petroleum fractions such as refined or semirefined petroleum products, for example, gasoline, naphtha, stove oil, diesel and gas oil; residual products obtained by the distillation of lower boiling fractions from a crude petroleum, such as bunker fuel oil and other residual products; low value refinery byproducts, such as catalytic cycle oil, lube oil extract, and the like; and liquefied normally gaseous hydrocarbons, such as propane, butane and LPG.

Stabilizing agents include partially oxygenated hydrocarbons such as monohydric aliphatic alcohols having 3 to 5 carbon atoms, dihydric alcohols containing 2 to 3 carbon atoms, aliphatic ketones containing 4 to 6 carbon atoms, glycol ethers containing 4 to 10 carbon atoms, polyhydric alkyl ethers such as dialkylene glycols containing 4 to 6 carbon atoms, and oxyalkylated alcohols containing 8 to 18 carbon atoms.

The surface-active material can be an oil-soluble anionic surfactant such as a higher alkyl sulfonate. These sulfonates are preferably used in the form of their sodium salts. However, other salts can be used. Superior micellar solutions can be prepared by employing a combination of preferentially oil-soluble organic sulfonates and preferentially water-soluble organic sulfonates. A readily available source of alkylaryl sulfonates are the natural petroleum sulfonates produced by sulfonating a relatively narrow boiling point range mixture of petroleum hydrocarbons.

The micellar solution compositions can be prepared by any of the conventional techniques. One suitable method is to prepare a substantially anhydrous soluble oil by admixing the hydrocarbon base stock, the stabilizing agent and the preferentially oil-soluble surface-active material. Thereafter, the preferentially water-soluble surface-active material, if used, is added. Water-in-oil microemulsions can be prepared by simply adding a desired amount of water to the substantially anhydrous soluble oil, or larger amount of water can be added to form an oil-in-water microemulsion.

The aqueous micellar surfactant solution can be prepared by simply admixing suitable proportions of water, the selected surfactant, and optionally, an organic stabilizing agent. Preferably, the water employed in forming the microemulsion is a salt-containing fresh water having a dissolved salt content of less than about 15,000 ppm, and more preferably less than about 5,000 ppm. Water-soluble salts of a monovalent metal can be added to obtain a water having a desired salt content.

Any of a wide variety of known selective plugging agents can be employed in the plugging step. These agents are injected in liquid form, i.e., the plugging agent is a liquid, or is dissolved or dispersed in water, so that when pumped into the well it will preferentially enter into the more permeable strata open to the well whereupon it reacts to form a plugging material.

Useful selective plugging agents include chemical agents that react with the reservoir rock or with connate reservoir or injected fluids to form a precipitate or plugging deposit in the reservoir, exemplary of which are alkali metal hydroxides, sodium silicate, and the like; two or more reactive chemical agents injected successively, such as various water-soluble salts of polyvalent metals which form a precipitate with separately injected aqueous solutions of sodium hydroxide, sodium carbonate, sodium borate, sodium bicarbonate, sodium silicate, sodium phosphate, etc.; various polymeric materials that form substantially permanent plugging deposits in the reservoir, such as crosslinked polyacrylamide; and reactive agents wherein the gelation or precipitation is delayed until the agent is placed in the reservoir, such as hydraulic cements and delayed action silica gels.

Also, various mixtures of these reactive plugging agents can be employed, such as admixtures of sodium silicate and polyacrylamide or crosslinked polyacrylamide, and admixtures of sodium silicates and hydraulic cement.

A preferred selective plugging agent is an aqueous solution of sodium silicate and a gelling agent such as an acid or an acid-forming compound, a water-soluble ammonium salt, a lower aldehyde, a polyvalent metal salt, or an alkali metal aluminate. Exemplary gelling agents are sulfonic acid, hydrochloric acid, ammonium sulfate, formaldehyde, calcium chloride, aluminum sulfate and sodium aluminate. Preferred gelling agents are those containing a polyvalent cation. The sodium silicate reacts in the presence of the gelling agent to form a silica or silica alumina gel, or to precipitate the silicate as insoluble silicate.

The volume of plugging agent solution injected and the concentration of plugging agent in the plugging agent solution can vary over wide ranges depending upon the particular agent selected, the specific characteristics of the reservoir and the connate reservoir fluids, the magnitude and extent of the heterogeneity, and the degree of fluid shutoff desired.

Where sodium silicate is employed as the plugging agent, the concentration of sodium silicate in the plugging solution can vary from about 1 to 30 wt %, weaker plugs being formed at the lower concentration and treating costs increasing at the higher concentrations. Thus, it is preferred that the concentration of sodium silicate in the plugging solution injected into the formation be between about 3 and 15 wt %, and preferably between about 3 and 10 wt %. The molar ratio of sodium oxide to silica (Na_2O/SiO_2) in the silicate can vary, but it is preferred that the sodium silicate solution contains 3 to 3.5 parts by weight of silica per part of sodium oxide.

Other types of silicates which can be used are the alkaline alkali metal silicates, i.e., an alkali metal silicate having a molar ratio of M_2O/SiO_2 of 1 and above, wherein M is an alkali metal atom, such as sodium, potassium, lithium, cesium and rubidium, exemplary of which are alkali metal orthosilicate, alkali metal metasilicate, alkali metal metasilicate pentahydrate, and alkali metal sesquisilicate. Particular agents useful in the process include sodium and potassium orthosilicate, sodium and potassium metasilicate, sodium and potassium metasilicate pentahydrate, and sodium and potassium sesquisilicate. Sodium orthosilicate is a particularly preferred alkaline alkali metal silicate because of its relatively high pH.

A wide variety of reagents can be employed to react with the alkaline alkali metal silicate to form the mobility adjusting precipitate, inclusive of which are acids and acid precursors such as chlorine, sulfur dioxide, sulfur trioxide; water-soluble salts of bivalent metals such as the halide and nitrate salts of iron, aluminum, calcium, cobalt, nickel, copper, mercury, silver, lead, chromium, zinc, cadmium and magnesium; and water-soluble ammonium salts. Preferred agents for reaction with the alkaline alkali metal silicate are those containing a polyvalent cation, such as calcium chloride.

Preferably, approximately the same volumetric quantities of each aqueous reactant solution are injected in each injection cycle, with the concentration of the water-soluble agent that reacts with the alkaline alkali metal silicate being adjusted to provide sufficient agent to stoichiometrically react with the silicate.

Each slug of reactant solution is injected at conventional flood water injection rates such as rates of about 100 to 2,000 barrels per day for a period of about 1 hour to about 7 days, and preferably for a period of about 4 hours to 1 day. The water slug injected intermediate the slugs of reactive solutions can be injected in smaller volume. The following is a typical injection cycle:

	...........Time...........	
Slug	**Broad Range**	**Preferred Range**
Alkaline alkali metal silicate solution	1 hr to 7 days	4 hr to 1 day
Water	1 hr to 1 day	1 hr to 8 hr
Aqueous solution of reactant	1 hr to 7 days	4 hr to 1 day
Water	1 hr to 1 day	1 hr to 8 hr

In a preferred mode of the process to recover oil from a subterranean reservoir, an aqueous solution of sodium orthosilicate is prepared having a sodium orthosilicate concentration selected to provide a pH sufficient to reduce the interfacial tension of the oil-water system to less than 5 dynes per centimeter and preferably to a value of less than 2 dynes per centimeter. The sodium orthosilicate solution is injected into the reservoir through an injection well for a period of about 4 hours to 1 day, followed by water injection for about 1 to 8 hours, then by a slug of an aqueous solution containing a stoichiometric quantity of a second reactant such as calcium chloride substantially equal in volume to the slug of alkaline sodium orthosilicate solution, and then by the injection of water for about 1 to 8 hours. This cycle is repeated throughout the flooding operation, and oil and other produced fluids are recovered from a spaced production well.

While the mobility-controlled caustic treatment can be followed by conventional water drive, it has been found in some instances that the subsequently injected flood water soon breaks through to the producing wells. Hence, it is preferred to maintain the abovedescribed chemical injections for substantially the entire recovery operation.

Example: This example illustrates the process of recovering oil from a heterogeneous reservoir having a highly permeable streak between an injection and a production well and wherein a small amount of oil is contained in the highly permeable streak. There is injected into the injection well 0.001 pore volume of an aqueous alkaline sodium orthosilicate solution prepared by admixing 0.958 pbw of a commercial low alkalinity sodium silicate solution containing 8.9 wt % Na_2O and 28.7 wt % SiO_2 (Na_2O/SiO_2 weight ratio of 0.31, Philadelphia Quartz Company) known as PQ Sodium Silicate N, with 1.35 pbw of 50 wt % sodium hydroxide solution. This is equivalent to 10 barrels of orthosilicate solution per vertical foot of interval treated.

Next there is injected 100 barrels water as a spacer followed by 1,000 barrels of an aqueous solution containing 0.07 wt % of calcium chloride gelling agent and finally 500 barrels water as a spacer. The well is then shut in for 24 hours to allow time for a silicate plug to form. The well is then produced at a rate of 500 barrels per day for 24 hours while the calcium ion content of the produced fluids is monitored to substantially completely remove the previously injected unreacted calcium ions from the reservoir.

Next there is injected 0.03 pore volume of an oil–external soluble oil prepared by admixing 33.7 vol % of a 37° API gravity Texas crude oil; 6.1 vol % of preferentially oil-soluble, surface-active, alkylaryl petroleum sulfonates known as Petronate RHL (Sonneborn Division of Witco Chemical Company, Inc.); 3.8 vol % of preferentially water-soluble, surface-active, alkylaryl petroleum sulfonate known as Petronate 30 (Sonneborn Division of Witco Chemical Company, Inc.); 1.4 vol % of ethylene glycol monobutyl ether known as butyl Cellosolve (Union Carbide Company); and 55 vol % of fresh water containing 1,000 ppm of sodium chloride.

Petronate RHL is an oil solution containing about 62 wt % of mixed preferentially oil-soluble alkylaryl sulfonates and not more than about 5 wt % water. Petronate 30 is an aqueous solution containing 30 wt % preferentially water-soluble alkylaryl sulfonates having an average molecular weight in the range of 330 to 350 and containing about 50 wt % water.

Finally, there is injected 0.8 pore volume reservoir brine containing polyacrylamide polymer thickener as a drive fluid. During the injection of the drive fluid into the injection well, substantial amounts of oil are produced at the production well indicating that a satisfactory plug of the highly permeable streak has been formed and that there has been no appreciable precipitation of the surface-active agent component of the soluble oil by reaction with previously injected calcium ions.

Preflush Fluid

G. Kalfoglou and K.H. Flournoy; U.S. Patents 4,143,716; March 13, 1979; and 4,157,306; June 5, 1979; both assigned to Texaco Inc. describe a surfactant flood oil recovery process whereby high salinity formation water is effectively displaced by injecting into a petroleum-bearing formation via an injection well a preflush of a thickened lower salinity aqueous fluid selected from the group comprising

(A) an aqueous fluid having dissolved therein a small amount of hydrophilic polymer; (B) an aqueous fluid containing a small amount of colloidal silica; and (C) an aqueous fluid having dissolved therein a small amount of hydrophilic polymer and a small amount of colloidal silica. Useful polymers include, for example, polyacrylamide, polysaccharide, methylcellulose, polyethylene oxide, or polyvinyl aromatic sulfonate, etc. In a second step an aqueous surfactant solution which may, if desired, also contain a polymeric thickening agent, is injected into the formation via the injection well and finally the petroleum displaced by the injections is recovered through a production well.

Type (A) Fluids: Type (A) fluids as described above generally will have dissolved therein from about 0.01 to about 0.10% by weight of a hydrophilic polymer such as polyacrylamide. Such aqueous fluids exhibit a viscosity of from about 6 to about 15 cp (measured at a shear rate of 300 reciprocal seconds) and are sufficient to effectively increase the efficiency of displacement of high salinity formation water by the preflush solution.

Type (B) Fluids: Colloidal silica in Types (B) and (C) fluids described above is different from precipitated silica or silica gel. The colloidal silica useful in this process is a fumed silica which is made up of chainlike formations sintered together. These chains are branched and have enormous external surface areas of from about 50 to about 400 m^2/g and each segment in the chain has many hydroxyl (OH) groups attached to silicon atoms at the surface. When the segments come into proximity to each other, these hydroxy groups will bond to each other by hydrogen bonding to form a three dimensional network. Colloidal silica acceptable for use will generally have a particle size ranging from about 7 to 15 millimicrons (mμ).

Generally the Type (B) fluids will contain from about 0.05 to about 0.6% by weight of colloidal silica and preferably they will contain about 0.05 to about 0.1% by weight.

If desired, the Type (B) fluid can contain from about 0.001 to about 0.01% by weight or more of a surfactant based on the weight of the fluid which can be, for example, a soap, the sodium salt of a high molecular weight sulfate or sulfonate, etc. Generally, the surfactant employed will be of the anionic type as exemplified by surfactants of the formula

where n is an integer of from 2 to about 10.

Type (C) Fluids: Type (C) fluids as described above will contain from about 0.01 to about 0.10% by weight of the hydrophilic polymer and from about 0.05 to about 0.60% by weight of colloidal silica.

Field Example: A petroleum-containing formation located at a depth of 5,600' is exploited by means of conventional waterflooding operations using an inverted five-spot pattern, until the water-oil ratio rises above 30. The formation thickness is 30' and the porosity is 26%. In this inverted five-spot pattern the center well is employed as an injection well while the four remaining wells serve as production wells. The dimensions of the square grid on which an inverted five-spot pattern is based is 500' and it is known that only 75% of the reservoir will be swept by the injected fluid using the standard five-spot pattern. The pore volume of the pattern swept by the injected fluid will be 500 x 500 x 30 x 0.26 x 0.75 = 1,462,500 ft^3. The salinity of the water contained in the formation is 225,000 ppm which is considerably above the tolerable salinity for petroleum sulfonate and other commonly used surfactants.

A total of 0.1 pore volume (146,000 ft^3) of a preflush solution having a salinity of only about 100,000 ppm and having dissolved therein 0.03 wt % of polyacrylamide, 0.01 wt % of a friction reducing copolymer which is random, linear water-soluble copolymer consisting essentially of about 1% by weight of (3-acrylamido-3-methyl)butyltrimethylammonium chloride and about 99 wt % acrylamide and having dispersed therein about 0.02 wt % of colloidal silica is injected into the formation via the injection well. This is followed by the injection into the formation of 0.1 pore volume (146,000 ft^3) of aqueous solution having dissolved therein 2.0% by weight of a sulfonate surfactant of the formula:

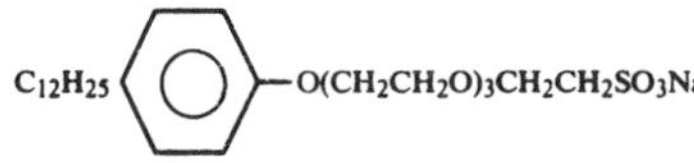

$$C_{12}H_{25}-C_6H_4-O(CH_2CH_2O)_3CH_2CH_2SO_3Na$$

and also having dissolved therein 0.02 wt % partially hydrolyzed polyacrylamide to increase the solution viscosity to about 9 cp. Next 0.1 pore volume (146,000 ft^3) of water containing 0.03 wt % of partially hydrolyzed polyacrylamide and having a viscosity of 9.4 cp is injected into the formation. This is followed by the injection of water having a salinity of about 100,000 ppm to displace the oil, surfactant solution and thickened water through the formation. Oil is produced through the associated production wells in the five-spot pattern, and the amount of oil produced is substantially in excess of that produced by waterflooding alone.

Waterflood Process for Two Zone Formation

J.E. Varnon, M.V. Kudchadker, A. Brown and L.E. Whittington; U.S. Patent 4,159,037; June 26, 1979; assigned to Texaco Inc. describe a process applicable to subterranean, petroleum-containing formations containing two or more zones, at least one of which has a permeability at least 50% greater than the other zone, which will permit application of enhanced oil recovery processes such as waterflooding or surfactant flooding in both zones.

The process involves first injecting water or other aqueous displacing fluid into the formation to pass through the more permeable zone, displacing petroleum therefrom, until the ratio of injected fluid to formation petroleum fluids being recovered from the formation reaches a predetermined or economically unsuitable level. This further increases the ratio of the permeability of the most permeable zone to the permeability of the less permeable zone or zones. Thereafter an aqueous emulsifying fluid is injected into the formation, which fluid flows substantially exclusively into and through the most permeable, previously waterflooded zone.

Injection into the zone may be by means of the well utilized as the injection well initially, or by means of the production well, or by both wells simultaneously or sequentially. The fluid has a viscosity not substantially greater than the viscosity of water, and contains a surfactant combination which readily emulsifies the residual oil present in the previously waterflooded zone.

The surfactant mixture present in the injected treating fluid must be one which forms an emulsion with the residual formation petroleum in the zone being treated at a salinity about equal to the salinity of the aqueous fluid present in the previously flooded, high permeability zone, and should also be relatively stable with changes in salinity since there are normally variations in water salinity from one point in a subterranean formation to another. The emulsion formed should also be stable for a long period of time at the temperature of the formation, in order to maintain the desired reduction of permeability within the treated zone.

The surfactant employed in the process comprises at least two components: (1) an organic sulfonate such as a water-soluble salt, preferably a sodium, potassium or ammonium salt of an alkyl or alkylaryl sulfonate having from 6 to 25 and preferably from 8 to 18 carbon atoms in the alkyl chain, which may be linear or branched, or a water-soluble salt, preferably a sodium, potassium or ammonium salt of petroleum sulfonate having a median equivalent weight from 325 to 475, and (2) a solubilizing cosurfactant, preferably a water-soluble salt of an alkyl or alkylarylpolyalkoxyalkylene sulfonate having the following formula:

$$RO(R'O)_nR''SO_3M$$

wherein R is an alkyl, linear or branched and having from 8 to 22 and preferably from 10 to 18 carbon atoms, or an alkylaryl such as benzene, toluene or xylene having attached thereto an alkyl, linear or branched, containing from 8 to 15 and preferably from 9 to 13 carbon atoms in the alkyl, R' is ethylene or a mixture of ethylene and higher alkylene such as propylene with relatively more ethylene than higher alkylene, preferably at least 65% ethylene, R'' is ethylene, propylene, hydroxy propylene or butylene, n is a number from 2 to 20 and preferably from 4 to 12, and M is a monovalent cation, preferably sodium, potassium or ammonium.

The equivalent weight of the organic sulfonate, the balance between R and n in the above formula and the weight ratio of solubilizing cosurfactant are all chosen based on experimentation using petroleum and brine from the formation into which the fluid will be injected, on the basis of exhibiting the best combination of the following:

(1) Optimum emulsifying property;

(2) Maximum emulsion viscosity;

(3) Emulsion brine tolerance;

(4) Maximum stability of emulsion at formation temperature and salinity;

(5) Tolerance for changes in aqueous fluid salinity within the range of variations expected in the formation and which may be caused by subsequently injecting fluids having greater or less salinity than the salinity of the aqueous phase of the emulsion.

A low molecular weight alcohol, e.g., a C_2 to C_7 alkanol or an alkyl-substituted phenol having from 1 to 5 carbon atoms in the alkyl chain, may be required for use or used advantageously in the emulsifying fluid, to enhance the emulsification-forming properties or to enhance the stability of the emulsion.

In a slightly different preferred embodiment of the process, a small amount of hydrocarbon is added to the emulsifying liquid to form an emulsion, micellar dispersion or microemulsion. The hydrocarbon may be crude oil, as well as kerosene, naphtha, gasoline or other commercially available mixtures, as well as C_6 to C_{24} hydrocarbons, preferably saturated hydrocarbons.

The amount of hydrocarbon which is incorporated in the emulsion is determined experimentally from the aqueous emulsifying fluid by preparing samples containing various concentrations of hydrocarbon over the range of 0.05 to about 15.0 and preferably 0.5 to 5.0% by volume, and determining the viscosity of each sample. The maximum amount of hydrocarbon which can be incorporated in the fluid without causing the viscosity of the fluid to exceed the viscosity of water or brine at formation conditions by more than 500% and preferably by no more than 200%, is selected. By incorporating this small, critical volume of hydrocarbon in the fluid, the desired viscosity development in the formation to be treated will be obtained much more rapidly than if an oil-free liquid were injected.

Maintaining the level below the abovedescribed limit ensures that the fluid may be injected easily and into the same flow channels of the formation as would be invaded by water.

The concentration of the alkyl or alkylaryl sulfonate or petroleum sulfonate surfactant will ordinarily be in the range of from about 0.01 to about 10 and preferably from about 0.5 to about 4.0% by weight. The concentration of the alkyl or alkylarylpolyalkoxyalkylene sulfonate surfactant will ordinarily be from about 0.1 to about 5.0 and preferably from about 0.4 to about 2.0% by weight. The ratio of organic sulfonate surfactant to the alkyl or alkylarylpolyalkoxyalkylene sulfonate will ordinarily be from about 0.5 to about 5.0, depending on the salinity of the fluid in which it is formulated, which in turn is usually about equal to the salinity of the fluid present in the subterranean formation.

The volume of treating fluid to be injected into the formation is ordinarily from about 1.0 to about 100 and preferably from 10 to 50 pore volume percent, based on the pore volume of the high permeability zone or zones to be contacted by the treating fluid. It is important to note that the pore volume on which these numbers are based relate to the pore volume of the high permeability zone to be treated, not the pore volume of the whole formation.

Sequential Injection of Two Surfactant Fluids

V.H. Schievelbein; U.S. Patent 4,165,785; August 28, 1979; assigned to Texaco Inc. describes a surfactant waterflooding supplemental oil recovery process, particularly one employing as the only surfactant or as a component in the surfactant fluid, an aliphatic or alkylarylpolyalkoxyalkyl sulfonate having the following formula: $R(OR')_nR''SO_3M$ wherein R is an aliphatic, preferably an alkyl group, linear or branched, having from 9 to 25 and preferably from 12 to 18 carbon atoms, or an alkylaryl group such as benzene, toluene or xylene having attached thereto at least one alkyl group, linear or branched, having from 9 to 15 and preferably from 10 to 13 carbon atoms; R' is ethylene or a mixture of ethylene and higher molecular weight alkylene with relatively more ethylene than higher molecular weight alkylene; n is a number including fractional numbers from 2 to 10 and preferably from 3 to 7; R'' is ethylene, propylene, hydroxy propylene, or butylene and M is a monovalent cation such as sodium, potassium, lithium or ammonium.

In applying the process, at least two separate surfactant fluids are injected sequentially into the formation. The first surfactant fluid employs a surfactant which produces a significant amount, preferably the surfactant capable of producing the maximum amount, of an emulsion containing the formation petroleum and the aqueous fluid, e.g., brine, present in the flow channels. In lab tests performed for the purpose of identifying the preferred emulsifying surfactant, the volume of emulsion formed should be at least 5% and preferably 40% based on the initial volume of surfactant solution.

The second surfactant fluid employs a surfactant which produces essentially no emulsion phase between the formation petroleum and the aqueous fluid present in the formation, but which exhibits the optimum effectiveness for the purpose of oil recovery, which may be identified by determining the particular surfactant which reduces the interfacial tension between the formation petroleum and the aqueous fluid present in the flow channels of the formation to a value less than 100 and preferably less than 20 millidynes per centimeter.

In the preferred embodiment, in which the sole surfactant or one of the surfactants present in each fluid is an aliphatic or alkylarylpolyalkoxyalkyl sulfonate, the surfactants employed in the two fluids will usually differ only in the average number of mols of ethylene oxide per mol of surfactant. Ordinarily the surfactant contained in the first surfactant fluid injected into the formation will contain from 0.2 to 0.8 fewer mols of alkylene oxide (e.g., ethylene oxide) per mol of surfactant than the surfactant employed in the second fluid injected into the subterranean formation.

For example, if it is determined that the minimum interfacial tension obtained in a series of tests employing samples of field brine and crude oil is 18 millidynes per centimeter, using 2.0% dodecylbenzenepolyethoxypropane sulfonate containing an average of 3.3 mols of ethylene oxide per mol of surfactant, this is the surfactant employed in the second fluid and the first fluid contains a like concentration of dodecylbenzenepolyethoxypropane sulfonate containing from 2.5 to 3.1 and preferably from 2.7 to 3.0 mols of ethylene oxide per mol of surfactant.

The volume of the first and second fluids injected into the formation will ordinarily be from 0.02 to 0.40 and preferably from 0.05 to 0.25 pore volume based on the pore volume of the formation to be exploited.

The concentration in each surfactant within the two fluids employed in the process will primarily be from 0.1 to 5.0 and preferably from 0.5 to 3.0% by weight (1 to 50 and preferably 5 to 30 kg/m^3).

Method of Preparing Microemulsions

A microemulsion is a dispersion of two immiscible liquids (one liquid phase being dispersed and the other being continuous) in which the individual droplets of the dispersed phase have an average radius less than about ¼ of the wavelength of light. Typically, in a microemulsion the dispersed phase droplets are less than about 1400 Å radius, and preferable in the order of 100 Å to 500 Å.

H.L. Rosano; U.S. Patent 4,146,499; March 27, 1979 describes a process for preparing microemulsions of a water-immiscible liquid in an essentially aqueous phase. For brevity in the following description, these will be referred to as oil and water phases. The process comprises four basic steps:

(1) An amphiphatic surfactant, herein referred to as the primary surfactant, is selected having a hydrophilic-lipophilic balance (herein HLB) not substantially less than required to make the primary surfactant soluble in the oil phase. The HLB depends upon the relative size and strength of the hydrophilic and lipophilic moieties of the amphiphatic surfactant (i.e., relatively lipophilic will be soluble in the oil while surfactants of a high HLB will be insoluble).

Solubility may be determined by a simple test. A fraction of a gram of surfactant is added to several milliliters of oil. If the surfactant is insoluble, a cloudy suspension will result; if it is soluble the solution will be clear. Using the simple test, a compound will be selected in which the HLB is just low enough to render the compound soluble.

(2) A solution of the primary surfactant selected in the first step is prepared in the oil which is to become the oil phase. That solution should contain sufficient primary surfactant to form an essentially monomolecular layer of surfactant on the dispersed oil phase droplets after the oil has been dispersed as a microemulsion. In the absence of a secondary surfactant as hereinafter described, the primary surfactant at this stage of the process should be capable of producing a finely-dispersed emulsion of the oil in the water. Typically, the oil will contain approximately 20% by volume of the primary surfactant but may contain as little as 10% of the primary surfactant.

(3) After dissolving the primary surfactant in the oil, the oil is then dispersed in the aqueous phase. If necessary, mild agitation may be provided to insure thorough dispersion. At this point, the resulting emulsion is a finely-dispersed oil-in-water emulsion usually appearing as a milky, or lactescent dispersion.

(4) Finally, a secondary surfactant (or cosurfactant) is provided in the aqueous phase to convert the lactescent dispersion into a microemulsion. In this step, the lactescent dispersion prepared in the third step may be titrated with a secondary surfactant or cosurfactant. In the alternative, a secondary surfactant may be dissolved in the aqueous phase prior to dispersion. Generally the cosurfactant has a lower HLB than

the primary surfactant selected in the first step. The cosurfactant may be selected from the same general series of compounds as the primary surfactant, but it is not necessary to do so.

Among the primary surfactants found particularly useful are nonylphenolpolyoxyethylene condensates, the sorbitan and sorbitol monoesters C_{12} to C_{18}; fatty acids, and their ethylene oxide condensates.

For secondary surfactants not only may the common water-soluble surface-active agents (such as those identified above as primary surfactants) be employed but also substances such as short-chain alcohols, for example, pentanol or hexanol.

The water-immiscible liquid may have a dominant portion of a light mineral oil. A representative oil is n-hexadecane, although obviously mixed oils can be used. The mineral oil can be used either alone or as a solvent for a number of lipophilic substances such as siloxane, lanolin, and hair or skin conditioners, or similar emollients.

Such systems can generally be emulsified by primary surfactants such as the glycerol fatty acid partial esters (such as glycerol monooleates) or ethoxylated alkyl phenols such as nonylphenol having 1.5 to 6 ethylene oxide units. Where the fatty group of the alkyl phenol is increased or decreased in size, the degree of ethoxylation will change to provide the correct HLB.

For this system the secondary surfactants are generally ethoxylated alkyl phenols having 4 to 20 carbon atoms, ethoxylated fatty alcohols or alkyl amino or amido betaines having an HLB greater than the HLB of the primary surfactant. Nonylphenol·15 EO is one such suitable secondary solvent.

The water-immiscible liquid may have a dominant portion of volatile aliphatic petroleum solvents such as pentane or hexane. Such volatile solvents are desirable, for example, when the water-immiscible phase is a solution containing an active ingredient to be emulsified, and from which the solvent is to be removed.

For such systems, a suitable primary solvent would be a sorbitan or sorbitol partial ester, such as sorbitan monolaurate. Because of the solubility relationships no ethoxylation is needed when sorbitan monolaurate is used as the primary surfactant. However, if the lipophilic moiety is increased in size, some small amount of ethylene oxide addition to the ester may be needed. A suitable secondary surfactant is sorbitan or sorbitol ester condensed with ethylene oxide, such as sorbitan monolaurate·20 EO.

Naphthenoyl Sulfobetaine Surfactants

H. Wagner; U.S. Patent 4,259,191; March 31, 1981; assigned to Th. Goldschmidt AG, Germany has discovered a process for the preparation of sulfobetaines of the general formula

$$(1) \qquad R^1{-}NH_x{-}R^2{-}\overset{\overset{\large R^3}{|}}{\underset{\underset{\large R_x^4}{|}}{N^+}}{-}R^5{-}SO_3^-$$

In this general formula, the substituents have the following meaning:

R^1 is the naphthenoyl residue derived from naphthenic acid. The naphthenic acids are defined as natural acids, obtained from crude oils and their products by extraction with caustic solution and subsequent acidification. Thus, they are essentially acid mixtures in which, in addition to linear carboxylic acids, especially alkylated cyclopentane carboxylic acids and cyclohexane carboxylic acids predominate. Those naphthenic acids which have an acid number of 80 to 350, preferably 120 to 250, are especially suitable.

R^2 is an alkylene residue with 2 to 6 carbon atoms. Preferably the alkylene residue is linear. Residues with 2 to 3 carbon atoms are particularly preferred.

The substituents R^3 and R^4 may be the same or different and represent a lower, linear alkyl residue with 1 to 4 carbon atoms. The methyl residue is particularly preferred.

R^5 is an alkylene residue with 1 to 4 carbon atoms, whereby an alkylene residue with 3 or 4 carbon atoms is especially preferred.

The index x has the value of 0 or 1. If x = 0, the quaternary nitrogen atom is connected through an additional R^2 group and by ring formation with the first nitrogen atom, which is linked to the carbonyl group of the naphthenic acid. At the same time, R^2 preferably is an alkylene group with 2 carbon atoms. In this case, the two nitrogen atoms are constituents of a piperazine ring.

The preparation of these compounds can be carried out by reacting aminoamides of naphthenic acid of the general formula (2) $R^1NH_xR^2NR^3R^4_x$

(a) With halogenalkylsulfonic acids of the general formula (3) XR^5SO_3H wherein X is a halogen atom, especially the chlorine atom, in the presence of at least equimolar amounts of an alkali hydroxide or alkali carbonate or

(b) With alkali salts of the abovementioned halogenalkylsulfonic acids, or

(c) Propane or butane sultone at temperatures of 50° to 150°C by known procedures. The aminoamides of naphthenic acids of the general formula (2) above can be prepared by reacting naphthenic acids with an alkylenediamine. The R^3 and R^4 groups can be introduced into the molecule before or after this reaction.

Example 1: 3,190 g of a crude naphthenic acid (acid number 193), oil content 2.9%, corresponding to a corrected molecular weight of 282 are converted to the amide with 1,380 g of dimethylaminopropylamine (20% excess) at 200°C within a period of seven hours. Towards the end of the reaction, when the acid number is less than 8, the product is freed from excess amine by distillation under a vacuum of 10 torr. The yield of the naphthenic acid amide of dimethylaminopropylamine, with a molecular weight of about 366, is about 4,100 g.

In order to prepare a 30% solution of the betaine, 1,464 g of this amide were emulsified in 4,556 g of water and heated with stirring to 50°C. 488 g of 1,3-propane sultone (3-hydroxy-1-propanesulfonic acid sultone) were added dropwise over a period of two hours. After the dropwise addition of propane-sultone is completed, the mixture is reacted for a further five hours at 70°C to form the betaine.

The product, naphthenyl-1,3-amidopropyldimethylaminopropanesulfonic acid betaine, can be characterized by the following formula:

$$R^1{-}CO{-}NH{-}(CH_2)_3{-}\overset{\overset{\large CH_3}{|}}{\underset{\underset{\large CH_3}{|}}{N^+}}{-}(CH_2)_3{-}SO_3^-$$

Example 2: 282 g of a crude naphthenic acid (acid number 193, oil content 2.9%; corresponding to a corrected molecular weight of 282) were converted to the amide with 139.5 g of 2-amino-1-diethylaminoethane (about 20% excess) at 210°C during a period of five hours. Towards the end of the reaction, when the acid number is less than 5, the product is freed from excess amine by distilling under a vacuum of 12 torr. The yield of naphthenic acid amide of 2-amino-1-diethylaminoethane with a molecular weight of about 380, is about 375 g.

In order to prepare a 30% solution of betaine, 113.5 g of this amide are emulsified in 350 g of water and heated with stirring to 50°C. 36.5 g of 1,3-propane-sultone (3-hydroxyl-1-propanesulfonic acid sultone) are added dropwise within a period of 30 seconds. After the dropwise addition of propanesultone is completed, the mixture is converted to the sulfobetaine by heating for five hours at 70°C.

The product, naphthenyl-1,2-amidoethyldiethylaminopropanesulfonic acid betaine, can be characterized by the following formula:

$$R^1{-}CO{-}NH{-}(CH_2)_2{-}\overset{\overset{\large C_2H_5}{|}}{\underset{\underset{\large C_2H_5}{|}}{N^+}}{-}(CH_2)_3{-}SO_3^-$$

The prepared compounds have excellent surface-active properties. Because of the naphthenoyl residue, the prepared sulfobetaines have a special affinity for mineral oil and mineral oil products.

The prepared sulfobetaines enable very stable mineral oil-water emulsions to be prepared and partially solubilize the mineral oil, while at the same time, because of their betaine structure, they are largely insensitive to the presence of any salt in the water.

In addition, when present in only very slight concentrations, they lower the interfacial tension between mineral oil and water, especially salt-containing water, for example, between crude oil and water in contact with such oil. Both the good emulsifying and solubilizing properties as well as the great reduction in interfacial tension between water and crude oil at very low concentrations in the ppm range make the prepared compounds into very effective, surface-active substances for the so-called tertiary oil recovery.

POLYMERIC WATERFLOODING COMPOSITIONS

Surfactant Addition to Improve Filterability

Enhanced recovery of hydrocarbons from subterranean formations by injection of water into an injection well and recovery of the hydrocarbons from a production well is well known. Also well known is addition of certain organic polymers to at least a portion of the aqueous fluid which is injected to thicken the waterflood and to further enhance recovery of hydrocarbons. Such thickened aqueous floods are commonly known as polymer floods.

Also well known is combination of such polymer flooding techniques with surfactant flooding and the like to further enhance recovery of hydrocarbons. One problem with injection of such aqueous floods containing organic polymers is plugging of the injected formation. This problem can be particularly troublesome with biopolymers of which the xanthan gums or heteropolysaccharides produced by fermentation with *Xanthomonas campestris* are a very commercially important example.

To mitigate the problem of plugging of the formation thus injected, aqueous solutions of such organic polymers are commonly filtered prior to injection. However, the filters employed to remove the plugging components also rapidly lose capacity and often become plugged.

H.H. Ferrell, D. Conley, B.M. Casad and O.M. Stokke; U.S. Patent 4,212,748; July 15, 1980; assigned to Conoco, Inc. describe a process whereby the filterability of aqueous solutions of polymers employed to thicken waterfloods such as biopolymers (e.g., heteropolysaccharides) is improved by addition of a surfactant (e.g., a sulfated alkoxylated alcohol). Filterability is further enhanced by addition of an ethoxylated alcohol surfactant and/or an alcohol.

According to a preferred mode of operation, a sulfated alkoxylated alcohol surfactant is employed and preferably is a sulfated ethoxylated alcohol. Examples of commercially available surfactants of this type include Alfonic 1412A (Continental Oil Company), Neodol 25-35 (Shell Chemical Company), and Tergitol 15-5-3.0 (Union Carbide).

According to another preferred mode of operation, a hydrocarbon sulfonate surfactant is employed, and is preferably a sodium hydrocarbon sulfonate surfactant, a potassium hydrocarbon sulfonate surfactant, an ammonium hydrocarbon sulfonate surfactant, or a substituted ammonium hydrocarbon sulfonate surfactant. Examples and disclosure of suitable hydrocarbon sulfonate surfactants are provided in U.S. Patents 3,874,454 and 4,058,467.

The sulfonates employed in this process when the sulfonate mode is employed can be either mixtures of oil-soluble sulfonates and water-soluble sulfonates, or individual sulfonate constituents. However, the average equivalent weight of sulfonates used will range from 300 to 600. Desirable results are obtained when the sulfonate constituent has an average equivalent weight of from 350 to 500. Especially desirable results are often obtained when the average molecular weight of the sulfonate employed is from 400 to 450. Examples of some ethoxylated alcohols which are suitable include polyoxyethylene lauryl, cetyl, stearyl, and oleyl ethers; ethoxylated alkyl phenols, e.g., octyl and nonylphenoxypoly(ethylenoxy) ethanol, and the like.

Examples of suitable alcohols which can be employed include water-soluble alcohols such as methanol, ethanol, isopropanol, methyl Carbitol, hexyl Carbitol, and the like.

Filtration can be by any conventional method such as by a diatomaceous earth filter, a revolving drum filter, sand filters, paper filters, combinations thereof, or the like. Filtration is enhanced such that backwashing or other cleaning, regeneration, or reconstitution of the filters is greatly reduced.

When the surfactants of the process are employed to enhance filtration, any amount that results in decreased capacity of the filter system or increased time between reconditioning of the filter system can be employed. When sulfated ethoxylated alcohol surfactants are employed, amounts of such surfactants above 0.01 g of surfactant per gram of polymer are suitable. Amounts in the range of 0.04 to 0.12 have exhibited good effects, and amounts in the range of 0.02 to 0.20 are believed to be a practical working range. The surfactant can be added to the polymer solution at any time and in any manner prior to filtration.

Improved Filterability by Caustic-Enzyme Treatment

O.M. Stokke; U.S. Patent 4,165,257; August 21, 1979; assigned to Continental Oil Company describes an improved hydrocarbon recovery process wherein subterranean petroliferous formations are flooded with aqueous mixtures of biopolymers to increase hydrocarbon recovery, the biopolymer being treated with bacillus-produced enzyme, Esperase (Novo Enzyme Corporation) to solubilize cell debris prior to filtration, the improvement comprising treating the polymer prior to filtration and use in a petroliferous formation with Esperase enzyme at a pH of 12.5 to 13.0 to improve filterability, then filtering prior to use if necessary.

The method for obtaining the caustic solution can be by any means well known in the art, such as by the addition of sodium hydroxide, potassium hydroxide, and so forth. Normally the treatment is carried out at a temperature of from 76° to 120°F but most preferred is at a temperature of from 100° to 120°F. The enzyme is normally added to the biopolymer at levels of from 2.5 to 10 wt % based upon the weight of the biopolymer. Additions of enzyme above levels effective to decrease the proteinaceous material contained in the polymer are wasteful and no additional benefit is seen.

Filtration following the treatment of the enzyme may or may not be necessary depending upon the particular formation and the particular starting biopolymer chosen. However, in most cases some filtration will be necessary, but filter plugging will be greatly reduced because of the decrease of debris in the biopolymer because of the treatment. Such filtration can be by any method well known to those skilled in the art but normally will be by methods such as diatomaceous earth (DE) filters.

Increasing Molecular Weight of Vinylpyrrolidone Polymer

D.H. Lorenz, E.P. Williams and H.S. Schultz; U.S. Patent 4,190,718; February 26, 1980; assigned to GAF Corporation describe a process for increasing the molecular weight of poly-N-vinylpyrrolidone over that normally obtained by suspension polymerization of N-vinylpyrrolidone in a saturated hydrocarbon suspension medium, by adding to the suspension medium a controlled amount of water, specifically an amount between 1 and 35 wt % based on the total weight of N-vinylpyrrolidone monomer. The significantly higher molecular weight of the resulting polymeric product can be still further increased by additionally incorporating in the suspension medium between 0.001 and 1 mol % of a difunctional vinyl cross-linking agent per mol of N-vinylpyrrolidone monomer. The polymeric products are preferably water-soluble granules having a porous structure, although the process also concerns the poly-N-vinylpyrrolidone having a number average molecular weight of 400,000 to 2,000,000, preferably between 450,000 and 1,500,000 in any physical form.

The preferred solvents for the process are those alkanes having boiling points between 50° and 100°C which are more easily separated from the polymeric product by boiling or evaporation. Most preferred of this group are normal and branched chain hexane, heptane, octane, and cycloalkanes such as cyclopentane and cyclohexane. The preferred amount of solvent employed in the polymerization is between 1.5 and 2 pbw per part of monomer.

Maintenance of the suspension is enhanced and prolonged during polymerization by the addition of small amounts of suspension stabilizers of the type used to prevent agglomeration of the suspended droplets or globules which become more viscous and sticky as polymerization progresses.

Such stabilizers or suspension aids which are found to be useful in this process are the poly-N-vinylpyrrolidone alkyl-modified polymers (e.g., N-vinylmethylpyrrolidone polymers and copolymers with methyl-substituted monomer) and the N-vinylpyrrolidone C_{10-20} olefin copolymers of Antaron P-904, Antaron P-804, Antaron V-816, Antaron V-516, Ganex P-904, Ganex V-220, Ganex V-516, etc., which are substantially insoluble in the solvents of this process. Generally, the stabilizer is employed in an amount between 0.5 and 5 wt %, preferably between 1 and 3 wt % based on the weight of monomer. Conveniently, the above stabilizers are polymers having a number average molecular weight between 8,000 and 20,500.

The catalyst or polymerization initiators found to be most effective in this suspension polymerization are those of the free radical type which are soluble in the monomer. Accordingly, suitable initiators for the process include azodiisobutyronitrile (Vazo-64), azodiisovaleronitrile (Vazo-52), dimethyl azodiisobutyrate, benzoyl peroxide, t-butyl hydroperoxide, t-butyl peracetate, acetyl peroxide, di-t-butyl peroxide, di-cumyl peroxide, cumyl hydroperoxide and generally any of the oil-soluble free radical initiators conventionally employed for polymerization.

The particular initiator may be selected in accordance with the temperature of polymerization so that the catalyst remains in the liquid phase during reaction. In an oxygen-free system, the proportion of initiator employed is between 0.02 and 2 wt %, preferably between 0.05 and 1.0 wt % based on total weight of monomer. However, when oxygen contaminant is present, an additional amount of initiator is employed, e.g., the above amount plus an equimolar amount for every mol of oxygen in a free state or reducible to a free state. Instead of, or in addition to, the initiators listed above, actinic light with or without the aid of photosensitizer, e.g., benzophenone, fluorescein, eosin, etc., may be used but is less desirable.

Example: Into a steel autoclave, containing a double 6-bladed turbine stirrer and a thermowell, which has been rinsed with distilled water and thoroughly purged with nitrogen, is added the following mixture: 2,125 g N-vinylpyrrolidone monomer, 750 g distilled water (30 wt % based on total weight vinylpyrrolidone), 50 g Ganex V-516 (a suspending agent by GAF Corporation and comprising a 50/50 graft copolymer of vinylpyrrolidone and C_{16} olefin), and 4,000 g n-heptane.

The order of addition for the above components involves introducing the N-vinylpyrrolidone to the mixture of heptane and suspending aid during continuous stirring and subsequently adding water. However, it is to be understood that any other order of addition in forming the suspension can be employed to equal advantage. Also, if desired, additional amounts of suspending agent may be added during the course of polymerization.

In a separate cylinder, 375 g of N-vinylpyrrolidone monomer is mixed with 1 g of azobisisobutyronitrile (a polymerization initiator of Du Pont, Vazo-64). Both the reactor and the cylinder are evacuated to 10" Hg vacuum gage and the vacuum released with dry grade nitrogen. This procedure is repeated 12 times.

After contents of the stirred reactor is brought to 75°C, the contents from the cylinder is charged to the reactor with continuous stirring. The reactor is then sealed and held at a temperature of between 74° and 76°C for a period of 8 hr after which it is cooled to about room temperature and the contents discharged to a filter. The resulting filter cake is then tray dried at 50°C under mild vacuum to provide a dry weight of 2,205 g poly-N-vinylpyrrolidone.

The relative viscosity of the poly-N-vinylpyrrolidone is determined by preparing 1% (w/v) solution of the dried polymer in CP ethanol. Viscosity of the above polymer is found to be 7.63 which corresponds to K value of 109.6 and a number average molecular weight of approximately 550,000.

The product of this example is particularly useful in the polymer flooding oil recovery process [fully described in the *Journal of Petroleum Technology*, pp 33-41 (January 1974)]. A solution of the above-dried product can be substituted for Pusher 700 (hydrolyzed polyacrylamide) in the same concentration and under the same conditions as set forth in the publication to provide greatly improved oil recovery.

Vinylpyrrolidone Polymer Solution Followed by Surfactant

W.C. Haltmar and E.S. Lacey; U.S. Patent 4,207,946; June 17, 1980; assigned to Texaco Inc. describe a tertiary hydrocarbon-recovery process in which the formation is first treated with a vinylpyrrolidone polymer followed by injection of an aqueous surfactant solution and finally recovering hydrocarbons via the production well. Optionally, an aqueous drive fluid is injected into the formation. Water-soluble vinylpyrrolidone polymers useful in the aqueous compositions utilized in the recovery process include those having recurring units of the formula:

$$\left[-\underset{R}{\overset{N(\text{ring, } C=O)}{C}}-\underset{R'}{CH}- \right]$$

wherein R and R' are independently selected from the group consisting of hydrogen and alkyl radicals having from 1 to 5 inclusive carbon atoms. Examples of alkyl radicals of 1 to 5 inclusive carbon atoms include methyl, ethyl, propyl, butyl, pentyl, and isomeric forms thereof. The sole limitation on the cited structure is that it be sufficiently soluble in the aqueous medium. Likewise, mixtures of these polymers may be employed. When R and R' of the above formula are both hydrogen, the resulting compound is polyvinylpyrrolidone, i.e., poly-N-vinyl-2-pyrrolidone, which is an especially useful polymer.

Water-soluble copolymers useful in the treating compositions are prepared by copolymerizing: (a) a compound of the formula:

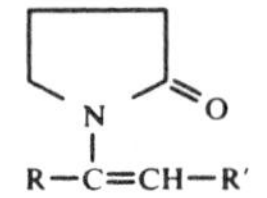

N (ring with C=O)
|
R—C=CH—R'

wherein R and R' are independently selected from the group consisting of hydrogen and alkyl radicals having 1 to 5 inclusive carbon atoms, and (b) a material selected from the group consisting of acrylamide, acrylic acid and vinyl sulfonic acid. Generally 75 to 98 wt % of the copolymer will comprise recurring units derived from (a) above. The useful water-soluble copolymers may be prepared by a variety of polymerization techniques well known in the art such as solution copolymerization, slurry copolymerization, etc., utilizing a wide variety of catalysts such as sodium lauryl sulfate, sodium metabisulfite, ammonium persulfate, azobisisobutyronitrile, ferrous sulfate heptahydrate, hydrogen peroxide, etc. The vinylpyrrolidone polymers can be prepared in the same manner as described above for the copolymers.

The number average molecular weights of the vinylpyrrolidone polymer and copolymers useful in this process will vary from 10,000 to 1,000,000 or more and, preferably will range from 100,000 to 400,000.

A wide variety of surfactants including the water-soluble petroleum sulfonates may be utilized in the process. Other useful surfactants include: (a) polyethoxylated alkyl benzene sulfonates having the formula:

$$R\text{-}C_6H_4\text{-}O(CH_2CH_2O)_nCH_2CH_2SO_3A$$

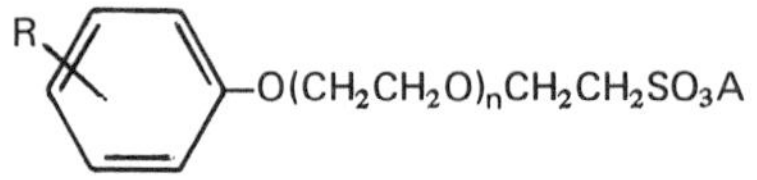

wherein R is alkyl of 8 to 22 carbon atoms, n is an integer of 2 to 10 and A is selected from the group consisting of hydrogen, sodium, potassium and the ammonium ion; (b) polyethoxylated alkyl benzene sulfates having the formula:

$$R\text{-}C_6H_4\text{-}O(CH_2CH_2O)_nSO_3A$$

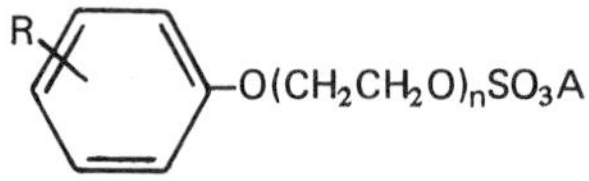

wherein R, A and n have the same meaning as previously described; (c) polyethoxylated alcohol sulfates having the formula: $RO(CH_2CH_2O)_mSO_3A$, wherein R and A have the same meaning as previously described and m is an integer of from 2 to 18; and (d) polyethoxylated alcohol sulfonates having the formula: $RO(CH_2CH_2O)_{m-1}CH_2CH_2SO_3A$, wherein R, m and A have the same meaning as previously described.

Preferably, the aqueous solution of the vinylpyrrolidone polymer injected into the hydrocarbon-bearing formation will contain dissolved therein 0.01 to 10 wt % or more of the polymer. Preferably, the aqueous polymer solution will contain 0.10 to 4.0 wt % of the vinylpyrrolidone polymer. Likewise, the weight percent of surfactant dissolved in the aqueous surfactant solution employed in the process will be from 0.05 to 5.0 wt % or more by weight.

The process can be carried out with a wide variety of injection and production systems which will comprise one or more wells penetrating the producing strata or formation.

Acrylic-Acrylate Polymers

D.F. Klemmensen and R.G. Bauer; U.S. Patent 4,148,746; April 10, 1979; assigned to The Goodyear Tire & Rubber Company describe a method wherein a fluid medium is introduced into a borehole in the earth and into contact with a nonfractured porous subterranean formation penetrated by the borehole, the fluid medium being an aqueous medium comprising water to which has been added a water-thickening amount of a water-thickening polymer which is prepared by polymerizing at least one alkyl acrylate monomer and at least one carboxylic polymerizable monomer selected from the group consisting of acrylic acid and methacrylic acid which are present from 20 to 95 pbw/100 parts of total monomer being polymerized and utilizing as an ionic emulsifier, an emulsifier selected from the group consisting of disodium or diammonium nonylphenoxy polyethoxy sulfosuccinate having the general formula:

$$CH_3(CH_2)_8C_6H_4O(CH_2CH_2O)_{8-16}COCH_2CH(COOM)SO_3Q$$

wherein M is hydrogen or one equivalent of a cation and Q, sodium or ammonium and a sodium or ammonium lauryl polyethoxysulfate having the general formula:

$$CH_3(CH_2)_{11}O(CH_2CH_2O)_{8-16}SO_3Q$$

wherein Q is sodium or ammonium, the polymer in latex form being neutralized to a pH ranging between a pH of 5.5 to 11.5 and containing from 1 to 200 wt % of a water-soluble compound of a polyvalent metal in which the valence of the metal is capable of being reduced to a lower valent state and from 1 to 200 amount of a water-soluble reducing agent, based on the weight of the polymer, which is effective to reduce at least a portion of the metal to its lower valent state.

From 0.05 to 2 wt % of polymer based on the weight of water may be employed. A more preferable range is from 0.1 to 0.5 wt %. Metal compounds which can be used include potassium permanganate, sodium permanganate, ammonium chromate, ammonium dichromate, the alkali metal chromates, the alkali metal dichromates, and chromium trioxide. Sodium or potassium dichromate is preferred, however, because of low cost and ready availability.

As a general guide, the amount of the starting polyvalent metal-containing compound used in preparing the aqueous gels will be in the range of 1.0 to 200 wt %, with 3.0 to 100 wt % based on the weight of the polymer used in the formation of the gel being preferred.

Suitable reducing agents which can be used include sulfur-containing compounds, such as sodium sulfite, sodium hydrosulfite, sodium metabisulfite, potassium sulfite, sodium bisulfite, potassium metabisulfite, sodium sulfide, sodium thiosulfate, ferrous sulfate, thioacetamide, hydrogen sulfide, and others; and non-sulfur-containing compounds such as hydroquinone, ferrous chloride, p-hydrazinobenzoic acid, hydrazine phosphite, hydrazine dichloride, and others. Some of the above reducing agents act more quickly than others, for example, sodium thiosulfate usually reacts slowly in the absence of heat, e.g., heating to 125° to 130°F. The most preferred reducing agents are sodium hydrosulfite or potassium hydrosulfite.

As a general guide, the amount of reducing agent used will generally be within the range of 1.0 to 200 wt %, preferably from 3 to 100 wt % of the stoichiometric amount required to reduce the metal in the starting polyvalent to the lower polyvalent valence state, e.g., +6 Cr to +3 Cr.

Various methods can be used for preparing the aqueous gels. Either the metal-containing compound or the reducing agent can be first added to a solution or dispersion of the polymer in water or brine, or the metal-containing compound and the reducing agent can be added simultaneously to the solution or aqueous medium containing the polymer. Generally, when convenient, the preferred method is to first disperse the polymer in the water or other aqueous medium, such as brine. The reducing agent is then added to the dispersion of polymer, with stirring. The metal-containing compound is then added to the solution or aqueous medium containing the polymer and the reducing agent, with stirring.

Gelation starts as soon as reduction of some of the higher valence metal in the starting polyvalent metal-containing compound to a lower valence state occurs. The newly formed lower valence metal ions, for example, +3 Cr obtained from +6 Cr, effect rapid crosslinking of the polymer and gelation of the solution or aqueous medium containing same.

A preferred procedure is to prepare a relatively concentrated or high viscosity gel and dilute same to a viscosity or concentration suited for the actual use of the gel. In many instances, this procedure results in more stable gel.

When employing the dilution technique, a starting solution or dispersion of polymer containing, for example, 1,000 to 10,000 ppm (0.1 to 1 wt %) or more of polymer can be used. This solution or dispersion is then gelled by the addition of suitable amounts of polyvalent metal compound and reducing agent. After gelation has proceeded to the desired extent, the resulting gel can be diluted with water to the concentration or viscosity most suited for its intended use. For example, if the gel is to be used in a waterflood operation, it could be diluted to a nominal 4,000, 2,500, 1,000, 500, 250 ppm or less gel by the addition of a suitable amount of water.

Addition of Salt in Polymerization Process

C.J. Phalangas, A.J. Restaino and H. Yun; U.S. Patent 4,137,969; February 6, 1979; assigned to Hercules Incorporated have found that improved flooding of oil-bearing formations is obtained with very high molecular weight, substantially linear, water-soluble polymers prepared in very short reaction time and at conversion of monomer to polymer levels of up to substantially 100% by a process which comprises irradiating an aqueous monomer solution containing at least one water-soluble salt of the class hereinafter defined, under carefully controlled conditions of monomer concentration, radiation intensity, total radiation dose, and monomer to polymer conversion as set forth hereinafter. The presence of the salt in the aqueous solution increases the rate of polymerization, and thereby shortens reaction times. More importantly, unexpectedly, the salt increases significantly the molecular weight of the polymer formed at any given level of conversion of monomer to polymer. The salt also gives a more linear polymer, as shown by its low Huggins value, at any given molecular weight. The presence of the salt also permits the formation of ultrahigh molecular weight polymers wherein substantially all of the monomer is converted to polymer. Such polymers obtain improved oil recovery results.

The water-soluble salts which may be used in the process are those water-soluble salts which are sufficiently soluble in the aqueous reaction medium used herein to furnish a solution containing at least 3 wt % of dissolved salt, based on the total weight of solution, and which are incapable of forming free radical scavengers under the influence of high energy ionizing radiation. In this latter category nitrate and nitrite salts are specifically excluded.

A preferred class of salts includes sodium chloride, sodium sulfate, sodium bisulfate, potassium chloride, potassium sulfate, potassium bisulfate, ammonium chloride, ammonium sulfate, and ammonium bisulfate. The sodium, potassium, lithium and ammonium salts may be used to polymerize any of the ethylenically unsaturated monomers described hereinafter, whereas the aluminum salts may be used only to polymerize monomer compositions consisting of 5 to 100% of cationic monomer and 0 to 95% of nonionic monomer.

The amount of salt used depends on the particular salt and reaction conditions used and on the molecular weight and reaction rate desired. In general, the amount of salt employed is from 3%, based on the total weight of solution, to the saturation point of the salt in the monomer solution at 35°C, and preferably from 4% to saturation. A particularly preferred amount of salt is from 6% to saturation. The polymers may be prepared from water-soluble monomers conforming to the formula:

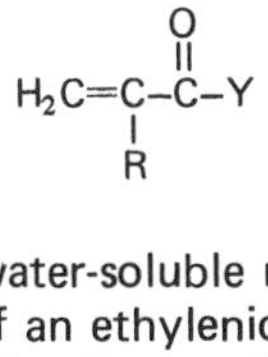

$$H_2C{=}\underset{\displaystyle R}{\underset{|}{C}}-\overset{\displaystyle O}{\overset{\|}{C}}-Y$$

mixtures of such monomers, or water-soluble mixtures of at least one of such monomers with up to 50 wt % of an ethylenically unsaturated monomer selected from the group consisting of vinyl sulfonic acid, alkali metal salts of vinyl sulfonic acid, diacetone acrylamide, and mixtures thereof. In the above formula, R represents hydrogen or methyl and Y represents $-NH_2$, $-OM$,

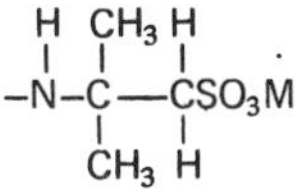

$$-\overset{\displaystyle H}{\overset{|}{N}}-\underset{\displaystyle CH_3}{\underset{|}{\overset{\displaystyle CH_3}{\overset{|}{C}}}}-\underset{\displaystyle H}{\underset{|}{\overset{\displaystyle H}{\overset{|}{C}}}}SO_3M \quad \text{or} \quad -OC_2H_4-\overset{R_1}{\underset{R_3}{N^+}}-R_2 \cdot X^-$$

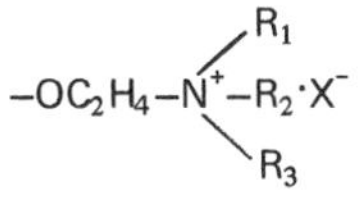

wherein M is hydrogen, H_4N^+, alkali metal, or any other cation yielding a water-soluble, polymerizable compound, R_1, R_2 and R_3 are 1 to 4 carbon alkyl radicals and X is an anion. Illustrative examples of monomers conforming to the formula include acrylamide, methacrylamide, 2-acrylamido-2-methylpropanesulfonic acid, acrylic and methacrylic acids and their water-soluble salts, such as their ammonium and alkali metal salts, quaternary salts of dimethylaminoethyl acrylate and dimethylaminoethyl methacrylate. Preferred nitrogen-bearing monomers within the above formula are acrylamide, dimethylaminoethyl acrylate quaternized with methyl chloride, dimethyl sulfate, or diethyl sulfate and dimethylaminoethyl methacrylate quaternized with methyl chloride, dimethyl sulfate, or diethyl sulfate.

Preferred combinations of monomers include mixtures of acrylamide and sodium acrylate and mixtures of acrylamide and quaternary salts of dimethylaminoethyl acrylate.

Irradiation of the monomer is carried out in an aqueous solution containing 10 to 40 wt %, and preferably 10 to 30 wt % of dissolved monomer.

It is usually desirable to employ radiation intensities of at least 1,000 rads and preferably at least 5,000 rads per hour. In order to obtain significantly higher molecular weight polymers, values below 200,000 rads per hour are desirable; and for producing polymers having molecular weights in the highest range, it is preferred that values below 100,000 rads per hour be employed.

The total radiation dose may influence the water solubility of the polymer, as it has been found that too high a radiation dose may render the resulting polymer water-insoluble. The upper limit of radiation dose is that which produces substantial amount of water-insoluble products. However, for most practical purposes, dosages up to 30,000 rads and preferably up to 15,000 rads are employed.

Although the polymerization reaction can be stopped at any monomer to polymer conversion, the advantages of this process are better realized by conversions from 80 to 100% and preferably from 85 to 98%. At these high monomer to polymer conversions the polymerization raw product is less vulnerable to crosslinking by postirradiation polymerization of unreacted monomer and, therefore, can be safely stored at room or lower temperature or used as such if it is economically preferred.

Example: 240 ml of deionized water are added to a 500 ml beaker. 0.6 g of glacial acrylic acid, 30 g of acrylamide, and 30 g of sodium sulfate are dissolved in the water. The pH of the resulting solution is adjusted to 10.0 with aqueous sodium hydroxide solution. The solution is then added to an irradiation vessel and flushed with nitrogen for 20 min. The solution is irradiated with gamma rays from a cobalt 60 source for 16 min at a radiation intensity of 20,000 rads per hour.

The reaction product is removed from the radiation source and allowed to stand until it has cooled to room temperature. The monomer to polymer conversion is 94.7%. A portion of the gel is extruded and a weighed amount of the gel strands are added to a beaker containing methanol. The polymer strands are allowed to stand in the methanol overnight. The strands are then ground on a Wiley mill using a 20 mesh screen. The ground product is slurried in methanol, filtered by vacuum filtration, washed three times on a filter with fresh methanol, and partially dried on the filter. The semidried powder is then dried in a vacuum oven for 24 hr at 36°C. The dried polymer powder is soluble in distilled water and has an intrinsic viscosity of 14.8 dl/g and a Huggins constant of 0.27.

Polyisocyanate Crosslinked Alkylene Oxide-Linear Nonionic Polysaccharide Reaction Product

S. Stournas; U.S. Patent 4,174,309; November 13, 1979; assigned to Mobil Oil Corporation describes an oil recovery process wherein an aqueous solution of a viscosifier comprising the water-soluble reaction product of an organic polyisocyanate and the addition product of an alkylene oxide and a linear, nonionic polysaccharide is injected to an oil-containing subterranean formation to decrease the mobility ratio between the injected water and oil and to improve the efficiency of the waterflood.

The process is practiced in a subterranean reservoir which is penetrated by spaced injection and production systems defining a recovery zone of the reservoir. In carrying out the process, an aqueous driving agent containing the viscosifier is injected into the reservoir through the injection system. The viscosifier is employed in the flooding water in an amount sufficient to increase its viscosity. In some cases, a total concentration of as small as 0.005 wt % will be satisfactory. Generally, a total concentration from 0.01 to 1%, and preferably from 0.05 to 0.5% is employed.

In general, any organic polyisocyanate may be used in the production of the viscosifier compositions. The polyisocyanates employed may be aliphatic or aromatic isocyanates having two, three or more reactive isocyanato groups. Examples of these isocyanates are hexamethylene diisocyanate, ethylene diisocyanate, cyclohexylene-1,2-diisocyanate, m-phenylene diisocyanate, 2,4-toluene diisocyanate, 2,6-toluene diisocyanate, 3,3'-dimethyl-4,4'-biphenyl diisocyanate, 1,5-naphthalene diisocyanate, p-phenylene diisocyanate, p,p'-methylene diphenyl diisocyanate, 3,3'-dichloro-4,4'-biphenylene diisocyanate and naphthalene triisocyanates. Other polyisocyanates useful in the process will readily occur to those skilled in the art.

If desired, a mixture of two or more different polyisocyanates may be used in the production of the viscosifiers. A mixture of 2,4- and 2,6-toluene diisocyanate may advantageously be used because of availability and relatively low cost.

In accordance with the process, the organic polyisocyanate is reacted with the addition product of an alkylene oxide and a linear, nonionic polysaccharide, the addition product having a molecular weight of at least 100,000 and preferably of at least 300,000. The alkylene oxides useful in the production of such addition products will generally include compounds of the following Formula 1:

$$\text{(1)} \qquad R_1-\overset{\overset{O}{\diagup\;\diagdown}}{\underset{R_2}{\underset{|}{C}}-\underset{R_3}{\underset{|}{C}}}-R_4$$

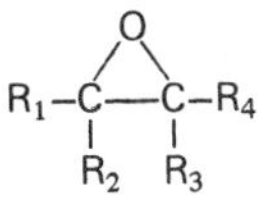

wherein R_1, R_2, R_3 and R_4 are the same or different and are selected from the class consisting of hydrogen and lower alkyl groups containing 1 to 5 carbon atoms, such as methyl, ethyl, propyl, butyl and pentyl groups. Preferably, at least one of R_1, R_2, R_3 and R_4 is hydrogen and the total number of carbons comprising R_1, R_2, R_3 and R_4 attached to the oxirane ring does not exceed five.

Examples of suitable alkylene oxides are ethylene oxide, propylene oxide, butylene oxide, pentylene oxide, 1-methyl-2-pentene oxide, trimethylene oxide and tetramethylene oxide. Other suitable alkylene oxides will readily occur to those skilled in the art. If desired, a mixture of two or more different alkylene oxides may be used to provide the viscosifiers. The amount of alkylene oxide to be used in the preparation of the viscosifiers may vary within wide limits and will to a large extent depend upon the polysaccharide used. Generally, enough alkylene oxide is reacted with the polysaccharide to make the resulting addition product as well as the polyisocyanate crosslinked addition product water-soluble.

As used herein the expression water-soluble means that the viscosifiers are sufficiently soluble in water to provide aqueous solutions including at least 1 wt % of the dissolved viscosifier. Preferably, the alkylene oxides will comprise ethylene and propylene oxides, and mixtures thereof, and the alkylene oxide content of the addition product will vary from 30 to 70 wt % of the total addition product.

The polysaccharides useful in the process are the linear, nonionic polysaccharides. Suitable polysaccharides include cellulose and starches, such for example, as amylose. Generally, the polysaccharide will be formed of repeating structural units of the following Formula 2.

```
                         CH2OH
                          |
                          CH-----O
                         /        \
(2)                  -CH           CH-O-
                         \        /
                          CH-----CH
                          |      |
                          OH     OH
```

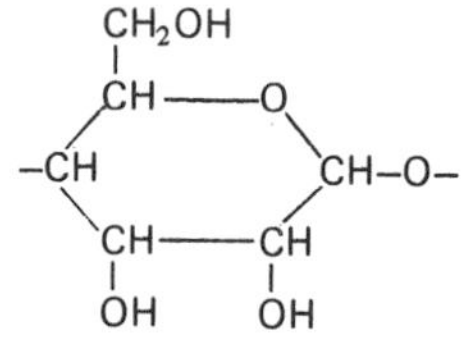

The useful polysaccharides include those which are not per se water-soluble, but which are rendered water-soluble by reaction with the alkylene oxides as described above to provide nonionic polysaccharide ethers which may then be crosslinked with the polyisocyanate to provide the viscosifier components. In this connection, the amount of polyisocyanate used to crosslink the polysaccharide ether may vary within wide limits provided that the resulting crosslinked product is water-soluble.

Generally, an amount of polyisocyanate is employed such that there is a minimum of an average of one crosslink derived from the polyisocyanate for every 2,000 repeating structural units of the type described in Formula 2, up to a maximum of an average of one crosslink for every 20 of such repeating structural units. Stated otherwise, the molar amount of polyisocyanate that should be used to crosslink the alkylene oxide-polysaccharide addition product will range from a value of 0.0005 to 0.05 n/m where n is the number of repeating units of Formula 2 in the polysaccharide, and m is the average number of isocyanato groups per polyisocyanate molecule.

Examples of suitable polysaccharide ethers which are the product of the reaction of a linear nonionic polysaccharide and an alkylene oxide are hydroxyalkyl celluloses, such as hydroxyethyl cellulose and hydroxypropyl cellulose, alkylhydroxyalkyl celluloses such as methylhydroxypropyl cellulose, and mixed hydroxyalkyl cellulose ethers such as hydroxyethyl-hydroxypropyl cellulose ethers. Other water-soluble nonionic polysaccharide ethers useful in the process will readily occur to those skilled in the art.

Polysaccharide ethers useful in the preparation of the viscosifiers may be prepared by reacting the linear, nonionic polysaccharide with the alkylene oxide under basic conditions in the presence of a suitable solvent by procedures well known to those skilled in the art, such as, for example, the procedures used in the production of hydroxyethyl cellulose by reacting ethylene oxide with cellulose. Generally, such polysaccharide ethers will contain an average of 2 to 5 alkylene oxide units per repeating unit of Formula 2 in the polysaccharide chain.

The subsequent crosslinking of the polysaccharide ether with the polyisocyanate may also be achieved by known procedures such as, for example, by dissolving the polysaccharide ether adduct in a suitable solvent, such as dimethylformamide, to provide a dilute solution (e.g., 1 wt %) to which may be added the polyisocyanate in the desired amount to obtain the required degree of crosslinking.

In general, the polyisocyanate will react with the hydroxyl groups of the polysaccharide ether at room temperature to provide intermolecular carbamate linkages. The resulting crosslinked product may then be precipitated from the solution by the addition of a nonsolvent, such as, for example, isopropyl alcohol. The resulting crosslinked polymer products may then be filtered and dried under vacuum at room temperature to provide the compositions in powdered form.

Copolymers of Amphoteric Polyelectrolytes

D.E. Byham; E.W. Sheppard and C.S.H. Chen; U.S. Patent 4,222,881; September 16, 1980; assigned to Mobil Oil Corporation describe a waterflooding process employing an amphoteric polyelectrolyte which is an effective thickening agent at high temperatures and in saline aqueous media which include the presence of significant quantities of divalent metal ions. The process is carried out in a subterranean oil-containing reservoir penetrated by spaced injection and production systems. In accordance with the process, at least a portion of the injected fluid is a thickened aqueous liquid containing a water-soluble copolymer including at least 20 wt % quaternary pyridinium sulfonate monomers of the formula:

$$-CH-CH_2-$$

$$N^+-R-SO_3^-$$

wherein R is a C_{1-4} alkylene group. The quaternary pyridinium sulfonate is copolymerized with a water-insoluble alpha-olefin or hydrogenated diene. A preferred application of the process is in cases where the formation waters or the injection waters or both contain divalent metal ion concentrations of at least 0.1 wt %.

In a preferred embodiment of the process, the amphoteric polymer employed as a thickening agent is a vinylpyridinium sulfonate-styrene block copolymer characterized by the following formula:

$$\left[-C-C- \; N^+-R-SO_3^- \right]_{1-m} \left[-C-C- \right]_m$$

wherein R is a C_{1-4} alkylene group and m is a mol fraction within the range of 0.2 to 0.9.

It normally will be desirable to employ the amphoteric polyelectrolyte in a concentration such that the viscosity of at least a portion of the mobility control slug is equal to or greater than that of the reservoir oil. Typically, the mobility control slug will be injected in an amount within the range of 0.2 to 0.6 pore volume.

Copolymer of Acrylamide and Vinyl Sulfonic Acid Alkoxylated with 2,3-Epoxy-1-Propanol

W.D. Hunter; U.S. Patents 4,228,016; October 14, 1980 and 4,228,017; October 14, 1980; both assigned to Texaco Development Corp. describes a process for recovering hydrocarbons from a subterranean hydrocarbon-bearing formation penetrated by an injection well and a production well which comprises:

(a) Injecting into the formation via an injection well a drive fluid comprising water having dissolved therein a small amount of a copolymer of acrylamide and vinyl sulfonic acid or the sodium, potassium or ammonium salt thereof alkoxylated with 2,3-epoxy-1-propanol,

(b) Forcing the fluid through the formation, and

(c) Recovering hydrocarbons through the production well.

Prior to practicing this process it is sometimes desirable to open up a communication path through the formation by a hydraulic fracturing operation. The copolymers of acrylamide and vinyl sulfonic acid utilized in preparing the alkoxylated copolymers employed in this process comprise recurring A-type units of the formula:

$$\left[-CH_2-\underset{\underset{NH_2}{|}}{\overset{}{\underset{C=O}{\underset{|}{CH}}}}- \right]$$

and recurring B-type units of the formula:

$$\left[-CH_2-\underset{\underset{SO_3M}{|}}{CH}- \right]$$

wherein M is selected from the group consisting of hydrogen, sodium, potassium and ammonium and wherein the copolymer and the weight percent of the A-type units range from 65 to 95 with the balance being B-type units. Generally the number average molecular weights of the acrylamide-vinyl sulfonic acid copolymers will range from 10,000 to 2,000,000 or more. The copolymers of acrylamide-vinyl sulfonic acid and the salts thereof are known materials which can be prepared by the usual vinyl compound polymerization methods. The preparation of acrylamide-vinyl sulfonic acid polymers is described in detail in Norton et al, U.S. Patent 3,779,917.

The alkoxylated copolymers of acrylamide and vinyl sulfonic acid or the sodium, potassium or ammonium salt thereof employed in the process of this method comprise the copolymer alkoxylated with from 2 to 100 wt % of 2,3-epoxy-1-propanol (i.e., glycidol).

The alkoxylation of the acrylamide-vinyl sulfonic acid copolymers can be conveniently conducted using methods well known in the art. For example, an aqueous solution of the copolymer comprising 10 to 30 wt % or more of the copolymer in water along with 0.5 wt % or more of powdered potassium hydroxide or sodium hydroxide is charged to an autoclave and the autoclave and contents heated to a temperature of 125° to 200°C after which the required weight of 2,3-epoxy-1-propanol is pressured with nitrogen into the autoclave over a period of 1 to 3 hr or more following which the autoclave is allowed to cool to room temperature and then vented. The reaction product remaining after being stripped to remove volatile materials yields the water-soluble, alkoxylated copolymer.

Example: A total of 400 cc of water, 6 g of powdered potassium hydroxide and 55 g of an acrylamide-vinyl sulfonic acid copolymer (number average molecular weight of about 520,000) are added to an autoclave which is then heated to a temperature of 120°C. Glycidol in the amount of 38 g is added to the autoclave under nitrogen pressure over a 1.2 hr period during which time the temperature of the autoclave is maintained at 120°C. Next, the autoclave and contents are allowed to cool to room temperature after which the autoclave is vented. The reaction mixture is then stripped of volatiles using a nitrogen purge. The resulting water-soluble product is the copolymer of acrylamide and vinyl sulfonic acid alkoxylated with about 39 wt % of glycidol.

In this secondary recovery process, generally the aqueous drive fluid will contain from 0.01 to 5.0 wt % or more of the alkoxylated acrylamide-vinyl sulfonic acid copolymer. Optionally, the aqueous drive fluid may be saturated with carbon dioxide and/or natural gas at the injection pressure which generally will be from 300 to 3,000 psig or more.

If desired, the aqueous drive fluid having dissolved therein the abovedescribed polymeric thickening agent may be made alkaline by adding of an alkaline agent. The advantageous results achieved with the aqueous alkaline medium used in the process are believed to be derived from the wettability improving characteristics of the alkaline agent.

Useful alkaline agents include compounds selected from the group consisting of alkali metal hydroxides, alkaline earth metal hydroxides, and the basic salts of the alkali metal or alkaline earth metals which are capable of hydrolyzing in an aqueous medium to give an alkaline solution. The concentration of the alkaline agent employed in the drive fluid is generally from 0.005 to 0.3 wt %. Also, alkaline materials such as sodium hypochlorite are highly effective as alkaline agents. Examples of these especially useful alkaline agents include sodium hydroxide, potassium hydroxide, lithium hydroxide, ammonium hydroxide, sodium hypochlorite, potassium hypochlorite, sodium carbonate and potassium carbonate.

A wide variety of surfactants such as linear alkylaryl sulfonates, alkyl polyethoxylated sulfates, etc. may also be included as a part of the aqueous drive fluid composition. Generally 0.001 to 1.0 wt % or more of the surfactant will be included in the drive fluid.

Vinyl Sulfonic Acid Alkoxylated with 2,3-Epoxy-1-Propanol

W.D. Hunter; U.S. Patent 4,217,230; August 12, 1980; assigned to Texaco Development Corporation describes a related process using a drive fluid such as water thickened with a polymer comprising repeating units of vinyl sulfonic acid alkoxylated with 2,3-epoxy-1-propanol.

The alkoxylation of vinyl sulfonic acid can be conveniently conducted using methods well known in the art, such as described in the previous process.

Vinyl Sulfonic Acid Alkoxylated with Ethylene Oxide

Similarly, *W.D. Hunter; U.S. Patent 4,216,098; August 5, 1980; assigned to Texaco Development Corp.* describes a process using a drive fluid such as water thickened with a polymer comprising repeating units of vinyl sulfonic acid alkoxylated with ethylene oxide or a mixture of ethylene oxide and propylene oxide.

The alkoxylated vinyl sulfonic acid monomers employed in preparing the polymers of alkoxylated vinyl sulfonic acid useful in this process comprise vinyl sulfonic acid alkoxylated with from 2 to 100 wt % of ethylene oxide or with a mixture of ethylene oxide and propylene oxide and wherein the weight percent of ethylene oxide in the mixture is about 60 to 95. The method of alkoxylation is the same as described previously using 2,3-epoxy-1-propanol.

Copolymer of Acrylamide and Vinyl Sulfonic Acid Alkoxylated with Ethylene Oxide

W.D. Hunter; U.S. Patents 4,228,019; October 14, 1980 and 4,228,018; October 14, 1980; both assigned to Texaco Development Corp. describes another related process using a drive fluid such as water thickened with a copolymer of (a) vinyl sulfonic acid alkoxylated with ethylene oxide or a mixture of ethylene oxide and propylene oxide, and (b) acrylamide.

Sulfonated Polyphenols

W.D. Hunter; U.S. Patent 4,226,731; October 7, 1980; assigned to Texaco Development Corp. describes a process for injecting into a hydrocarbon-bearing formation via an injection well a fluid comprising water containing a small amount of a water-soluble, sulfonated, ethoxylated polyphenol, forcing the fluid through the formation and recovering hydrocarbons through a production well. The fluids employed may, if desired, contain an alkaline agent such as sodium hydroxide.

The water-soluble, sulfonated, ethoxylated polyphenols useful in preparing the driving fluids comprise recurring A-type units of the formula:

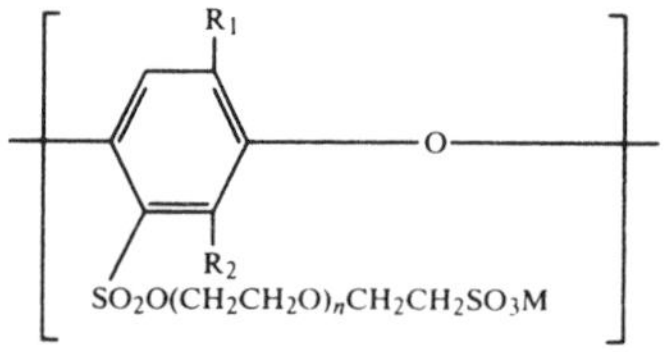

R_1

O

R_2

$SO_2O(CH_2CH_2O)_nCH_2CH_2SO_3M$

wherein R_1 and R_2 are independently selected straight chain alkyl groups of 1 to 3 inclusive carbon atoms, n is an integer of 3 to 30, and M is a cation selected from the group consisting of hydrogen, ammonium, sodium and potassium, and recurring B-type units of the formula:

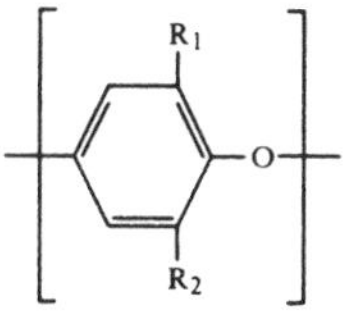

R_1 — O — R_2

wherein R_1 and R_2 have the same meaning as previously described and wherein in the sulfonated, ethoxylated polyphenol the weight percent of the A-type units range from 20 to 60 with the balance being B-type units.

The number average molecular weights of the sulfonated, ethoxylated polyphenols utilized in preparing the driving fluids of this process will range from 5,000 to 250,000 or more and preferably will be from 5,000 to 50,000. Generally, the driving fluid will contain dissolved therein about 0.01 to 2.0 wt % of the abovedescribed water-soluble, sulfonated, ethoxylated polyphenol.

The abovedescribed sulfonated, ethoxylated polyphenols can be conveniently made by a number of processes well known in the art.

In a similar process described by *W.D. Hunter; U.S. Patent 4,226,730; October 7, 1980; assigned to Texaco Development Corp.*, the recurring A-type units are of the formula:

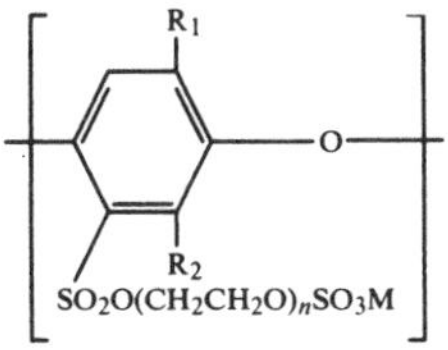

R_1 — O — R_2

$SO_2O(CH_2CH_2O)_nSO_3M$

wherein R_1, R_2, n, and M are as described previously.

Thermal Stabilization of Biopolymers with Alcohol

M.K. Abdo; U.S. Patent 4,141,842; February 27, 1979; assigned to Mobil Oil Corporation describes a waterflooding process in which a polysaccharide produced by action of bacteria of the genus *Xanthomonas* on a carbohydrate is employed as a thickening agent and stabilized against thermal degradation by the use of certain aliphatic alcohols, either alone or in combination with alkali metal carbonates. *Xanthomonas* polysaccharides and their methods of preparation are well known to those skilled in the art.

The alcohol employed to impart thermal stability to the *Xanthomonas* polysaccharide is a water-soluble C_{3-5} aliphatic alcohol. Thus, the stabilizing alcohol may be selected from the group consisting of propyl, butyl, and amyl alcohol. Unsaturated alcohols such as allyl alcohol and crotyl alcohol may also be employed although the saturated alcohols normally will be used. Also, while the propyl alcohols may be employed in carrying out the process, they are relatively volatile and, depending upon the reservoir pressure, may exhibit a tendency to vaporize at the relatively high reservoir temperatures encountered. A similar consideration applies to tertiary butyl alcohol. Thus, the preferred alcohols for use in carrying out the process are the C_{4-5} aliphatic alcohols and particularly isobutyl alcohol and n-butyl alcohol.

While any alkali metal carbonate may be employed, sodium carbonate usually will be preferred from the standpoint of economy and availability. The biopolymer solution should be buffered at a pH above 7 and preferably above 8.5. On the other hand, the pH of the solution normally should not exceed 10. Within these constraints, the alkali metal carbonate may be employed in any suitable concentration. A preferred concentration range is 0.0001 to 0.01 wt % (10 to 100 ppm). While greater amounts of alkali metal carbonate may be employed in the stabilized polysaccharide system, it usually will be desirable to limit the alkali metal carbonate concentration to a value not greater than 0.1 wt % (1,000 ppm).

When the alcohol is employed in conjunction with the alkali metal carbonate, it is preferred to add the alcohol to the system in an amount within the range of 0.1 to 1.0 vol %. Thus, in a preferred embodiment, the alcohol is present in concentration within the range of 0.1 to 0.5 vol % and the alkali metal carbonate in a concentration within the range of 0.001 to 0.005 wt %.

While the thickened aqueous solution of stabilized *Xanthomonas* polysaccharide may be employed as the sole displacing liquid, it usually will be injected as a discrete slug and then followed by a suitable driving fluid. Typically the thickened aqueous liquid will be injected in an amount within the range of 0.1 to 0.5 pore volume and then followed with a driving fluid.

Stabilization of Biopolymers with Antioxidants and Alcohol

A related process is described by *S.L. Wellington; U.S. Patent 4,218,327; August 19, 1980; assigned to Shell Oil Company* whereby in an aqueous solution thickened with a water-soluble anionic polysaccharide polymer (xanthan gum polymer), injected into a subterranean reservoir, the stability of the solution viscosity is improved by deoxygenating the aqueous liquid and then adding a sulfur-containing antioxidant, a readily oxidizable water-soluble alcohol or glycol and the xanthan gum polymer.

Numerous types of materials and techniques for treating aqueous solutions to remove dissolved oxygen are known to those skilled in the art. In general, such treatments are effected by or completed by dissolving a strong reducing agent (or oxygen scavenger) in the solution.

The aqueous liquid used in this process can be substantially any fresh or saline water but is preferably a relatively soft and not extensively saline water. Such a water preferably has a total dissolved salt content of not more than 50,000 ppm and a hardness (in terms of parts per million of calcium ions) of not more than 5,000 ppm. When deoxygenated for use in the process, such a water is preferably substantially completely free of dissolved oxygen and its total dissolved salt content includes from 10 to 100 ppm SO_3-group-containing oxygen scavenger (in terms of SO_3-group equivalent).

Water-soluble inorganic compounds that contain or form ions that contain an SO_3-group are particularly suitable oxygen-scavengers for use in this process.

Such compounds include water-soluble alkali metal sulfites, bisulfites, dithionites, etc. As known to those skilled in the art, such an oxygen scavenger is preferably used in a slight stoichiometric excess (relative to the amount needed to remove substantially all of the dissolved oxygen in the solution being treated). Such an excess is preferably from 10 to 500% more than stoichiometric; and, where a significant excess is used, the oxygen-scavenger is preferably an alkali metal sulfite or bisulfite.

The sulfur-containing antioxidant used in the process can comprise substantially any such water-soluble antioxidant composition (transfer agent, terminator, peroxide decomposer) which is effective with respect to decomposing peroxides in aqueous solutions, and is capable of protecting a xanthan gum polymer solution from drastic loss of viscosity due to being boiled at atmospheric pressure for about 5 min. Examples of such compounds include relatively water-soluble mercaptans, thioethers, thiocarbanols, and the like. Particularly suitable examples are thiourea, thiodiacetic acid (thiodiglycolic acid), 3,3'-thiodiacetic acid (dithiodiglycolic acid) and their water-soluble homologues.

Suitable readily oxidizable alcohols or glycols for use in this process include substantially any water-soluble primary and secondary alcohols or glycols that are easily oxidized, and are capable of protecting a xanthan gum polymer solution from drastic loss of viscosity due to being boiled at atmospheric pressure for about 5 min. Examples of such compounds include methanol, ethanol, allyl alcohol, isopropyl alcohol, isobutyl alcohol, ethylene glycol, and the like.

The anionic polysaccharide polymers, or xanthan gum polymer, suitable for use in this process, can be substantially any such materials produced by the fermentation of carbohydrates by bacteria of the genus *Xanthomonas*. In general, the anionic polysaccharide B-1459 is preferred. Examples of polymers comprise the xanthan biopolymers from Pfizer Chemical Company, the General Mills xanthan biopolymers, the Kelzan or Xanflood anionic polysaccharides from Kelco Company and the like.

The anionic polysaccharides used in this process (and/or the fermentation broth in which they are made) are preferably treated with enzymes such as a proteinase to ensure the removal of (or destruction of) bacterial cells which may impede the flow of a solution into fine pores within subterranean earth formations.

As known to those skilled in the art, in an oil recovery process in which fluids are displaced within a subterranean reservoir by injecting a viscosity enhanced aqueous solution, the effective viscosity (or reciprocal mobility within the reservoir) should be at least substantially equal to and preferably greater than that of the fluid to be displaced. In this process, the concentration of anionic polysaccharide in such a solution should be in the order of 100 to 2,000 pbw of polysaccharide per million parts by weight of aqueous liquid. Such concentrations generally provide viscosities in the order of from 2 to 50 cp at room temperature, in a water containing about 400 ppm total dissolved solids.

In this process, the concentration of antioxidant can be relatively low, in the order of about 50 ppm (w/w of aqueous liquid) and preferably from 200 to 800 ppm. The readily oxidizable alcohol or glycol concentration can be from 50 to 2,000 ppm, and preferably from 500 to 1,000 ppm. In general, the concentrations of the readily oxidizable alcohol or glycol and the polysaccharide polymer are preferably kept at least nearly equal (e.g., at least within 10% of each other). Substantially fresh water solutions containing 800 ppm Kelzan polysaccharide, 200 to 800 ppm thiourea, and 500 to 1,000 ppm isopropyl alcohol have retained 70 to 90% of their original viscosity after 8-mo storage at 97°C (207° F). In such storage tests, the best and most consistent results were obtained when the isopropyl alcohol and Kelzan concentrations were about equal and the thiourea concentration was about half the isopropyl alcohol concentration.

Reduction of Relative Water/Petroleum Movement

Relatively low viscosity brine generally moves more readily through a reservoir containing both oil and brine than does relatively high viscosity oil. In water drive reservoirs, water is almost unavoidably produced along with the petroleum. Normally, as production of a well continues, the degree of water saturation of the reservoir increases and the relative oil permeability is reduced. Thus, a well tends to produce more and more water and less and less petroleum.

In wells with scale problems, it is common to inject into the surrounding reservoir a scale inhibitor contained in an aqueous medium. When such wells are returned to production, it is desired that the inhibitor mix with reservoir fluids and be produced back at a slow rate over an extended period of time. In fact, the inhibitor is often produced back at a fast rate and is soon depleted from the reservoir.

A.M. Sarem; U.S. Patent 4,191,249; March 4, 1980; assigned to Union Oil Company of California describes a method for treating a subterranean reservoir penetrated by one or more wells wherein the flow of an aqueous fluid through the reservoir is reduced without a substantial reduction in the flow of petroleum therethrough by first injecting a slug of petroleum liquid containing both an oil-soluble thickening agent and a suspension of a solid particulate water-soluble thickening agent, and subsequently producing the well. If the method is preceded by injection into the reservoir of an aqueous solution of a scale inhibitor, the inhibitor is produced from the reservoir at a slower rate than normal.

The treating fluid which controls the water/petroleum ratio in the process is a petroleum base liquid, such as crude oil or a refined petroleum material, for example, kerosine, diesel oil and the like. The amount of the treating fluid to be injected depends on the nature of the reservoir which is to be treated, the particular reservoir fluids and/or previously injected treating fluids which will be contacted and the reservoir temperature and pressure conditions. Generally the treating fluid is injected in a slug of 10 to 100 gal/ft of reservoir treated. Additional slugs of a similar size can be employed later in the productive life of the well.

The oil-soluble thickening agent component can be any of the well-known oil thickeners including polymethyl laurate, polyalkyl styrene, polybutadiene, polyisobutylene, the bivalent or trivalent metallic soaps of monocarboxylic acids having 14 or more carbon atoms per molecule and powdered colloidal silicas which are fire-dried fumed silicas having a surface area between 200 and 480 m^2/g. The molecular weight of the abovedescribed polymers which serve as oil thickeners is preferably in the range of 50,000 to 1,000,000. The amount of oil-soluble thickening agent to be used will vary depending on which petroleum-base liquid is used and its initial viscosity as well as the particular thickening agent employed.

Generally, enough thickening agent is used to increase the viscosity of the petroleum base liquid about fourfold at reservoir temperature and pressure. Such a viscosity increase is achieved by using 0.2 to 2 wt % thickening agent.

The water-soluble thickening agent component can be any of the well-known water thickeners including water-soluble polymers such as polyacrylamide, partially hydrolyzed polyacrylamide, polyacrylic acid, polyvinyl alcohol, polyvinyl pyrrolidone, polystyrene sulfonate, polyethylene oxide and a heteropolysaccharide produced by bacteria of the genus *Xanthomonas*; cellulose derivatives such as methylcellulose, ethylcellulose, carboxymethylcellulose and carboxymethylhydroxyethylcellulose; an alkaline alkali metal silicate having a molar ratio of M_2O/SiO_2 of 1 or above wherein M is an alkali metal atom; and natural gums such as guar, xanthan, karaya and the like.

The molecular weight of the abovedescribed water-soluble polymers can vary over a wide range, e.g., 10,000 to 25,000,000. The preferred polymers have a molecular weight in excess of 1,000,000. The preferred partially hydrolyzed polyacrylamides have between 12 to 67% of carboxamide groups hydrolyzed to carboxyl groups. The amount of water-soluble thickening agent employed can vary depending on such factors as the nature and quantity of reservoir fluids and/or previously injected fluids which will be contacted in the reservoir by the treating fluid, the down-hole well conditions of temperature and pressure and the particular thickening agent selected. Generally 0.2 to 2 wt % water-soluble thickening agent in finely divided particulate form is suspended in the treating fluid.

The scale inhibitor can be any one of a number of well-known scale inhibitors including: water-soluble metal salts of relatively low molecular weight hydrolyzed polyacrylamides, wherein the metal is a polyvalent cation such as Ca^{++}, Zn^{++}, Pb^{++}, Fe^{+++}, Cr^{+++} and Al^{+++} and the molecular weight range is about 5,000 to 50,000; a sulfonated alkali lignin having a degree of sulfonation of 0.8 to 4.0; a polyamino polycarboxylic compound such as the tetrasodium salt of ethylenediaminetetraacetic acid; polyphosphate glasses such as alkali metal calcium polyphosphate, alkali metal magnesium polyphosphate and alkali metal calcium magnesium polyphosphate; organic phosphates such as 1-hydroxy, 1,1-diphosphonic acid ethane and ethylenediamine tetramethylene phosphonic acid; phosphate mixed esters of hydroxy amines containing less than 8 carbon atoms in hydrocarbon groups attached to the amino nitrogen and hydroxy hydrocarbons containing at least 6 carbon atoms in a hydrocarbon group; and various other inorganic or organic phosphate materials such as a phosphoric acid, e.g., orthophosphoric acid.

The scale inhibitors are usually injected as an aqueous solution or in suspension in an aqueous medium if they have limited water solubility. Usually a concentration of 0.5 to 5 wt % scale inhibitor in an aqueous medium is used. The aqueous medium can be either fresh water or brine. A slug of 2 to 20 gal/ft aqueous solution or dispersion of the scale inhibitor can be used.

Example: During the producing life of a well, the water/oil ratio increases requiring that an increasing volume of fluids be produced to maintain the same volume of oil production. The increasing amount of water produced creates additional problems of disposal. In an attempt to reduce the produced water/oil ratio, the producing interval of the well is treated by injecting therein a slug of

50 gal/ft of producing interval of lease crude containing 0.75 wt % aluminum stearate oil-soluble thickener and a suspension of 0.2 wt % Enjay B-9702 [a heteropolysaccharide produced by bacteria of the genus *Xanthomonas* (Enjay Chemical Company)] water-soluble thickener. This composition is displaced out into the reservoir by injecting into the well a slug of 20 gal/ft of producing interval of lease crude. The well is then returned to production. At the same production rate as used before the treatment, the volume of oil produced increases substantially and the volume of water produced decreases sharply. Thus, the produced water/oil ratio is reduced as desired.

Production of a Heteropolysaccharide from *Klebsiella pneumoniae*

K.S. Kang, G.T. Veeder, III and D.D. Richey; U.S. Patent 4,186,025; January 29, 1980; assigned to Merck & Co., Inc. describe a process for producing a heteropolysaccharide by bacterial fermentation of a selected carbon source under controlled conditions. The heteropolysaccharide of this process is a high molecular weight polysaccharide containing primarily carbohydrate residues and a minor amount of protein and will be referred to as Heteropolysaccharide 10.

This compound may be prepared by fermentation of a suitable nutrient medium with a strain of *Klebsiella pneumoniae* which does not grow at 37°C in the absence of iron. A deposit of this organism was made with the American Type Culture Collection on August 11, 1971 under Accession No. ATCC 21711.

A restricted deposit of a double-blocked mutant of this organism which requires iron, insufficient of which is available in the human body, for growth at 37°C has been deposited with the American Type Culture Collection on July 25, 1977, under Accession No. ATCC 31311.

In practicing the process, a suitable nutrient fermentation medium is inoculated with heteropolysaccharide-producing strain of *Klebsiella pneumoniae* and permitted to incubate at a temperature of 33° to 37°C, preferably about 35°C, or in the case of the mutant at a temperature of 28° to 32°C, preferably about 30°C for a period of 45 to 60 hr. The bacteria are quite fastidious in their nutritional characteristics in that they require a fairly specific carbon source in order to produce massive amounts of the heteropolysaccharide. The carbon source required by the bacteria in order to produce the polysaccharide is an oligosaccharide containing from 3 to 10 monomer units at a concentration of 1 to 5 wt % and preferably 2 to 4 wt %.

The most suitable carbon source is starch that has been hydrolyzed with an α-amylase enzyme followed by heating at an elevated temperature. The type of starch used is not critical. Representative examples of starches that may be employed are corn starch (which is preferred), wheat starch, rice starch, potato starch and tapioca starch. Also acceptable are the commercial corn syrups which are starch hydrolysates.

A further ingredient which is present in the fermentation medium is a source of magnesium ions. The magnesium salt content of the fermentation medium may range from 0.005 to 0.02 wt %. Suitable sources of magnesium ions include water-soluble magnesium salts, such as magnesium sulfate heptahydrate, magnesium acetate, magnesium chloride, magnesium nitrate and magnesium acid phosphate which may be deliberately added or present as an impurity in the carbon source or the water used.

The pH of the fermentation medium is important to suitable growth of the bacteria. It was found that the optimum pH for production of Heteropolysaccharide-10 is in the range of 6.0 to 7.5, and preferably 6.0 to 6.5. Control of the pH can generally be obtained by the use of a buffer compound such as dipotassium acid phosphate at a concentration from 0.4 to 0.6 wt % of the fermentation medium. Any of the various sodium and potassium salts of phosphoric acid may be used as buffer, e.g., KH_2PO_4, K_2HPO_4, K_3PO_4, NaH_2PO_4, Na_2HPO_4 or Na_3PO_4. When the pH is adjusted to about 7, there will be present about equal amounts of mono- and dibasic phosphates.

At least a trace quantity of phosphorus, generally in the form of a soluble potassium salt, is also present in the fermentation medium. Larger quantities of phosphorus, such as about 0.65 wt % (calculated as dipotassium acid phosphate) of the fermentation medium, can, however, also be used.

The liquid medium should contain 5 to 10% of the amount of oxygen that can be dissolved in the medium, when the oxygen is added as air.

A source of nitrogen is also present in the fermentation medium. The nitrogen source may be organic in nature as, for example, soy protein; an enzymatic digest of soybean meal such as Soy Peptone, Type-T; Promosoy 100; a pancreatic hydrolysate of casein, such as N-Z Amine Type A; an enzymatic digest of proteins, such as Ferm Amine Type IV, or distillers' solubles, such as Stimuflav. When utilizing an organic nitrogen source in the fermentation medium, an amount ranging between 0.01 and 0.07 wt % of the fermentation medium is satisfactory.

Also, if desired, an inorganic nitrogen source, such as ammonium nitrate, ammonium chloride, ammonium sulfate, ammonium citrate or ammonium acetate may be present in the fermentation medium. The amount of such a salt which may be employed can range from 0.02 to 0.15 wt % and preferably from 0.045 to 0.1 wt % of the fermentation medium.

On completion of the fermentation, the desired Heteropolysaccharide-10 may be recovered by treatment of the fermentation beer with a miscible solvent which is a poor solvent for the heteropolysaccharide and does not react with it. In this way the heteropolysaccharide is precipitated from solution. The quantity of solvent employed generally ranges from 2 to 3 v/v of fermentation beer.

Among the various solvents which may be employed are acetone and lower alkanols such as methanol, ethanol, isopropanol, n-butanol, sec-butanol, tertiary butanol, isobutanol, and n-amyl alcohol. Isopropanol is preferred. Precipitation of the desired heteropolysaccharide is facilitated when the fermentation beer is first heated to a temperature of 70° to 90°C for a short time, e.g., 5 to 10 min, and then cooled to about 30°C or lower before addition of the solvent. Thus, this is a preferred method of precipitating the heteropolysaccharide from the fermentation beer. The solid is recovered by separating it from the liquid, as by filtering or straining, and then drying at elevated temperature.

Heteropolysaccharide-10 is useful as a fluid loss control agent in drilling muds, completion fluids and similar aqueous media from which fluid losses to subterranean strata have to be controlled. In waterflooding compositions, it is used as a thickening agent to impart sufficient viscosity to the aqueous medium so that the crude oil may be effectively displaced from the reservoir. In drilling muds and in waterflooding operations there will normally be present other materials such as weighting agents and metal salts.

In these applications, the heteropolysaccharide is added in low concentration, i.e., from 0.3 to 3.0 wt %, using mixing and formulating techniques well known to those skilled in the particular art. The viscosity of the composition may be varied as desired by adjusting the amount of Heteropolysaccharide-10 employed.

Modified *Xanthomonas* Heteropolysaccharide

L.A. Naslund and A.I. Laskin; U.S. Patent 4,182,860; January 8, 1980 describe a process for producing modified heteropolysaccharides derived from a bacteria of the genus *Xanthomonas*, comprising the steps:

(a) Preparing an aqueous solution containing from 200 to 30,000 ppm, by weight, of a heteropolysaccharide fermentation product produced by the action of bacteria of the genus *Xanthomonas* and at least 0.5 wt % of at least one salt, preferably from 0.5 to 10 wt % inorganic salts selected from the group consisting of sodium chloride, calcium chloride and mixtures thereof to obtain a saline heteropolysaccharide solution;

(b) Heating the saline heteropolysaccharide solution to a temperature of at least 100°C, preferably in the range from 100° to 180°C;

(c) Maintaining the saline heteropolysaccharide solution at a temperature of at least 100°C, preferably for a period of time sufficient to improve the filterability characteristics but not so long that the viscosity-imparting properties of the saline-heat-treated heteropolysaccharide are substantially impaired, e.g., from 1 to 300 min.

The heteropolysaccharide that has been subjected to the abovedescribed saline heat treatment is more readily separated from debris in the solution, such as bacterial cells and proteinaceous material, by filtration or other separation means. The saline-heat-treated heteropolysaccharide is physically different from a heteropolysaccharide which has not been subjected to such treatment. Tests demonstrate that the treatment results in a reduction in average particle size.

The heteropolysaccharide which has been subjected to the saline heat treatment and separated from the debris in the heteropolysaccharide solution can be employed in aqueous solutions to recover oil from subterranean formations. It can be injected into such formations without serious plugging of the porous medium.

GAS DISPLACEMENT AND THERMAL RECOVERY METHODS

Injection of Nitrogen-Generating Liquid

E.A. Richardson and R.F. Scheuerman; U.S. Patent 4,178,993; December 18, 1979; assigned to Shell Oil Company describe a process for treating a gas well from which production is prevented by the hydrostatic pressure of liquid contained within the well.

An aqueous liquid which contains or forms a nitrogen-gas-forming mixture of reactants is injected into a well conduit to displace enough liquid to reduce the hydrostatic pressure within the well to less than the fluid pressure in the near-well portion of the reservoir. The injected aqueous liquid is substantially inert to the well conduits and reservoir components and forms or contains a mixture of:

(a) At least one water-soluble compound which contains at least one nitrogen atom to which at least one hydrogen atom is attached and is capable of reacting with an oxidizing agent within an aqueous medium to yield nitrogen gas and by-products which are substantially inert to the well conduits and reservoir components,

(b) At least one oxidizing agent which is capable of reacting with the nitrogen-containing compound to form the nitrogen gas and by-products, and

(c) An aqueous liquid capable of dissolving those reactants and reaction by-products.

The composition of the nitrogen-gas-forming mixture is correlated with the pressure, temperature and volume properties of the reservoir and well conduits so that the pressure and volume of the generated nitrogen gas is capable of displacing sufficient liquid from the well to reduce the hydrostatic pressure to less than the fluid pressure in the adjacent portion of the reservoir and cause fluid to flow from the reservoir to the well.

In a preferred embodiment, the nitrogen-gas-forming mixture is injected into a production tubing string at a rate such that the gas is formed within and accumulated at the top of the tubing string. The gas is subsequently released to initiate the production of gas from the well and reservoir. Alternatively, the composition of the gas-forming mixture and its rate of injection can be adjusted so at least some of the gas is formed within the pores of the reservoir formation.

In another preferred embodiment, an aqueous solution or dispersion of a foam-forming surfactant is injected, before, during or after the injection of the nitrogen-gas-forming mixture, so that a release of gas from the well induces foaming and a foam-transporting of liquid out of the well.

Water-soluble amino nitrogen compounds, which contain at least one nitrogen atom to which at least one hydrogen atom is attached and are capable of reacting with an oxidizing agent to yield nitrogen gas within an aqueous medium, which are suitable for use in the process can comprise substantially any water-soluble ammonium salts of organic or inorganic acids, amines, amides and/or nitrogen-linked hydrocarbon-radical substituted homologs of such compounds as long as the substituted compounds react in a manner substantially equivalent to the parent compounds with respect to the production of nitrogen gas and by-products which are liquid or dissolve to form aqueous liquid which are substantially inert relative to the well conduits and reservoir formation.

Examples of such compounds include ammonium chloride, ammonium nitrate, ammonium acetate, ammonium formate, ethylenediamine, formamide, acetamide, urea, benzylurea, butylurea, hydrazine, phenylhydrazine, phenylhydrazine hydrochloride, and the like. Such ammonium salts, e.g., ammonium chloride, ammonium formate or ammonium acetate are particularly suitable.

Oxidizing agents suitable for use in this process can comprise substantially any water-soluble oxidizing agents capable of reacting with a water-soluble nitrogen-containing compound such as an ammonium salt or a urea or hydrazine compound as described above to produce nitrogen gas and the indicated types of by-products. Examples of such oxidizing agents include alkali metal hypochlorites (which can, of course, be formed by injecting chlorine gas into a stream of alkaline liquid being injected into the well), alkali metal or ammonium salts of nitrous acid such as sodium or potassium or ammonium nitrate and the like. The alkali metal or ammonium nitrites are particularly suitable for use with nitrogen-containing compounds such as the ammonium salts.

Aqueous liquids suitable for use in the process can comprise substantially any relatively soft fresh water or brine. Such aqueous liquid solutions preferably have a total dissolved salt content of from 1 to 100 ppm, and a total hardness in terms of calcium ion equivalents of no more than 50 ppm.

E.A. Richardson, R.F. Scheuerman and D.C. Berkshire; U.S. Patent 4,219,083; August 26, 1980; assigned to Shell Oil Company have found that:

(a) The gas-generating reaction in the previous process can operate in the presence of both a reaction-delaying buffer and a buffer-overriding reactant,

(b) The so-modified solution can be injected through a well bore and into a reservoir to there provide a very fast-rising pulse of both fluid pressure and heat within a near-well portion of the reservoir, and

(c) Such a provision of such a pulse can initiate a perforation-cleaning backsurge of fluid through the perforations in a well casing.

Alkaline buffer compounds or systems suitable for initially retarding the rate of gas generation can comprise substantially any water-soluble buffer which is compatible with the gas-forming components and their products and tends to maintain the pH of an aqueous solution at a value of at least 7. Examples of suitable buffering materials include the alkali metal and ammonium salts of acids such as carbonic, formic, acetic, citric, and like acids.

Reactants for reducing the pH of the aqueous solution and overriding the buffer can comprise substantially any water-soluble, relatively easily hydrolyzable materials which are compatible with the gas-forming reactants and their products and are capable of releasing hydrogen ions at a rate slow enough to allow the buffered solution of the gas-generating reactants to be injected into the reservoir formation before the pH is reduced to a value less than 7. Examples of suitable reactants include: lower alcohol esters of the lower fatty acids such as the methyl and ethyl acetates, formates and the like; hydrolyzable acyl halides, such as benzoyl chloride; relatively slowly hydrolyzable acid anhydrides; relatively slowly hydrolyzable phosphoric or sulfonic acid esters; and the like.

Because of its versatility, the process is often more advantageous than a mechanically induced backsurging treatment for cleaning well casing perforations. For example, it can be used where the well depth and/or reservoir pressure is inadequate for the provision of a low pressure chamber of sufficient volume to cause an inflow pressure gradient that induces an adequate rate and extent of backsurge.

With this process, an adequate inflow pressure gradient can be chemically induced by generating a localized pulse of pressure and heat in the portion of the reservoir immediately adjacent to the well. And, when the viscosity of the reservoir oil tends to reduce the flow rate of a mechanically induced backsurge, such a flow rate can be increased by the heating and thermal mobilizing effect of this chemically induced backsurge.

Attic Oil Reservoir Recovery Method

As the petroleum reservoir is developed as in an oil field, the oil and gas are drawn off and the oil-water contact in the reservoir rises. This process continues until the oil-water contact reaches the structurally highest producing well in the oil field. Normally, this will mark the economic limit to the further development of the oil field and its subsequent abandonment. Nevertheless, in many reservoirs there remains a substantial volume of oil in that portion of the reservoir above the uppermost well and below the permeability barrier which forms a seal at the top of the reservoir. This untapped portion is called the attic oil reservoir.

The normal technique to produce this attic oil is to inject gas into the reservoir. Injection of the gas causes a downward displacement of the oil-water contact corresponding to the volume of gas injected. The injected bubble of gas rises to the top of the attic reservoir and displaces a corresponding volume of oil downward toward the producing interval in the uppermost well. This in turn allows resumed production of oil from this well. If this well begins to produce too much water again, signifying that the oil-water contact has again risen to the level of producing interval in the well, the gas injection process can be repeated.

This gas injection method is not without problems, however. As the uppermost well is produced, the water level beneath the well, marked by the oil-water contact, commonly will not remain as a relatively flat surface, but will instead form a cone with the apex at the producing interval in that well. This effect is commonly referred to as water coning. It is caused by the difference in relative mobility between the connate water and the oil. When the uppermost well is put back on production, a low-pressure area is created in the reservoir at the point of the producing interval in that well.

The initial fluid to be produced will be oil, of course, since the oil should now surround the producing interval in the well. However, as this oil is produced, the oil-water contact beneath the well begins to be distorted upward due to the higher relative mobility of the formation water until the formation water again breaks through into the producing well.

A.O. Clauset, Jr.; U.S. Patent 4,205,723; June 3, 1980; assigned to Texaco, Inc. describes a method of recovering oil from an attic oil reservoir comprising:

(a) Injecting a volume of gas into the uppermost producing well,

(b) Waiting a short period of time until the gas concentration at the well bore is less than 10%,

(c) Pumping a water-excluding agent into the well bore,

(d) Waiting for the injected gas volume to migrate to the top of the attic oil reservoir, and

(e) Placing the well on production.

In order for an attic oil recovery program to be successful, several conditions must be satisfied. First, the formation water in the vicinity of the well must be flushed away by the injected gas sufficiently to reduce the relative water permeability to the lowest possible value in that region. This will facilitate the flow of the displaced oil back into this region. Second, the oil-water contact must be lowered sufficiently to prevent the early formation of a water cone. Third, the volume of the injected gas must be adequate to displace the oil layer downward to the level of the producing interval, yet not so large as to displace the oil layer below the producing interval, resulting in gas production rather than oil production when the well is put back on production. Fourth, the injected gas must have sufficient time to migrate upward to displace the oil down to the producing well.

The first condition can be satisfied in part by injecting the gas at a high injection rate. This will serve to flush the formation water out of the zone around the producing well and in turn lower this region's relative permeability to water to a minimum. To this end, the gas should be injected at rates above 0.5 Mscf per day and preferably at a rate between 1 and 5 Mscf per day.

The second and third conditions can be satisfied by determining the amount of gas needed to be injected per foot of depression of the oil-water contact for a particular reservoir. This can be calculated by using the following formula:

$$\Delta Q_I = 0.178 \Phi E_{DOW} hL/Bg \sin \theta$$

where ΔQ_I is gas volume injected per foot of depression of the oil-water contact (Mscf), Φ is porosity, E_{DOW} is displacement efficiency of oil driving water downward (typical value is 0.8), h is true reservoir thickness at well (ft), L is width of reservoir at oil-water contact (ft), B_g is gas conversion factor RB/Mscf where RB is reservoir barrels and Mscf is thousands of standard cubic feet, and θ is dip angle of reservoir measured from oil-water contact. Use of this formula is demonstrated in the following Example.

Example: For a reservoir with the following parameters B_g is 0.7 RB/Mscf, h is 34.2 ft, L is 1,000 ft, Φ is 0.25, θ is 20°, and E_{DOW} is 0.8.

$$\Delta Q_I = \frac{0.178 \times 0.25 \times 0.8 \times 34.2 \times 1{,}000}{0.7 \times 0.342}$$

$$= 5{,}100 \text{ Mscf of gas per foot of oil-water contact depression}$$

If the operator desires to lower the oil-water contact 20 ft, ΔQ_I total is: 20 x 5,100 is 102,000 Mscf.

The fourth condition requires that the operator wait a sufficient length of time for the injected gas bubble to migrate upward before resuming production in order that the opening of the well to production not pull the injected gas bubble back into the well before it has had time to migrate to the top of the reservoir. This time interval will depend on the particular parameters of the individual oil reservoir and is best determined by the operator in the field. Generally, it can be assumed that sufficient time has elapsed when the concentration of gas in the vicinity of the well bore has decreased to a level below 5%. Computer simulations for a medium porosity reservoir having a permeability of greater than 500 md and containing relatively light oil of a viscosity of about 5 cp or less show that waiting times can be expected to be relatively short, a period of ten days or less.

This process comprises the step of adding a water-excluding agent to the well bore during the waiting period. As the injected gas bubble moves towards the top of the reservoir, formation liquids are displaced downwards along with the injected water-excluding agent. By the time the well is put back on production, the water-excluding agent was moved to a position around and beneath the producing well. In this manner a selective permeability barrier to the flow of water is created in precisely the position necessary to inhibit the formation of a water cone.

Several water-excluding agents are useful in the process. Among them are partially hydrolyzed polyacrylamides and copolymers of acrylic acid and acrylamide such as Zone Control P (Dow) and WC-500 (Calgon). Their use is well known in the art.

Two-Step Recovery System Using Combustible Gases

J.C. Allen; U.S. Patent 4,203,853; May 20, 1980; assigned to Texaco Inc. describes a method for recovering hydrocarbons from an underground reservoir penetrated by an injection well and a production well which comprises:

(a) Establishing a burning zone in the reservoir at the face of the injection well,

(b) Introducing a combustion-supporting gas into the reservoir via the injection well to propogate the zone toward the production well,

(c) Terminating the injection of the combustion-supporting gas into the reservoir,

(d) Injecting into the reservoir via the injection well a combustible gaseous mixture comprising a mixture of a combustible gas and a combustion-supporting gas,

(e) Effecting combustion of the mixture in the formation thereby establishing a second burning zone which moves toward the production well and displaces hydrocarbons from the formation, and

(f) Recovering the displaced hydrocarbons via the production well.

In another embodiment of this process the heat stored in the formation at the end of the second combustion step may be recovered by injecting water as a drive fluid into the heated formation. The water injected into the hot formation in this step is all or partially converted into steam which displaces more of the in-place oils through the formation and results in the recovery of additional oil via the production well.

If desired, the water injected via the injection well may contain from 0.001 to 0.50 wt % or more of an interfacial tension reducer in order to increase the oil recovery. Alkaline fluids may also be injected via the injection well in the process. The drive water is made alkaline by the addition of sodium hydroxide or potassium hydroxide to the water in an amount sufficient to give concentration of 0.01 to 0.1 wt % based on the total drive water weight. Interfacial tension reducers which are highly useful in the process include sulfated compounds of the formula:

$$\text{(quinolyl)}-SO_2-(OC_3H_6)_r-(OC_2H_4)_sOSO_3M$$

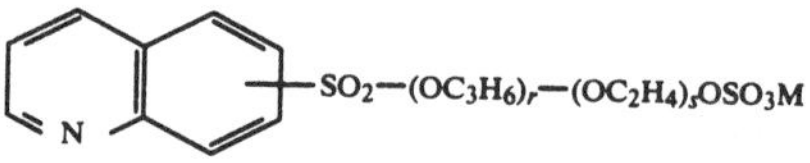

wherein r is an integer of 2 to 5, s is an integer of 8 to 60 and M is selected from the group consisting of hydrogen, sodium, potassium and the ammonium ion and compounds of the formula:

$$\text{(quinolyl)}-SO_2-(OC_2H_4)_tOSO_3M$$

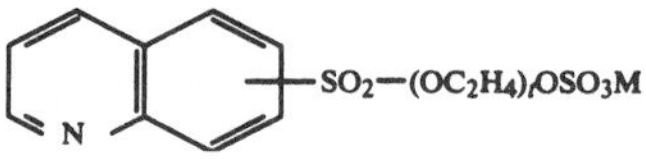

wherein t is an integer of 8 to 40, and M has the same meaning as previously described. Interfacial tension reducers of this type can be formed by sulfating compounds of the formula:

$$\text{(quinolyl)}-SO_2-(OC_3H_6)_r-(OC_2H_4)_sOH$$

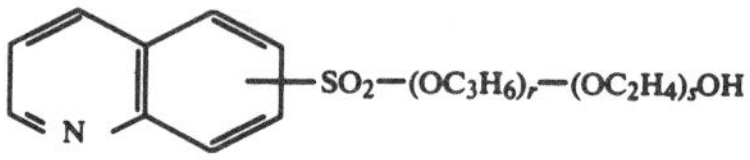

where r and s have the same meaning as before and compounds of the formula:

$$\text{(quinolyl)}-SO_2-(OC_2H_4)_tOH$$

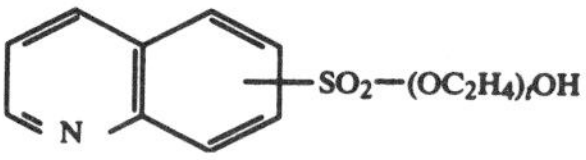

where t has the same meaning as before, batchwise with, for example, chlorosulfonic acid in a glass-lined kettle at about 30°C followed by reaction with the corresponding base, if desired.

This process will be more fully understood by reference to the following description of one embodiment thereof. A hydrocarbon-bearing formation is penetrated by an injection well which is spaced apart from a production well. The wells are of a suitable type for carrying out a procedure of forward in situ combustion for recovering hydrocarbons from the formation. The injection well and the production well each has a casing which extends from the earth's surface down into the lower portions of the formation. The bottom of the casing of each well is sealed by a casing shoe.

The injection well is equipped with tubing which extends through the wellhead downward to adjacent the lower extremity of the casing. A packer is positioned on the tubing in the injection well at a point opposite the producing formation and between two separate sets of casing perforations, i.e., an upper and a lower set of perforations, both opposite the hydrocarbon-bearing formation and in communication therewith. The production well has a single set of perforations through the casing wall opposite the hydrocarbon-bearing formation. The injection well by virtue of the packer previously described and the two separate sets of casing perforations provide two segregated fluid entry avenues. Fluids introduced through the wellhead and into the annulus between the tubing and casing are in communication with the upper area of producing formation via the upper set of casing perforations while fluids introduced through the wellhead via the tubing are in communication with the lower area of the producing area via the lower set of casing perforations. An in situ combustion front is begun by injecting air into the injection well via the annulus between the tubing and casing and then into the formation via the upper casing perforations and via the tubing and into the formation through the lower casing perforations.

The hydrocarbons in the reservoir are ignited by conventional techniques such as by using electrical igniter. Air injection is continued for about 40 days and at the end of that time the air injection is terminated. In the next step air is injected via the injection well into the annulus between the tubing and the casing and into the formation through the upper set of casing perforations and a mixture of carbon monoxide and hydrogen (about 38 vol % carbon monoxide) at a temperature of about 400°F is injected into the formation via the tubing end of the injection well through the lower set of casing perforations.

Ignition is instantaneous and injection of the air and mixture of carbon monoxide is continued for 30 days. In the next step, water is injected into the formation via the annulus in the injection well and through the upper set of casing perforations for a period of about 45 days during which time hydrocarbons displaced through the formation enter the well bore of the casing of the production well through the casing perforations and are recovered via the production well.

Two-Step Recovery System Using Steam and Gases

J.C. Allen; U.S. Patent 4,156,462; May 29, 1979; assigned to Texaco Inc. describes a method of recovering hydrocarbons in which in a first step the formation is heated by injection of steam via an injection well and in a second step a mixture of carbon monoxide and hydrogen is pressured into the formation via the injection well and hydrocarbons are recovered from the formation via a production well.

In the steam injection step of the process saturated steam, wet steam or superheated steam having a temperature of 500° to 1500°F may be utilized. As this first steam heating operation proceeds, the heat from the operation lowers the viscosity of the in-place hydrocarbons which are moved toward the production wells where they are produced.

The steam injection step is generally conducted for a period of 25 to 140 days or more and after the formation has been heated for some distance away from the well bore, i.e., 10 to 100 ft or more, injection of steam is terminated. The temperature of the heated portion of the formation at the conclusion of the steam injection step is preferably 300° to 1200°F.

In the next step of the process a mixture of hydrogen and carbon monoxide, for example, as obtained from a synthesis gas generator is injected into the formation via the injection well. If a hydrogen-carbon monoxide mixture obtained from a synthesis gas generator is employed, the ratio of hydrogen to carbon monoxide by volume will depend on the type of fuel employed as feed for the generator. During the step in which the hydrogen-carbon monoxide mixture is injected, additional hydrogen and carbon dioxide are formed by reaction between the carbon monoxide and steam injected into the formation during the initial heating step. The advantages of carbon dioxide in recovering oil from hydrocarbon-bearing formations are well known and its use as a displacement medium has been demonstrated. Carbon dioxide having a high solubility in oil causes the oil to swell and substantially reduces the oil viscosity.

If desired, saturated steam, wet steam or superheated steam in an amount of from 5 to 50 vol % based on the volume of the carbon monoxide-hydrogen mixture may be introduced into the formation along with the carbon monoxide-hydrogen mixture. The step of heating the formation by injection of steam and the step in which a mixture of hydrogen and carbon monoxide is injected into the formation may be repeated in a cyclic manner, as desired, in operating the process.

In a preferred method of operating the initial steam heating step of the process, steam is injected at a temperature of 800° to 1200°F for a period of 30 to 70 days or more followed by injection via the injection well of a mixture of carbon monoxide and hydrogen for a period of 20 to 60 days or more.

In another embodiment of this process, the heat stored in the formation at the end of the reaction period during which the carbon monoxide injected into the formation in the hydrogen-carbon monoxide mixture reacts with the steam in the formation may be recovered by injecting water as a drive fluid into the heated formation via the injection well. The water injected into the hot formation in this step is in turn partially converted into additional steam which displaces more of the in-place oil through the formation and results in the recovery of additional oil via the production well.

If desired, the water injected via the injection well may contain from 0.001 to 1.0 wt % or more of an interfacial tension reducer in order to increase the oil recovery. Alkaline fluids may also be injected via the injection well. The drive water is made alkaline, if desired, by the addition of sodium hydroxide or potassium hydroxide to the water in an amount sufficient to give a concentration of 0.01 to 0.2 wt % in the drive water.

Interfacial tension reducers which are highly useful in the process include sulfated compounds of the formula:

$$NO_2-C_6H_4-(OC_3H_6)_r-(OC_2H_4)_sOSO_3M$$

wherein r is an integer of 2 to 5, s is an integer of 8 to 60 and M is selected from the group consisting of hydrogen, sodium, potassium and the ammonium ion and compounds of the formula:

$$NO_2-C_6H_4-(OC_2H_4)_tOSO_3M$$

wherein t is an integer of 8 to 40, and M has the same meaning as previously described. Interfacial tension reducers of this type can be formed by sulfating compounds of the formula:

$$(NO_2)C_6H_4-(OC_3H_6)_r-(OC_2H_4)_sOH$$

where r and s have the same meaning as before and compounds of the formula:

$$(NO_2)C_6H_4-(OC_2H_4)_tOH$$

where t has the same meaning as before, batchwise with, for example, chlorosulfonic acid in a glass-lined kettle at about 30°C followed by reaction with the corresponding base, if desired.

The process will be more fully understood by reference to the following description of one embodiment thereof. The description of the production and injection wells are the same as in the previous process.

The initial heating step is begun by injecting steam (90% quality) at a temperature of 800°F into the tubing of the injection well and then into the formation via the casing perforations. Steam injection is continued for about 50 days and at the end of that time the measured temperature of the formation at a point adjacent the casing perforations is about 780°F. In the next step, a mixture of carbon monoxide and hydrogen (about 35 vol % carbon monoxide) at a temperature of about 600°F is injected via the tubing of the injection well and through the casing perforations into the formation over a period of about 75 days during which time the carbon monoxide in the mixture reacts with the steam present in the formation thus forming additional hydrogen and carbon dioxide and during this time hydrocarbons displaced through the formation enter the well bore of the production well through the casing perforations and are recovered via the production well.

Aqueous Solution of Ammonium Bisulfite

J.H. Estes and E.P. Buinicky; U.S. Patent 4,222,439; September 16, 1980; assigned to Texaco Inc. describe a process whereby an aqueous liquid solution of ammonium bisulfite is injected into an oil-bearing earth formation through one or more injection wells, preferably at a temperature less than about 120°F. At temperatures below 120°F, the aqueous ammonium bisulfite solution is stable and not particularly corrosive, thus is easily handled without special equipment or procedure.

Concentration of ammonium bisulfite in the solution is within the range of 0.01 to 1.0 mol/ℓ, and preferably 0.1 mol/ℓ. The aqueous solvent is preferably selected from water or brine, and may contain other additives, such as surface active agents, oil-soluble components, thickening agents, etc., which do not adversely affect the action of the ammonium bisulfite for enhanced recovery of oil from the oil-bearing earth formation.

Within the earth formation, the ammonium bisulfite solution is heated to a temperature above 120°F preferably 200° to 300°F. In the case where the earth formation is hot, the ammonium bisulfite solution may be heated by direct heat exchange with the hot earth formation. In other cases, wherein the earth formation is not sufficiently hot, the ammonium bisulfite may be heated by injection of hot fluids such as steam, hot water, combustion gases, etc. In appropriate cases, the ammonium bisulfite may be heated by in situ combustion of a portion of the hydrocarbons in place within the earth formation. Maximum temperature of the injected ammonium bisulfite solution should be maintained below the boiling point, at formation pressure, to ensure presence of a liquid for enhancing oil recovery.

The hot ammonium bisulfite solution is forced through the earth formation by injection of a drive fluid into the earth formation via the injection well. The drive fluid may be additional ammonium bisulfite solution, or it may be the same fluid employed for heating the ammonium bisulfite, e.g., steam, hot water, air, etc. In cases where the ammonium bisulfite solution is maintained at a selected temperature above 120°F, by direct heat transfer from the earth formation, the drive fluid may be selected from other fluids such as carbon dioxide, light hydrocarbon gases, water, brine, etc. The drive fluid forces the ammonium bisulfite solution into the earth formation wherein the ammonium bisulfite solution acts to displace oil from the pores and interstitial spaces of the earth formation.

In a preferred embodiment of the process, the oil displaced by the ammonium bisulfite solution is moved through the earth formation to a recovery well, from which the displaced oil is produced to the surface. Production of oil in this manner is continued until such production is no longer economically justified.

In a second embodiment, ammonium bisulfite solution is driven into the earth formation, under pressure, from the injection well for a period of time, displacing oil from the pores and interstitial spaces of the earth formation. At the end of the time period, pressure upon the injection well is relieved, and petroleum in admixture with injected fluids is produced from the injection well. This process may be repeated until the production rate of petroleum is no longer economically justified.

Recovery of Heavy Oil

P.S. Buckley and D.M. Grist; U.S. Patent 4,191,252; March 4, 1980; assigned to The British Petroleum Company Limited, England describe a process for the treatment of a viscous oil reservoir to increase the water permeability of the reservoir prior to steam injection which process comprises injecting a solvent for the oil down a well in the reservoir wherein the solvent is steam distillable and contains a surfactant which is either:

(a) Steam distillable so that the solvent and surfactant are caused to penetrate the reservoir by the steam and be present in the zone where the steam condenses, the surfactant also being adsorbable on the reservoir rock so as to be adsorbed before reaching the production well, or

(b) Thermally labile under steam injection conditions so that the surfactant properties can be lost before the end of the subsequent steam injection.

The steam injection and the oil recovery steps can be those of a conventional cyclic steam recovery process or of a steam drive process.

By the term steam-distillable solvent is meant a solvent which has a vapor pressure curve close to that of water. It is preferred that the solvent employed shall be capable of dissolving the asphaltenes present in the crude oil and, therefore, preferred solvents are those having an aromatic content of at least 25%. By aromatic content is meant that determined by fluorescent indicator absorption by ASTM D1319. The suitability of a solvent may be readily determined by shaking a sample of the crude oil with an approximately equal volume of the solvent to see whether or not solution occurs. It is advisable to study the test solution under a microscope to ensure that no small insoluble particles, i.e., less than 1 are present.

Preferred solvents may vary according to the nature of the viscous oil but suitable steam-distillable solvents are toluene, refined oils, e.g., kerosine, naphtha, gasoline and steam-cracked gasoline. Other solvents such as pyridine, cyclohexane and chloroform may also be used.

The addition of the solvent greatly increases the amount of steam that can be injected into the well and the quantity chosen is determined empirically in connection with each well so that the desired amount of steam is injected.

Steam under pressure can be at temperatures up to 350°C and, therefore, it is possible to heat the oil in the reservoir up to this temperature. By choosing a surfactant that decomposes below this temperature and then operating the steam heating step so that those parts of the reservoir in which the surfactant is injected are heated above the decomposition temperature of the surfactant, the latter can be destroyed.

Examples of such thermally labile surfactants are the petroleum sulfonates which are sometimes referred to as alkaryl sulfonates or alkaryl naphthenic sulfonates. They may be employed as sodium, potassium, ammonium or substituted ammonium salts. They can be obtained by sulfonating at least a portion of a sulfonatable hydrocarbon (e.g., gas oils) and then neutralizing the product with the appropriate alkali. Typical commercially available sulfonates may contain 5 to 100% active sulfonate. Most commercially available petroleum sulfonates will break down at the elevated temperatures and pressures induced in a viscous oil reservoir by steaming (e.g., a petroleum sulfonate of molecular weight 460 was found to decompose above 210°C). Their ability to break down may readily be assessed by thermal gravimetric analysis in the laboratory.

Preferred materials have an average molecular weight of 360 to 520 and preferably from 430 to 480. The sulfonate can be a mixture of low and high molecular weight materials. Synthetic alkyl benzene sulfonates and alkyl benzene disulfonates can also be employed provided they have the required thermal instability.

Other anionic surfactants that can be employed are the alkyl ethoxy sulfates and the alkyl phenol ethoxy sulfates. Generally such materials are thermally unstable and can, therefore, be used in the process.

The steam-distillable surfactant which is adsorbable on the reservoir rock can be amine, for example, a primary alkylamine (particularly butylamine, hexylamine and dodecylamine), branch chain alkylamines (e.g., polyisobutylamines) and secondary and tertiary alkylamines. These materials are readily adsorbed on the rock/sand surfaces present in a viscous oil reservoir and hence removed from the fluids therein.

The quantity of surfactant employed can be varied considerably but clearly it is undesirable to employ more surfactant than is necessary to attain the desired improvement in the ability of the solvent to clear oil from the porous structure at the opening of the well into the reservoir and thus increase the quantity of steam that can be injected. The presence of the more expensive surfactant increases the efficiency of the relatively cheaper solvent and thus enables a small quantity of a solvent to be employed. Generally an economical amount of surfactant to employ lies in the range of 0.1 to 5.0 wt % of the solvent.

Example: A bed of an oil-sand mixture was prepared by saturating sand with reservoir water and then pumping a Cold Lake heavy crude oil through the bed to displace the water and saturate the bed with oil. The bed was maintained at 276°C and was then treated with a 2 wt % solution of ammonium petroleum sulfonate in toluene. The volume of solution employed was equivalent to half the pore volume of the bed. The temperature of the solution injected was 22°C. Before treatment the water permeability of the bed was 0.8 darcy and after treatment it was 2.8 darcys.

Production of Hydrocarbons from Igneous Sources

I.C. Bechtold and H.R. Emmerson; U.S. Patent 4,140,184; February 20, 1979 describe a method for employing subterranean heat to effect hydrocarbon production from carbon-containing material and water. As will be seen, the method basically involves the use or provision of a reaction chamber in a subterranean formation and includes the following steps:

(a) Forming an aqueous slurry containing the material in divided, flowable form,
(b) Passing the slurry into the subterranean chamber for decomposition of the material and reaction with water to form the hydrocarbon, and
(c) Recovering the hydrocarbon.

Referring to Figure 4.2, a carbon-containing material is obtained or delivered from a source, as for example, the limestone layer or bed **10** with or without other carbon-containing materials. In general, the bed may consist of calcium and/or magnesium carbonate, with other usable carbon-containing materials including lignite, oil shale, tar sand, gilsonite, and graphite or coal which may or may not lie in intimately bonded or in adherent combination with rock.

The feed may typically be delivered as at **11** to a mill or crusher **12** producing a comminuted feed stream at **13**. Water, metallic compounds, clay and carbon may also be delivered to the mill at **14**, **15** and **15a** whereby the feed stream **13** may consist of slurry of carbon-containing material and clay and metallic compounds, these being in the correct proportion to act as a catalyst, as will be described.

Figure 4.2: Production of Hydrocarbons from Igneous Sources

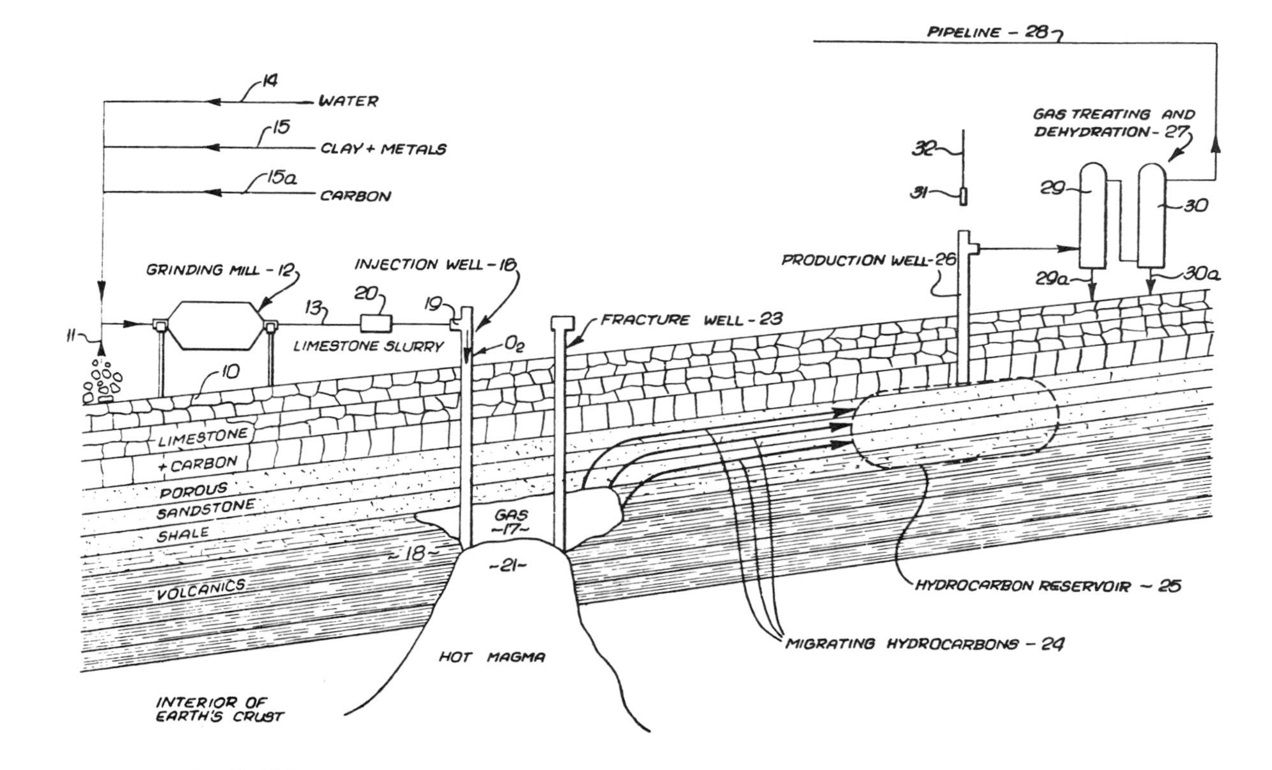

Source: U.S. Patent 4,140,184

An injection well or down passage **16** is formed in the earth to communicate between the surface and a subterranean chamber **17** located in a hot formation **18**. The slurry is introduced into the upper head end of the well at **19**, and passes downwardly to chamber **17**. If necessary, the slurry may be pumped into the wellhead, as via pump **20**. The well may be cased and the wellhead may be enclosed, as shown, to contain pressure which can enhance the desired reactions and provide for additional forces to extend fracturing.

Slurry in chamber **17** is heated to elevated temperatures, to undergo hydrocarbon-forming reactions. Subterranean heat is transferred to the chamber **17** typically as from a hot magma source **21** near which the chamber **17** may be formed. In this regard, the chamber may comprise the lower end of the well **16** drilled into the earth into proximity to the hot magma. The latter may be either molten or solid, so long as the required heating of the slurry is realized.

The produced CO_2 reacts with carbon and H_2O in the chamber typically to form hydrogen and ultimately hydrocarbons. The catalyst clay may, for example, be selected from the group consisting of montmorillonite, kaolinite and illite. The metallic compounds may be supplied by iron or other heavy metal oxides or combinations of same or other compounds having catalytic activity. By utilizing available heat and temperatures for a sufficient time, reactions may occur, and form polymers, hence liquid hydrocarbons. Also, it is possible in this way to form unsaturated hydrocarbons such as those containing ethylene, propylene and acetylene. Carbon-containing materials such as lignite, oil shale, tar sand, and gilsonite, may enhance production.

Recovery of the hydrocarbon gases typically includes the step of providing an up passage extending from the porous rock formation **25** to the earths' surface, for passing the gases to the surface, typically a well **26**. The gas flowing upwardly in passage **26** is collected and passed through a dehydration and treating station **27**, and then fed to a pipeline **28**. Heavier hydrocarbons may be withdrawn from the treating vessels **29** and **30** at **29a** and **30a**. Liquid hydrocarbons in zone **25** may be upwardly removed as by pumping, if necessary. For this purpose, a pump is shown at **31** suspended by tubing **32**, to be lowered into well **26**.

At the injection well **16** other components may be inserted, such as oxygen, nitrogen, ammonia, or other reactants which with the hydrocarbons will react to form compounds other than hydrocarbons, such as alcohols, ketones, esters, amines, etc.

FORMATION PRETREATMENT COMPOSITIONS

Treatment of Subterranean Formations with Chelated Polyvalent Metal Ions

It is well recognized by persons skilled in the art of oil recovery that rock properties are of prime importance in supplemental oil recovery techniques involving injection of water. Maintaining injectivity becomes a problem in cases where producing formations are water sensitive. Damage to permeability is caused by small amounts of montmorillonite and other types of clay present in the formation. When clay particles in the portion of the formation adjacent to pore spaces or flow channels of the formation are exposed to fresh water, the clays swell and disperse, moving downstream and forming blockages in the flow channels of the formation.

G. Kalfoglou; U.S. Patent 4,230,183; October 28, 1980; assigned to Texaco, Inc. has found that permeable earth formations including petroleum-containing formations, which contain water-sensitive clay minerals including sodium montmorillonites, illites, and other clay minerals which swell or expand on being contacted with fresh water, can be treated so as to render the water-swellable clay fractions of formation insensitive to subsequent contact with fresh water, or to reverse the permeability-decreasing swelling phenomenon which has already occurred as a consequence of prior contact with low salinity water, by a method which is relatively inexpensive, and which treatment is very persistent with time and subsequent passage of fluids through the formation, and which is not affected by contact with high pH fluids in subsequently applied processes.

The method comprises injecting into the formation an aqueous solution of a chelated, polyvalent transition metal ion. The specific metal ions used include magnesium, vanadium, chromium, manganese, iron, cobalt, nickel, copper, zinc, ruthenium, rubidium, silver, cadmium, iridium, platinum, silver, mercury and lead. The organic ligand which is the chelating agent are those which form five- or six-membered ring structures with the metal cations, and which preserve the positive charge on the cation.

Amines which can be used as the chelating ligands are ethylenediamine; 1,2-diaminopropane; 1,3-diaminopropane; 2,3-diaminobutane; diethylenetriamine; triethylenetetramine; tetraethylenepentamine; pentaethylenehexamine; tris(aminoethyl)amine; 1,2,3-triaminopropane; 1,2-diamino-2-aminoethylpropane; 1,2-diamino-2-methylpropane; 2,3-diamino-2,3-dimethylbutane; 2,2'-bipyridine; 2,2'-dipyridylamine; 1,10-phenanthrolamine; 2-aminoethylpyridine; terpyridine; biguanide and pyridine-2-aldazine. The aqueous treating fluid contains from 0.01 to 10 and preferably 0.1 to 5.0 wt % of the chelated polyvalent metal ion.

The treatment may be applied prior to contact with fresh water in order to prevent loss of permeability in the flow channels of the earth formation resulting from subsequent contact with fresh water, or it may be applied to a formation which has already experienced permeability loss due to contact between fresh water and the water-sensitive clay minerals. The treatment fluid may be injected into the formation for near well bore treatment of the portion of the formation immediately adjacent to an injection well or a production well or both, and displaced away from the formation by subsequently injected oil displacement fluids in an enhanced recovery process.

When used for near well bore treatment, the quantity of treating fluids should be from 50 to 1,000 and preferably 100 to 700 gallons of fluid per foot of formation thickness. When injected into the formation for in-depth treatment of water-sensitive clays in combination with an enhanced recovery process, the quantity of fluid required is from 0.01 to 1 and preferably from 0.05 to 0.5 pore volume of treating fluid.

Reservoir Stabilization Using Nitrogen-Containing Compounds

D.W. Nooner; U.S. Patent 4,227,575; October 14, 1980; assigned to Texaco, Inc. describes a method of mineralogically transforming a formation matrix by injecting thereinto an aqueous solution of nitrogen-containing compounds at elevated temperatures, continuing the injection until the formation is heated to an elevated temperature for a desired length of time when stabilization of the formation is effected to a desired radial distance from the well bore.

With this method, clay-containing formations are effectively transformed to stabilize their swelling upon subsequent exposure to aqueous media, such as water, hot water, or steam. The method has utility as a pretreatment operation for any operation, such as well stimulation or secondary recovery, wherein a hydrothermal treatment is to be undertaken. For example, a subsequent steam flooding operation, by means well known in the art, may be accomplished without the normally associated clay swelling and/or dispersion. Such pretreatment permits more effective recovery of the hydrocarbons contained in the formation, since blocking and plugging of the formation by the flooding fluid is precluded or retarded, thereby allowing the easier penetration of the formation by the flooding fluids.

The process provides an improved method for the stabilization of subterranean formations. Exposure of the formation to temperatures of from 260° to 310°C and an aqueous solution of a nitrogen-containing compound having a concentration of about 0.5 to 3.0 N produces a transformation of the water-sensitive montmorillonite clays to a more stable illitic-type clay.

Treatment with guanidine hydrochloride, guanidine carbonate, guanidine acetate, urea and formamide at 300°C for seven days caused transformation of the swelling clay (montmorillonite) to nonswelling clay (illite) and lesser swelling clays (mixed-layer).

Reservoir Stabilization Using Potassium Salts

D.W. Nooner; U.S. Patent 4,164,979; August 21, 1979; assigned to Texaco Inc. describes a similar process whereby use is made of an aqueous solution of a potassium salt of an organic acid at elevated temperatures, continuing the injection until the formation is heated to an elevated temperature for a desired length of time when stabilization of the formation is effected to a desired radial distance from the well bore.

Treatment with potassium formate, potassium biphthalate and potassium acetate at 300°C for seven days caused transformation of the swelling clay (montmorillonite) to nonswelling clay (illite) and lesser swelling clays (mixed-layer).

Oxygenated Polyamine Stabilizing Agent

R.W. Anderson and B.G. Kannenberg; U.S. Patent 4,158,521; June 19, 1979; assigned to The Western Company of North America have found that a stabilizing agent in the form of a solution of an oxygenated polyamine, and specifically one that is a copolymer product of epichlorohydrin and dimethylamine, can be employed to substantially permanently stabilize clay formations. Preferred copolymers of this type have a mol ratio of epichlorohydrin to dimethylamine in the range of 0.79 to 0.93. A preferred intrinsic viscosity of the copolymer is about 0.04.

Any suitable diluent which is a polar solvent, such as water, ethyl alcohol, methanol, glycols, or mixtures of these in water, for example, can be used to carry the copolymers. Aqueous salt brines are preferred diluents. The method of treatment can comprise the injection of an aqueous solution of the polymer into a bore hole in a manner which insures sufficient contact with the clay. Aqueous treating solutions of the polymer can be prepared simply, and are of low viscosity, facilitating easy handling. Further, such treating solutions are water miscible, and are readily soluble in brines and acids. Furthermore, the treating solutions are effective over a wide pH range and are, therefore, useful in strong acids, weak acids, low pH fluids, neutral waters, brines or basic solutions. Formations containing clay which are treated according to this method remain substantially unaffected by subsequent well treatments or contact with formation fluids. The amount of clay stabilization agent present in a treating solution thereof is dependent upon the types and amounts of clay present in the formation, the relative size and permeability of the formation to be treated, and the type of solvent employed.

Acylated Polyether Polyol Pretreatment

C.M. Blair, Jr.; U.S. Patent 4,216,828; August 12, 1980; assigned to Magna Corporation describes a process to provide a pretreating flood of a thin film spreading agent composition (TFSA) having present therein an acylated polyether polyol, to improve the recovery of oil by subsequent flooding with water or other aqueous systems such as viscous, aqueous polymer solutions, caustic solutions and detergent solutions.

Effective TFSA may be derived from a wide variety of chemical reactants and may contain numerous different groups or moieties. Particularly effective products are those which are described as:

$$R \begin{cases} (OA_jH)_n \\ (NR'A_kH)_m \end{cases}$$

wherein:

A is an alkylene oxide group, $-C_iH_{2i}O-$;
O is oxygen;
i is a positive integer no greater than about 10;
j is a positive integer no greater than about 100;
k is a positive integer no greater than about 100;
N is nitrogen;
R^1 is one of hydrogen, a monovalent hydrocarbon group containing less than C_{11}, or (A_LH);
L is a positive integer no greater than about 100;
R is a hydrocarbon moiety of a polyol, a primary or secondary amine, a primary or secondary polyamine, a primary or secondary amino alcohol, or hydrogen; and
m + n is no greater than about 4 when R is other than hydrogen and one of m and n is zero and the other is unity when R is hydrogen.

The acylated polyether polyol should be the reaction product of the polyether polyol and a member selected from the class consisting of mono- and polybasic carboxylic acids, acid anhydrides, and iso-, diiso-, and polyisocyanates, the acylated polyether polyol at about 25°C: (a) being less than 1 vol % soluble in water and in isooctane; (b) having a solubility parameter in the range of 6.9 and 8.5; and (c) spreading at the interface between distilled water and refined mineral oil to form a film having a thickness no greater than 20 Å at a film pressure of 16 dynes/cm. These compositions must conform to the physical property parameters set forth above.

Alternatively, the TFSA compositions may be described as acylated polyether polyols wherein the polyether polyols are derivable by reaction of an alkylene oxide containing less than 10 carbon atoms with a member of the group consisting of polyols, amines, polyamines and amino alcohols containing from 2 to 10 active hydrogen groups capable of reaction with alkylene oxides and the acylating agent being a member selected from the class consisting of mono- and polybasic carboxylic acids, acid anhydrides and iso-, diiso-, and polyisocyanates.

Representative of these compositions is polypropylene glycol, having an average molecular weight of 1,200 to which about 20 wt % of ethylene oxide has been added. Such a polyether glycol is theoretically obtainable by condensing about 20 mols of propylene oxide with about 1 mol of water, followed by addition of about 6 mols of ethylene oxide. Alternatively, one may condense about 20 mols of propylene oxide with a previously prepared polyethylene glycol of about 240 average molecular weight.

Silicate-Activated Lignosulfonate Gel

D.D. Lawrence and B.J. Felber; U.S. Patent 4,257,813; March 24, 1981; assigned to Standard Oil Company, Indiana describe a process for treating conductive zones in subterranean formations by injecting an aqueous fluid containing gel-forming amounts of a water-soluble lignosulfonate and a water-soluble silicate into the highly conductive zones. Suitable treating fluids contain about 2 to 10 wt % ammonium or sodium lignosulfonate and sodium silicate. The sodium silicate is included in the treating fluid at weight ratios of SiO_2 to lignosulfonate of 0.2 to 1. A satisfactory sodium silicate has a mol ratio of 3.33 SiO_2 to Na_2O.

Gelation time required for injecting these treating fluids, as is well known to those skilled in the art, depends largely upon the type of treatment to be performed. If the problem zone is at or near the well bore, a treating fluid having a short gelation or set time, such as 10 hr or less, may be used. On the other hand, if it is intended to treat a large volume of the formation extending from the well bore, a treating fluid having a longer gelation time, such as 100 or possibly 1,000 hr, may be used.

SAND CONSOLIDATION AND PLUGGING

PLUGGING COMPOSITIONS

Selective Plugging Using Debilitating Agent

In secondary or tertiary oil recovery operations, recovery of oil is maximized if the driven fluid is permitted to build up in a wide bank in front of the driving fluid which moves uniformly toward a producing well. To keep this bank of oil intact, and moving toward a producing well, a substantially uniform permeability must exist throughout the strata. If this uniform permeability does not exist, and it generally does not, the flooding will seek the areas of high permeability, and channeling occurs with the appearance of excess driving fluid at the producing well. Moreover, as the more permeable strata are depleted, the driving fluid has a tendency to follow these channels and further increase water production as reflected in an increased water/oil ratio to the point that the process becomes economically undesirable.

It is known in the art that more uniform flood fronts can be obtained in formations of nonuniform permeability by control or permeability correction of the more permeable strata of the formation. A number of methods for reducing the permeability of these permeable strata have been proposed, including the injection of plugging materials into these strata which at least partially plug the permeable zones so as to achieve more uniform permeability.

Some of these methods of permeability correction accomplish the plugging step by the in situ formation of plugging material in the formation by the injection of one or more reactant substances which chemically react to form a solid residue. These reactant substances include various hydraulic cements, precipitate-forming materials, and monomers or prepolymers which are polymerizable under formation conditions. Unfortunately, however, these plugging materials can also plug less permeable zones. Particularly with the polymerizable materials which may be relatively fluid when injected, the situation can even be made worse by completely shutting off the already less permeable zones.

H.M. Barton, Jr.; U.S. Patent 4,190,109; February 26, 1980; assigned to Phillips Petroleum Company describes the selective plugging of more permeable zones known as thief zones to improve the sweep efficiency of tertiary or secondary oil recovery processes.

The preferred plug forming materials are polymerizable monomers. Accordingly, the process preferably involves injecting into the formation a polymerization poison or inhibitor for the particular monomer to be polymerized. For instance, materials such as potassium ferricyanide, hydroquinone and the like may be injected into the formation. These are known inhibitors for the polymerization of monomer systems such as acrylamide and acrylic acid; acrylonitrile and acrylic acid; acrylonitrile and sodium styrene sulfonate; acrylic acid and methacrylic acid; and acrylic acid with diallylamine.

Examples of polymerization catalysts for the above systems are organic peroxides such as tert-butyl hydroperoxide, di(tert-butyl) peroxide, and methyl ethyl ketone peroxide as well as inorganic peroxides such as ammonium persulfate. Also, various accelerators may be included with the catalyst system such as N,N-dimethylaniline and β-dimethylaminopropionitrile.

Poisons such as p-benzoquinone which are known poisons for formation of popcorn polymer as disclosed in U.S. Patent 3,771,599, can be injected followed by the injection of the monomer system.

Other polymerization inhibitors useful in this process include chloranil and aromatic nitro compounds such as nitrobenzene, 2,4-dinitrochlorobenzene, o-nitrophenol, m-dinitrobenzene, picric acid, naphthalene picrate and the like.

In addition to the formation of material strictly classified as polymers, the process also encompasses the injection of a poison or inhibitor for any two substances that tend to react to form a solid in the presence of a catalyst or the presence of each other, or in the presence of materials inherent in the reservoir which act as catalysts. For instance, an acid can be introduced and allowed to permeate substantially the entire formation, after which the acid is flushed from the more permeable zones. Then a base such as NaOH is added followed by $MgCl_2$ which forms a $Mg(OH)_2$ precipitate in the more permeable zones but does not do so, or does so to a lesser extent in the less permeable zones, because the acid neutralizes the base.

Alternatively, the initial injection can be that of a material which causes depolymerization or decomposition or otherwise reduces the viscosity wherein the subsequent injection would be that of a polymer or other chemical compound formed above ground.

As a general proposition in heterogeneous systems, more than one pore volume of fluid is required to completely displace the connate fluids because higher permeability zones tend to accept the injected medium first, and thus an excess is required to permeate the less permeable zones. Consequently, a larger pore volume of the debilitating agent, i.e., poison or inhibitor, will be injected as compared with the subsequent injection of the flush. For instance, more than one pore volume of retarder (1.1 to 10, preferably 1.5 to 3, for instance) or poison may be injected followed by less than one pore volume (0.1 to 0.9, preferably 0.25 to 0.75, for instance) of the flush. Thereafter this is followed by the monomer which would generally be employed in amounts similar to that utilized for the flush, although lesser or greater amounts can also be used.

Following the injection of the monomer and whatever induction time is required for plugging to occur, normal secondary or tertiary operations are then employed. For instance, a surfactant system may be injected and thereafter a mobility buffer, followed by a drive fluid as is known in the art, or conventional water flooding can be injected immediately after a selective plugging operation is complete. In either case, with the more permeable zones selectively plugged there is formed a more uniform bank of oil which is driven from the injection well or wells toward the recovery well or wells where it is produced. The process is also applicable to other utilities where it is desired to achieve a more uniform permeability in a subterranean formation.

Example: Two pore volumes of a one weight percent aqueous solution of potassium ferricyanide are introduced through an injection well into an oil bearing formation which has been subjected to secondary water flooding operations until the ratio of water to oil in the production wells has become greater than 100:1 due to channeling of the water flood fluid through the more permeable zones. Then the formation is flushed with 0.5 pore volume of water containing 15,000 ppm total dissolved solids which are primarily sodium chloride so as to remove the potassium ferricyanide from the more permeable zones without removing same from the less permeable zones.

Thereafter there is injected a composition comprising 97 pbw water, 3 pbw acrylamide, 2 pbw sodium bicarbonate, 0.0075 pbw methylenebisacrylamide, 0.0025 pbw cobalt acetate, 0.1 pbw disodium salt of ethylenediaminetetraacetic acid, and 0.125 pbw of a 70% tert-butyl hydroperoxide/30% di-tert-butyl hydroperoxide mixture. After about a 4 hr latent time a normal tertiary oil recovery process is begun.

First, a surfactant system comprising 3.1% of a petroleum sulfonate having an equivalent weight of 407, 3% isopropanol, 1.7% oil and 92% freshwater containing 15,000 ppm NaCl is introduced in the amount of about 0.1 pore volume. This is followed by one pore volume of a 2,000 ppm polyacrylamide mobility buffer in freshwater and one final pore volume of freshwater driving fluid. Oil recovery from the recovery well is thereby increased as evidenced by a lower ratio of water to oil.

Amide Polymer Gel

R.J. Pilny and T.W. Regulski; U.S. Patent 4,199,625; April 22, 1980; assigned to Dow Chemical Company describe a method for rendering porous structures impermeable to the passage of liquids by treating them with an aqueous gelable reaction mixture comprising a water-soluble or water-dispersible aliphatic polyaldehyde, a hypohalite salt and a water-soluble polymer derived from an ethylenically unsaturated amide.

The amide polymers employed have polyethylenic backbones bearing pendant carboxamide moieties. Such amide polymers are normally addition polymers containing polymerized ethylenically unsaturated carboxamide monomers which may contain up to 50 mol percent of another ethylenically unsaturated monomer copolymerizable with the carboxamide monomer or monomers.

For ease of handling and placing in the porous structure, it is preferred to employ amide polymers characterized by viscosities of from about 100 cp to 15,000 cp for an aqueous 20% by weight solution thereof.

To form the desired gel capable of rendering the porous structure impermeable, the amide polymer preferably contains from about 50 to 100 mol percent of one or more ethylenically unsaturated carboxamide monomers, more preferably from 70 to about 100 mol percent, and most preferably from 90 to about 100 mol percent of amide monomers. Exemplary carboxamide monomers include acrylamide, methacrylamide, fumaramide, ethacrylamide, N-methylacrylamide and the like. It is understood that the finished polymer contains sufficient water-solvating carboxamide moieties to render the finished polymer soluble in water to the extent of at least 5% by weight and preferably to the extent of 20% or more by weight. Acrylamide, per se, is the preferred carboxamide monomer.

In general, any aliphatic polyaldehyde, having sufficient solubility or dispersibility in water to enable rapid, intimate mixing with an aqueous solution of amide polymer, may be employed. In practice, saturated aliphatic polyaldehydes are preferred. Suitable polyaldehydes include dialdehydes, such as glyoxal, succinaldehyde, glutaraldehyde and the like, as well as more complex chemicals such as water-soluble or water-dispersible polyaldehyde starch derivatives. For most purposes a dialdehyde, particularly glyoxal, is preferred.

The hypohalite salt is suitably any metal hypohalite, but is preferably an alkali metal hypochlorite or hypobromite, most preferably sodium or potassium hypochlorite.

In practicing the method, it is only necessary that the amide polymer, hypohalite salt and the polyaldehyde be thoroughly mixed in the proper proportions in an aqueous medium under conditions of suitable alkalinity to provide a gelable aqueous reaction mixture. One such procedure is carried out by thoroughly mixing an aqueous solution of the hypohalite salt with an aqueous solution of the amide polymer and the polyaldehyde under proper alkaline conditions. Alternatively, an aqueous solution of polyaldehyde is mixed with an aqueous solution of amide polymer and hypohalite salt under proper alkaline conditions.

In the gelable aqueous reaction mixture, the concentration of the amide polymer varies depending upon the molecular weight of the polymer and the firmness of the gel desired. If the amide polymer has a relatively low molecular weight, the concentration of amide polymer is advantageously from 0.25 to about 30, preferably from 3 to about 20, most preferably from 5 to about 15, weight percent based on the reaction mixture.

For the purposes of this process, an amide polymer has a relatively low molecular weight if a 40 weight percent aqueous solution of the polymer exhibits a viscosity in the range from 50 to about 15,000 cp as determined by a Brookfield LVT viscometer (No. 2 spindle, 15 rpm, 25°C). It is understood, however, that somewhat lower concentrations than the aforementioned can be employed with higher molecular weight amide polymers.

The concentration of polyaldehyde in the gelable reaction mixture is at least an amount sufficient to cause the reaction mixture to form a water-insoluble, three dimensional gel within 5 min after the mixture is subjected to gelation conditions, up to the saturation concentration of the polyaldehyde in the reaction mixture. Advantageously, however, the gelable reaction mixture contains enough polyaldehyde to provide from 1 to about 300, preferably from 5 to about 100, mmol of aldehyde moiety per mol of carboxamide moiety

$$-\overset{\displaystyle O}{\overset{\|}{C}}-NH_2$$

in the amide polymer. When the polyaldehyde is glyoxal, the concentrations of glyoxal sufficient to provide the aforementioned mol ratios are within the range from 0.004 to about 1.2, preferably from about 0.02 to about 0.4, weight percent based on the gelable reaction mixture. For sewer grouting and similar subterranean plugging applications, it is preferred to employ from 15 to about 60 mmol of polyaldehyde per mol of carboxamide moiety in the amide polymer.

The concentration of the hypohalite salt in the gelable reaction mixture is an amount sufficient to provide the resulting gel with alkaline stability. A gel possesses the requisite alkaline stability if it does not dissolve in an aqueous medium having a pH of 8 in a period of at least a week after the gel is placed in the aqueous medium. Gels having preferred alkaline stability do not dissolve in an aqueous medium having a pH of 8 to 10 for a period of six months, those gels which are insoluble in 5 N NaOH for a week being most preferred.

The concentration of hypohalite salt is in the range from 2 to about 600, preferably from 10 to about 250, most preferably from 15 to about 160 mmol of the hypohalite anion per mol of carboxamide moiety in the amide polymer.

The hypohalite salt is advantageously employed in the form of an aqueous solution prepared by dissolving the corresponding free halogen in a slight molar excess of alkali metal hydroxide or other relatively strong base with cooling to prevent the formation of halites or halates. As a result of this preferred practice, a solution is made which contains a mol of halide ion for each mol of hypohalite ion formed. A slight excess of the base is beneficially employed to stabilize the hypohalite solution and to provide an aqueous solution of hypohalite having a pH of at least about 12 and preferably a pH over 13 without containing such an excess of alkali as to cause undesired hydrolysis of the amide groups of the polymer when the hypohalite solution is mixed with the amide polymer solution.

For economical reasons, it is desirable to employ a commercial household bleach which is an aqueous solution containing about 5.25 to 5.5 weight percent of sodium hypochlorite, an approximately equimolar proportion of sodium chloride and sufficient excess of sodium hydroxide to provide a solution having a pH of 13.5 or slightly higher. In commercial bleach the stabilizing excess of NaOH corresponds to about 0.3 to 1% by weight of the solution.

Advantageously, the pH of the reaction mixture at the initiation of gelation should be at least 7.5 up to 14, preferably from 8 to about 13.5, most preferably from 10 to about 13. This desired alkalinity at the outset of gelation is accomplished by the addition of a relatively strong base to one of the aforementioned starting ingredients, preferably the polyaldehyde and/or the hypohalite salt, most preferably the hypohalite salt, or to the reaction mixture prior to gelation.

Generally any base capable of generating the needed alkaline pH which does not interfere with the gelation reaction is usefully employed. Examples of relatively strong bases advantageously employed to provide this desired alkalinity include alkali metal hydroxides such as sodium and potassium hydroxide; metal phosphates such as trisodium phosphate; metal carbonates such as sodium carbonate; alkylamines such as dimethylamine, methylamine and trimethylamine; and other organic bases such as ethylenediamine. Of the foregoing bases, those such as trisodium phosphate which provide a maximum pH in the range from 9 to about 14 are preferred.

In practice, when operating at temperatures from about 20°C up to a temperature at which the amide polymer or other reactants degrade prematurely, the gelation reaction is initiated rapidly when an aqueous solution of the amide polymer and polyaldehyde is brought together with the hypohalite salt dissolved in the alkaline solution, particularly when the reaction mixture has a pH in the preferred pH range from 10 to about 13.

Thus, for example, when an aqueous polyacrylamide/glyoxal solution at a pH below 7 and at a temperature from about 20° to 25°C is combined with an aqueous solution of sodium hypochlorite containing sufficient trisodium phosphate so that the pH of the resulting reaction mixture is in the range from about 10 to about 13.5, the resulting mixture sets to a firm water-insoluble gel within a matter of seconds while the pH falls to a value within the range of 7 to 11.

Example: An aqueous solution containing ~20% of a homopolymer of acrylamide is prepared and found to have a pH of about 5 and a viscosity of ~475 cp at 27°C as determined with a Brookfield LVT viscometer using a No. 2 spindle at 30 rpm. To 25 g of the foregoing solution is added with stirring 0.2 ml of a 40% solution of glyoxal in water. To this solution at 21°C is added with mixing a second solution consisting of 17 g of deionized water, 0.425 g of $Na_3PO_4 \cdot 12H_2O$ and 8 g of 5.25% aqueous solution of NaOCl. The resulting mixture has an initial pH of about 11.5 and forms a firm, nonpourable, water-insoluble gel in ~19 sec. The gel is self-supporting and does not exude water on standing. The gel is immersed in 5 N NaOH for several days without any noticeable deterioration.

When the gelable composition of this example is placed in a porous subterranean structure and gelled by the foregoing procedure, the structure is rendered impermeable to the passage of aqueous liquid.

For purposes of comparison, a gel is prepared according to the foregoing procedure except that no NaOCl is employed. While a nonpourable gel is formed by this procedure within about 15 sec, the gel is entirely destroyed when immersed in 5 N NaOH for about 1 hr.

Spheroidal Microgels

C.J. McDonald, J.V. Van Landingham and S.P. Givens; U.S. Patent 4,182,417; January 8, 1980; and M.L. Zweigle and J.C. Lamphere; U.S. Patent 4,172,066; October 23, 1979; both assigned to The Dow Chemical Company describe an improved oil recovery method whereby the fluid permeability of a subterranean geological formation is modified, preferably to restrict the passage of fluids (particularly aqueous fluids) therethrough. The desired control of fluid mobility is achieved by introducing into the subterranean formation, preferably via a well bore, a fluid medium containing discrete, spheroidal microgels of a water-swellable or water-swollen, crosslinked polymer in an amount sufficient to control the mobility of fluids in the subterranean formation.

The microgels have partly or totally water-swollen diameters which are generally within the range from 0.5 to about 200 μm and are very useful in treatment of porous subterranean strata that are commonly found in hydrocarbon-bearing formations. The microgels when dispersed in water or other aqueous media exist as discrete, spheroidal, water-swollen particles which can be separated from the aqueous media by filtration or other similar technique.

The fluid medium containing the microgels is most advantageously low in viscosity and therefore much more easily pumped into the desired porous formation than are conventional high viscosity water-soluble polymers or partially gelled polymers used hereinbefore. This advantageous property is believed to be due to the discrete, spheroidal characteristics of the microgels as well as to their relatively controlled particle size and crosslinked character. These characteristics enable the microgels to penetrate deeply into the pores of the porous structure to be restricted or plugged. Surprisingly, the microgels are not readily displaced even when the direction of liquid flow in the subterranean structure is reversed.

Accordingly, the method is most advantageously employed in enhanced oil recovery operations wherein a drive fluid is introduced through a borehole in the earth into a porous subterranean formation penetrated by the borehole thereby driving oil from oil bearing structures toward a producing well. In addition such fluid media containing the microgels are usefully employed as the fluid in well drilling operations, as packer fluids in well completion operations and as mobility control fluids in other enhanced oil recovery operations.

Methods using such media in drilling operations, in well completion operations, and as friction reduction aids in normal water-flooding operations as well as in fracturing processes are additional aspects of the process. It is further observed that fluid media containing the microgels are very useful in the treatment of subterranean structures containing substantial amounts of saltwater or brine which normally hinder or prevent the gelation reactions required in in situ gelation procedures.

Ethylenically unsaturated monomers suitable for use in preparing the microgels are those which are sufficiently water-soluble to form at least 5 weight percent solutions when dissolved in water and which readily undergo addition polymerization to form polymers which are at least inherently water dispersible and preferably water-soluble. By "inherently water dispersible", it is meant that the polymer when contacted with an aqueous medium will disperse therein without the aid of surfactants to form a colloidal dispersion of the polymer in the aqueous medium. Exemplary monomers include the water-soluble ethylenically unsaturated amides such as acrylamide, methacrylamide and fumaramide; N-substituted ethylenically unsaturated amides and quaternized derivatives thereof; ethylenically unsaturated carboxylic acids; ethylenically unsaturated quaternary ammonium compounds; sulfoalkyl esters or carboxylic acids and the like.

Of the foregoing water-soluble monomers, acrylamide, methacrylamide and combinations thereof with acrylic acids or methacrylic acid are preferred, with acrylamide and combinations thereof with up to 70 weight percent of acrylic acid being more preferred. Most preferred are the copolymers of acrylamide with from 5 to about 40, especially from 15 to about 30, weight percent of acrylic acid. The particle size of the microgels of these most preferred copolymers is more easily controlled than are the acid-free copolymers. For example, the addition of polyvalent metal ions such as calcium, magnesium and the like to aqueous compositions containing the microgels reduces the particle sizes of microgels by a highly predictable amount.

In the most preferred embodiments, it is desirable that the total monomer mixture contain a relatively small proportion (i.e., an amount sufficient to crosslink the polymer, thereby converting the polymer to a nonlinear polymeric microgel without appreciably reducing water swellability characteristics of the polymer) of a copolymerizable polyethylenic monomer. Exemplary suitable comonomeric crosslinking agents include divinylarylsulfonates; diethylenically unsaturated diesters; ethylenically unsaturated esters of ethylenically unsaturated carboxylic acids; diethylenically unsaturated ethers etc.

When a crosslinking comonomer is the means employed to provide the necessary crosslinking, any amount of such crosslinking agent in the monomer mixture is suitable provided that it is sufficient to crosslink the polymer to form a discrete, spheroidal, water-swellable microgel as defined herein. Preferably, however, good results have been achieved when the crosslinking agent is employed in concentrations from 5 to about 200, more preferably from 10 to about 100 pbw of crosslinking agent per million weight parts of total monomer.

The microgels are advantageously prepared by microdisperse solution polymerization techniques, e.g., the water-in-oil polymerization method described in U.S. Patent 3,284,393. In the practice of this method, a water-in-oil emulsifying agent is dissolved in the oil phase while a free radical initiator, when one is used, is dissolved in the oil or monomer (aqueous) phase, depending on whether an oil or water-soluble initiator is used. An aqueous solution of monomer or mixed monomers or a monomer per se is added to the oil phase with agitation until the monomer phase is emulsified in the oil phase. In cases where a crosslinking comonomer is used, it is added along with the other monomer to the

The reaction is initiated by purging the reaction medium of inhibitory oxygen and continued with agitation until conversion is substantially complete. The product obtained has the general appearance of a polymeric latex. When it is desirable to recover the microgel in essentially dry form, the polymer microgel is readily separated from the reaction medium by adding a flocculating agent and filtering and then washing and drying the microgel. Alternatively, and preferably, the water-in-oil emulsion reaction product is suitably employed as the fluid medium containing the microgels.

It is desirable to disperse the microgel in a fluid medium, preferably water or a water-in-oil emulsion, such that the resulting dispersion is reasonably stable.

In preferred embodiments utilized for fluid mobility control in enhanced oil recovery, it is desirable that the concentration of the microgel in the fluid medium be in the range from 100 to about 50,000 ppm of dry polymer based on the total weight of the fluid medium, more preferably from 250 to about 10,000 ppm, most preferably 250 to about 5,000 ppm.

Polymer, Acid-Generating Salt and Melamine Resin Gel Composition

G.T. Colegrove; U.S. Patent 4,157,322; June 5, 1979; assigned to Merck & Co., Inc. describes gel compositions consisting of water, a polysaccharide polymer, an acid-generating salt and a melamine resin.

The heat of underground formations catalyzes the reaction which crosslinks the polysaccharide to a gelled state.

The polymers that can be utilized are generally those polymers that contain at least one crosslinkable hydroxyl, carboxyl or amide group. It is desirable, of course, to utilize polymers wherein more than one crosslink is possible so that a more highly crosslinked gel can be formed.

Crosslinkable polymers that may be utilized include synthetic polymers such as partially hydrolyzed polyacrylamides. Also included among the polymers that can be utilized are cellulose and cellulose derivatives such as carboxymethylcellulose, methylcellulose, methylhydroxypropylcellulose, hydroxyethylcellulose or polysaccharides such as xanthan gums, guar gums, locust bean gums, gum tragacanths, alginates and their derivatives.

The polymer may be utilized over a rather broad range of concentrations from about 0.05% by weight based on the water to about 2.0%. However, a more preferred range is from 0.5 to about 1.5%.

The gelation reaction is catalyzed by a salt which is acid-generating upon the application of heat. The gelation reaction occurs under acidic conditions. An important advantage of this process is that the solution pH is close to neutral when made up, but when the heat from underground decomposes the salt, an acid is formed in situ, lowering the pH and causing the gelling reaction to occur.

Some salts that may be utilized are the ammonium salts of strong acids such as ammonium chloride, ammonium sulfate, ammonium nitrate, ammonium phosphate monobasic and the like.

The acid-generating salt can be utilized over a range from 0.5 by weight of water to about 1.0%. A more preferred range is from 0.1 to about 0.5%.

The melamine resin that is utilized can be a commercial product such as the reaction product of melamine and formaldehyde. Included among these melamine-formaldehyde (melamine) resins which are useful in this process are the partially methylated resins and the hexamethoxymethyl resins (i.e., American Cyanamid's Cymel 373, Cymel 370 and Cymel 380). The resin, however, has to be one that is soluble or dispersible in an aqueous medium. The resin concentration by weight based on water is from 0.3 to about 2.0%. A more preferred range is from about 0.5 to about 1.25%.

The gelation rate of the composition depends on the amount of each of the components and the temperature at which the reaction is conducted. Thus, one can tailor the gel rate and gel strength of the composition by adjusting the amount of the acid-generating salt, the resin amount and the temperature. The higher the temperature at given concentrations of resin, acid-generating salt and polymer, the faster will be the gelation time. If one desires a more gelled composition, he may increase the polymer and resin concentrations at a given temperature.

The rate of the crosslinking reaction can be accelerated by the addition of formaldehyde to the composition.

Example: A simulated core model packed with coarse gravel is utilized to demonstrate the water-diverting characteristics of the composition.

When plain water is passed through the core model packed with coarse gravel, a high flow rate indicates a permeability of about 6 darcies.

An aqueous solution containing 0.5% xanthan gum (Kelzan), 0.5% partially methylated melamine formaldehyde resin (Cymel 373), 0.25% monoammonium phosphate, and 0.2% formalin by weight of composition is then injected into the core model containing coarse gravel and the solution allowed to gel for 8 hr at 130°F. An attempt to pass water through the core after the gelled composition indicates a flow rate effectively reduced to about zero which indicates complete plugging of the porous channels.

Self-Foaming Aqueous Composition

E.A. Richardson, R.F. Scheuerman, D.C. Berkshire, J. Reisberg and J.H. Lybarger; U.S. Patent 4,232,741; November 11, 1980; assigned to Shell Oil Company describe a process for temporarily plugging pores or other openings in or along a subterranean reservoir by flowing through a well and into those openings a self-foaming aqueous solution which remains substantially unreactive until it has entered the openings.

That solution is compounded by dissolving in an aqueous liquid (a) a reactant for generating nitrogen gas at a rate affected by the temperature and pH of the solution, (b) surface active materials capable of causing the solution to be converted into a relatively immobile foam as the gas is generated within the solution (c) a pH-controlling system for initially maintaining a relatively high pH at which the reaction rate is relatively low at temperatures less than the reservoir temperature and subsequently maintaining a barely acidic pH at which the gas-generating reaction rate is moderately fast at temperatures near the reservoir temperature and, (d) a relatively slowly reactive acid-yielding compound in a proportion such that it relatively gradually changes the pH from the initial relatively high value to the subsequent barely acidic value.

The so-compounded solution is injected through the well conduits and into contact with the reservoir at a rate such that a significant portion of the solution (a) enters at least the largest pores or other openings in or along the reservoir before generating a significant fraction of the amount of gas it is capable of generating, and, (b) remains within those openings until it becomes a relatively immobile foam.

Nitrogen-containing gas-forming reactants which are suitable for use in the process can comprise water-soluble amino nitrogen-containing compounds which contain at least one nitrogen atom to which at least one hydrogen atom is attached and are capable of reacting with an oxidizing agent to yield nitrogen gas within an aqueous medium. Such water-soluble nitrogen-containing compounds can include ammonium salts or organic or inorganic acids, amines, amides, and/or nitrogen/linked hydrocarbon-radical substituted homologs of such compounds, as long as they react with an oxidizing agent to produce nitrogen gas and by-products which are liquid or dissolve in water to form liquids which are substantially inert relative to the well conduits and reservoir formations. Examples of such nitrogen-containing compounds include ammonium chloride, ammonium nitrate, ammonium nitrite, ammonium acetate, ammonium formate, ethylenediamine, formamide, acetamide, urea and the like. Such ammonium salts, e.g., ammonium chloride, ammonium formate or ammonium nitrate are particularly suitable.

Oxidizing agents suitable for use in the process can comprise substantially any water-soluble oxidizing agents capable of reacting with a water-soluble nitrogen-containing compound of the type described above to produce nitrogen gas. Examples of such oxidizing agents include alkali metal hypochlorites (which can, of course, be formed by injecting chlorine gas into a stream of alkaline liquid being injected into the well), alkali metal or ammonium salts of nitrous acid such as sodium or potassium or ammonium nitrite, and the like.

The alkali metal or ammonium nitrites are particularly suitable for use with nitrogen-containing compounds such as the ammonium salts. Since the reaction can occur between ammonium ions and nitrite ions, ammonium nitrite is uniquely capable of providing both the nitrogen-containing and oxidizing reactants in a single compound that is very soluble in water.

Aqueous liquids suitable for use in the process can comprise substantially any in which the salt content does not (e.g., by a common ion effect) prevent the dissolving of the desired proportions of N-containing and oxidizing reactants. In general, any relatively soft freshwater or brine can be used. Such aqueous liquid solutions preferably have a dissolved salt content of less than about 1,000 ppm monovalent salts and less than about 100 ppm multivalent salts.

Buffer compounds or systems suitable for moderating the rate of gas generation can comprise substantially any water-soluble buffer which is compatible with the gas-forming components and their products and tends to maintain the pH of an aqueous solution at a barely acidic pH at which the reaction rate is only moderately high at temperatures near the reservoir temperature. Such a pH is preferably from about 5.5 to 7. Examples of suitable buffering materials include the alkali metal and ammonium salts of acids such as carbonic, formic, acetic, citric, and the like, acids.

A small but significant proportion of alkali metal hydroxide is added to such a buffer system to provide an initially alkaline pH which is higher than that which will be maintained by the buffer and high enough to provide a low rate of reaction at temperatures lower than the reservoir temperature.

Reactants for reducing the pH of the aqueous solution to the barely acidic pH maintained by the buffer can comprise substantially any soluble, relatively easily hydrolyzable materials which are compatible with the gas-forming reactants and their products and are capable of releasing hydrogen ions at a rate slow enough to allow the buffered and pH-increased solution of the gas-generating reactants to be injected into the reservoir formation before the pH is reduced to a value less than about 7. Examples of suitable reactants include: lower alcohol esters of the lower fatty acids such as the methyl and ethyl acetates, formates and the like; hydrolyzable acyl halides, such as benzoyl chloride; relatively slowly hydrolyzable acid anhydrides; relatively slowly hydrolyzable phosphoric or sulfonic acid esters; and the like.

It is generally desirable to use nitrogen-containing compounds and oxidizing agents which are dissolved in substantially stoichiometric proportions and in relatively high concentrations within the gas-generating solution. Such reactants are typified by ammonium chloride and sodium nitrite and they are preferably used in substantially equimolar amounts of from about 3 to 6 mols per liter.

A number of tests of the capabilities of various solutions to generate nitrogen gas were conducted under conditions simulating those encountered within a subterranean reservoir formation. In such tests, the reactants were dissolved in an amount of aqueous liquid filling about one-fourth of the volume of a pressure resistant chamber or bomb. The bomb was maintained at a constant temperature within a circulating bath of liquid and measurements were made of the variations in gas pressure with time. The data obtained from typical tests in which the gas-producing reactants were ammonium chloride and sodium nitrite are listed in the table on the following page.

In tests 7, 8 and 9, sodium bicarbonate was used as a buffer which was capable of providing a relatively slow rate of reaction at a temperature of 210°F. Test 7 shows that, with that buffer by itself, the half-life was 681 minutes. Test 8 shows that the addition of sodium hydroxide and methyl formate reduced the half-life to 72 min. The reaction rate for the hydrolysis of the methyl formate proved to be too fast at the temperature of Test 8. Because of that, the pH was quickly lowered to one at which the gas was quickly produced. Test 9 shows that a mixture of the bicarbonate, sodium hydroxide, methyl formate and methyl acetate provided a half-life of 178 min.

The tests thus indicate that the slower hydrolysis of the methyl acetate continues the hydrogen ion-releasing after the exhaustion of the methyl formate, so that the pH is kept substantially constant. The data fits second order kinetics during about 80% of the generation of the nitrogen gas. And thus, it is now apparent that such a gas-generating reaction can be significantly delayed for a time and then allowed to proceed, by adding, to a solution which contains a rate-moderating buffering material, both a rate-reducing amount of alkali metal hydroxide and a rate-restoring amount of acid-generating material.

Foam-forming surfactants suitable for use in the process can comprise substantially any which are capable of being dissolved or dispersed in an aqueous liquid solution containing the nitrogen-containing compound and oxidizing agent and remaining substantially inert during the nitrogen-gas-producing reaction between the nitrogen containing compounds and the oxidizing agent. Examples of suitable surfactants comprise nonionic and anionic surfactants, (American Alcolac Company); sulfonate sulfate surfactant mixtures, e.g., those described in U.S. Patent 3,508,612; petroleum sulfonates (Bray Chemical Company); Petronates and Pyronates (Sonnoborn Division of Witco Chemical Company); fatty acid and tall oil acid soaps, e.g., Acintol heads (Arizona Chemical Company); nonionic surfactants, e.g., Triton X100; and the like surfactant materials which are soluble or dispersible in aqueous liquids.

Reaction Parameters for $NH_4^+ + NO_2^-$ Reaction

Test No.	Composition (mols/ℓ) NH_4Cl	$NaNO_2$	Other Components	Bath Temperature (°F)	pH at 70°F When Prepared	Maximum Pressure Attained (psi)	t½ (min)*
1	1.0	1.0	0.2 NaAc** + 0.2 HAc***	72.5	4.50	204	590
2	1.0	1.0	0.2 NaAc** + 0.2 HAc***	113	4.50	317	61
3	2.0	1.0	0.2 NaAc** + 0.2 HAc***	113	4.47	344	24
4	1.0	1.0	1 MeFor†	122	6.42	380	49
5	2.0	2.0	0.2 $NaHCO_3$ + 0.1 MeFor†	210	7.12	865	76
6	3.0	2.0	0.2 $NaHCO_3$	210	6.80	917	85
7	2.0	2.0	0.2 $NaHCO_3$	210	8.54	650	681
8	2.0	2.0	0.2 $NaHCO_3$ + 0.3 NaOH + 0.4 MeFor†	210	8.70	865	72
9	2.0	2.0	0.2 $NaHCO_3$ + 0.3 NaOH + 0.2 MeFor† + 0.2 MeAc	210	8.65	850	178
10	2.0	2.0	0.2 $NaHPO_4$ + 0.17 HCl	210	6.50	870	31
11	2.0	2.0	0.2 $NaHCO_3$ + 0.225 NaOH + 0.245 MeFor†	210	8.18	795	146
12	4.0	4.0	–	130	6.17	757	500
13	4.0	4.0	0.15 $NaHCO_3$	210	6.92	248	115
14	3.0	3.0	0.05 $NaHCO_3$	160	6.70	275	981
15	3.0	3.0	0.02 $NaHCO_3$	160	6.56	250	636
16	2.25	3.0	1 NH_4Ac	160	6.87	371	188
17	4.0	4.0	0.05 $NaHCO_3$	160	6.66	360	660
18	4.0	4.0	0.05 $NaHCO_3$ + 0.05 MeFor†	160	6.68	520	86
19	3.0	3.0	0.5 NaAc	135	–	–	559

*Half-life, i.e., the time by which the reactants have produced one-half of the stoichiometrically available N_2 gas.
**Sodium acetate.
***Acetic acid.
†Methyl formate.

Water-thickening agents suitable for use in the process can comprise substantially any water-soluble polymer or gel capable of dissolving in, and/or substantially any solids-free oil capable of being emulsified with, an aqueous liquid solution containing the nitrogen-containing compound and oxidizing agent and remaining substantially inert during the flowing of the resultant nitrogen-gas-producing solution or emulsion into the reservoir formation while also increasing the effective viscosity of the foam which is subsequently generated within the reservoir formation. Examples of suitable thickening agents which are water-soluble include xanthan gum polymer solutions such as Kelzan or Xanflood (Kelco Corporation); hydroxyethylcellulose, carboxymethylcellulose, guar gum and other such agents.

Such soluble thickening agents are particularly effective in relatively low temperature reservoirs having relatively high permeabilities. Suitable oils include relatively viscous refined oil or crude oil which have been freed of substantially all solids.

Soft Plug Material

J.U. Messenger; U.S. Patent 4,173,999; November 13, 1979; assigned to Mobil Oil Corporation describes a method of controlling or alleviating the loss of whole liquid drilling fluid, often referred to as drilling mud, into a lost circulation zone of a subterranean formation that is penetrated by a well during the drilling of a well.

In accordance with this process, when drilling a well and loss of drilling fluid into a lost circulation zone of an incompetent subterranean formation occurs, a nonaqueous slurry is formed of a nonaqueous liquid, a hydratable material and a solids suspending agent, which slurry may also include therein an oil wetting dispersing agent. The slurry is introduced into the well, normally through the drill string, and is displaced down the well into the vicinity of the incompetent formation.

An aqueous fluid, normally drilling mud where water-based drilling mud is being employed in the drilling of the well, is displaced down the well normally through the annulus formed about the drill string and is mixed with the nonaqueous slurry in the vicinity of the incompetent formation to form a gel which is referred to as a soft plug and the soft plug is then displaced or squeezed into the incompetent formation to prevent the loss of drilling fluid thereinto.

The ratios of the amounts of material used in forming the nonaqueous slurry are as follows: hydratable material, 300 to 800 lb/bbl of nonaqueous liquid; solids suspending agent, 0.1 to 5.0 lb/bbl of nonaqueous liquid; and dispersing agent, 0.5 to 3.0 lb/bbl of nonaqueous liquid. The most commonly used hydratable materials are bentonite, cement and mixtures thereof, though other hydratable materials such as attapulgite may also be used. A preferred ratio of bentonite to cement is 100 lb of bentonite to 188 lb of cement. A slurry formed using this preferred ratio of bentonite to cement has good flow properties which facilitates its placement in a well and yields a strong soft plug.

Preferred nonaqueous liquids for use are light hydrocarbon oils such as diesel oil and kerosene. Such light hydrocarbon oils are preferred because of their low cost and ready availability at most well sites. The preferred solids suspending agent is an oleophilic clay. Such a clay is a water-swellable clay which has undergone treatment to render it oleophilic. Examples of suitable oleophilic clays are the bentonites or other clays which have been treated with an oil wetting surfactant such as a long chain quaternary or nonquaternary amine. For a more detailed description of such clays and the method of preparation, reference is made to *Chemical Engineering,* March 1952, pp. 226-230; U.S. Patent 2,531,812; and U.S. Patent 2,675,353. Suitable oleophilic clays may also be prepared by dehydrating a hydrophilic clay such as bentonite and then treating the dehydrated clay with a glycol or glycol ether. For a more detailed description of this procedure, reference is made to U.S. Patent 2,637,692. Suitable oleophilic clays are "Geltone" and "Petrotone."

The oil wetting dispersing agents used may be any suitable "thinners" which are oil soluble and have the characteristics of oil wetting the surfaces of the hydratable materials employed. Examples of oil wetting dispersing agents which may be employed are given in the table below by trade name, distributor and chemical composition. Also given in the table are the density of the dispersing agent and practicable amounts of dispersing agents which may be used in pounds per final barrel of the nonaqueous slurry.

Trade Name of Dispersant	Distributor	Chemical Composition	Density (lb/gal)	Practical Amounts* (lb/final bbl)
EZ-Mul	Baroid	Half amide, salt terminated	8.1	0.5
Driltreat	Baroid	Lecithin	8.7	0.5
Surf-Cote	Milchem	Oil-soluble amine dodecyl benzene sulfonate	8.16	1.0
SA-47	Oil Base, Inc.	Aryl alkyl sodium sulfonate	8.12	0.5
Fazethin	Magcobar	–	–	1.0
Ken-Thin	Imco	Imidazolin	7.9	1.5
SE-11	Magcobar	Modified alkyl aryl sulfonate plus imidazoline	7.83	1.5
Carbo-Mul	Milchem	Oil-soluble alkanol amide	7.5	1.5

*To use in plug.

The problem of plugging the drill string can be avoided by including the solids suspending agent in the nonaqueous slurry along with the hydratable material. The inclusion of both a solids suspending agent and an oil wetting dispersing agent results in a slurry with better flow properties, greater solids suspension and a lower filter loss, all effective in preventing plugging of the drill string.

Sealing a Fracture Formation

A well-known and widely practiced technique for enhancing the production of hydrocarbons from a subterranean formation entails hydraulic fracturing of the formation with a pressurized fluid containing a particulate propping material. When the fracturing fluid is removed following development of the fracture, the propping material remains in place to mechanically prevent closure of the fracture.

Various conditions are sometimes encountered in fracturing which prevent optimization of the degree of production stimulation achieved thereby. At times, the fracture, or a portion of it, is intercepted by a zone which bears water, which therefore flows into the hydrocarbons entering the fracture and being produced therefrom. On other occasions, the fracture extends, in part, into a so-called thief zone, and because of the relatively high permeability of this zone, undesirable quantities of the fracturing fluid and/or the hydrocarbons may be lost to this zone.

G.R. Coulter; U.S. Patent 4,157,116; June 5, 1979; assigned to Halliburton Company describes a method by which undesirable fluid migration at a portion of a fracture boundary can be shut off, and in the course of the procedure, production of hydrocarbons into the fracture from a producing interval defining the remainder of the fracture can be stimulated. The method makes use of a plugging or sealing material and a removable diverting material placed in the fracture at particular times and places for accomplishing these primary objectives.

More specifically considered, the process is used where a producing well stimulated by fracturing is producing at a less than optimum rate or economy due to the fracture having, in part, extended into a subterranean zone which, because of its permeability or fluid content, has a deleterious effect upon the production of hydrocarbons originating at a portion of the fracture face contiguous to that zone.

The process comprises the steps of first introducing or emplacing a porous bed of solid particles against that face of the fracture which is defined by the zone which can be beneficially blocked or occluded to prevent or reduce undesirable flow of fluid across the fracture-zone interface at this location.

With the porous bed of solid particles so located within the fracture, a removable diverting material susceptible to pumping into the fracture is then placed in all or a portion of the fracture which is not occupied by the bed of solid particles, and is allowed to set up to a solid or semisolid state at that location. The essential aspect of the location of this removable diverting material is that it forms a barrier within the fracture at a location such that a sealing material subsequently passed from the well bore into the fracture will be diverted in its path of flow so as to substantially entirely enter the interstices between the solid particles in the porous bed, and will not contact the predominantly hydrocarbon-producing interval.

With the diverting material placed in the fracture and the porous bed of solid particles in place, a sealant material is then pumped through the well bore and into the bed of solid particles so as to enter the interstices of the solid particles and form a substantially continuous mass of material overlying the fracture-zone interface. As the sealant solidifies, a fluid-impermeable plug or barrier is formed at this location which blocks or substantially impedes the transfer of fluids across the interface and through this portion of the fracture. In the case of a water-bearing zone which supplies undesirable water to the fracture for admixture with the hydrocarbons under production, the sealant barrier thus formed will prevent such water flow.

Where the blocked zone is a thief zone, loss of hydrocarbons or fracturing fluid from the fracture to the thief zone will be prevented by the sealant barrier. The same advantage of thief zone blockage is provided where such zones are encountered adjacent an injection well in secondary and tertiary recovery situations.

After the sealant material has set up in the interstices of the solid particles in the porous bed to form a plug or barrier at the fracture-zone interface, the removable diverting material is removed from the fracture via the well bore. Removal of the diverting material, which may be a gel containing an internal breaker or other type of known diverting material, can be accomplished in accordance with techniques well understood in the art. Such removal of the diverting material opens the fracture to the flow of hydrocarbons from the interval, and the well can then again be placed on stream for production purposes. The quality and/or quantity of production is thus substantially enhanced by the prevention of water infiltration and admixture with the hydrocarbons, or by the prevention of undesirable hydrocarbon loss to the thief zone.

In a preferred method, the diverting material which is utilized in a portion of the fracture includes a solid proppant material. The particles of the proppant thus moved into the fracture as a constituent of the diverting material are left in position in the fracture when the diverting material is to be removed to provide a permeable bed of proppant which aids in hydrocarbon production by maintaining the fracture width against the closing propensities of overburden forces. The following example of the process wil aid in its understanding.

Example: The main pay interval of the San Andreas formation in Yoakum County, Texas, lies at a depth of from 5,000 to 5,200 ft. It is hydraulically fractured to yield a fracture having a height of 200 ft, an average width of 0.19 inch and an average distance of horizontal extension from the well bore of about 220 ft. Approximately ½ the total height of the fracture (the lower 100 ft) traverses a zone highly saturated with water. The bottom hole temperature of the well is 100°F and the bottom hole treating pressure is 4,000 psi. The formation has an average overall permeability of 5 md, and an average porosity of 12%.

For the purpose of placing a bed of solid proppant particles in the lower portion of the fracture adjacent the water zone, 10-20 sand is mixed into a low viscosity fracturing fluid to provide a sand concentration in the fluid of 2.0 lb/gal. The low viscosity fracturing fluid is water containing 1 weight percent potassium chloride, and the following additional additives per 1,000 gal of water: 30 lb of silica flour (as a fluid loss additive), 30 lb of guar gum (as a viscosifier), and 10 lb of sodium dihydrogen phosphate (for pH control). A small amount of ethoxylated alcohol is also added as a nonemulsifying agent.

The low viscosity fluid carrying the suspended sand is injected into the fracture at a rate of 15 bbl/min until 40,000 gal of the fluid has been placed. 690 sacks of the 10-20 sand are required for this volume of emplaced sand-carrying fracturing fluid. Pumping is then stopped and the sand is permitted to settle to the bottom of the fracture. Following this step 218 ft of the fracture are covered to a depth of 100 ft by the sand.

A temporary diverting material is next prepared using as a base liquid the potassium chloride-containing water described above. To the water (per each 1,000 gal) are added 80 lb of guar gum, 3 gal of potassium pyroantimonate (as a gelling crosslinker), 20 lb of sodium dihydrogen phosphate and small amounts of a nonemulsifying agent and cellulase to function as a gel breaker. To this composition a solid particulate proppant is then added at the rate of 0.77 lb/gal.

The temporary diverting material as thus constituted is pumped via the well bore into the upper portion of the fracture at a rate of 10 lb/min. 10,000 gal of the diverting material are pumped into the fracture, requiring a total of 40 sacks of the proppant. The volume of the fracture occupied by the diverting material, when in place, is 200 ft in length and 100 ft in height. After emplacement of the diverting material, the well is shut in for a period of 2 hr to permit the diverting material to set up to a semisolid, high-strength gel.

A sealant material composed of 16% Grade 40 sodium silicate containing a latent acid catalyst and having a viscosity of 1.5 cp is injected into the fracture from the well bore, and is diverted by the emplaced diverting material into the interstices of the 10-20 sand bed laid down in the lower portion of the fracture in the first step of the procedure. 3,000 gal of the sealant material are injected to completely fill the interstices in the sand bed. This volume is adjusted to account for fluid loss to the formation.

The well is then shut in and the sealant permitted to set up to a strong, solid gel. During this time, the internal breaker (the cellulase) in the temporary diverting agent commences to break the gelled diverting agent. After about 6 hr, breaking of the diverting agent is complete and the sealant material has set up to a semisolid state. The broken liquid portion of the diverting material is then pumped from the well, and the well is returned to production. A 3.8-fold increase in hydrocarbon production is realized after completion of the described procedure.

Magnetic Tape Particles

H.L. Hamilton; U.S. Patent 4,222,444; September 16, 1980 describes a method to close off areas of porosity in a well bore wall while the drilling operation continues by the addition of finely ground polyester tape to the well drilling fluid. If this procedure proves to be futile, larger polyester particles may be introduced into the well bore in a further attempt to close off the cracks or fissures, or the fissures may be located by the use of an electric well logging instrument which will positively locate the fissures through the aid of the magnetic properties of the tape lodged therein. At this time, cement slurry may be introduced in the vicinity of the cracks or fissures for positively sealing them off. Thus, the method provides a means for not only preventing against well fluid leak but also provides a means of disposing of discarded computer tape in an economical manner.

Method for Rigless Zone Abandonment

R.C. Martin; U.S. Patent 4,189,002; February 19, 1980; assigned to The Dow Chemical Company describes a method for permanently plugging a zone of a subterranean formation penetrated by a well bore. In carrying out the method, a resin system containing at least one thermosetting resin and at least one curing agent or catalyst for the resin is injected into the formation via the well bore. The resin system is displaced into the formation by a second liquid containing an effective amount of at least one chain stopping compound to react with at least one component of the resin system. Preferably, the second liquid also contains a fluid loss additive. The chain-stopping or set-inhibiting stage may be separated from the resin stage if desired, e.g., by a wiper plug, and is injected in a manner so that substantially none of the set-inhibiting stage is itself injected into the formation.

Reaction of any of the resin system remaining in the well bore is thereby capped before sufficiently high crosslink density is achieved to form a solid residue in the well bore. The liquid remaining in the well bore can be removed at a later time, e.g., by recirculation to the surface, bailing, displacement into another zone, and the like. The resin system injected into the formation cures to form a consolidated mass which is substantially impermeable.

The present process permits a zone to be plugged and abandoned without the need for drilling out residual material left in the well bore, making the process particularly attractive for use in wells having a plurality of completion zones and for wells in remote locations where rig time is extremely costly.

The resin system stage is conventional and may comprise any thermosetting resin and a suitable curing agent or catalyst appropriate for the resin employed, to provide a liquid system which can be injected into the formation to be treated in a single stage and which will cure in the formation to substantially reduce the permeability of the formation in the vicinity of the well bore, preferably to a permeability of about 0.01 millidarcy or less. Thus, suitable resin systems based on phenolic resins, furan resins, furfural alcohol resins, vinyl ester resins, and the like may be employed as those skilled in the art will appreciate. However, epoxy resin type systems are preferred.

Epoxy resins suitable for use in the process comprise those organic materials possessing more than one epoxy group. Examples of the polyepoxides include 1,4-bis(2,3-epoxypropoxy)benzene, 1,3-bis(2,3-epoxypropoxy)benzene, 1,4'-bis(2,3-epoxypropoxy)diphenyl ether, 4,4'-bis(2-methoxy-3,4-epoxybutoxy)diphenyl dimethylmethane, and 1,4-bis(2-methoxy-4,5-epoxypentoxy)benzene.

A number of curing agents are known which harden unset epoxy resins. Specific classes of curing agents include, for example, amines, dibasic acids and acid anhydrides. The preferred hardening agents are the amines, especially those having a plurality of amino hydrogen groups. Included are aliphatic, cycloaliphatic, aromatic or heterocyclic polyamines, such as diethylenetriamine, ethylenediamine, triethylenetetramine, dimethylaminopropylamine, diethylaminopropylamine, piperidine, methanediamine, triethylamine, benzyl dimethylamine, dimethylaminomethyl phenol, tri(dimethylaminomethyl)phenol, pyridine, and the like. Mixtures of various amines may be employed. The amines or other curing agents react rather slowly to convert the polyepoxides to an insoluble form.

A suitable curing agent and concentration thereof best suited for a particular well can easily be determined by a knowledge of temperature conditions and available working time, i.e., length of time between adding the curing agent and final positioning of the resin-containing mixture in the formation.

The curing agent can be employed in an amount ranging from about 40 to more than 125%, preferably 70 to 110% and more preferably about 85 to 100%, of that stoichiometrically required.

Preferably, the resin and curing agent are admixed in a mutual solvent to provide a solution which is easily pumped and which can be readily injected into the formation. Too much solvent, however, can result in incomplete plugging or prolonged cure times. Those skilled in the art will be able to select a suitable solvent and solvent ratio depending on the choice of resin and curing agent. The solvent may be, for example, an organic alcohol, ester, ether, ketone, acetate, etc., and mixtures thereof.

Amyl acetate, methyl ethyl ketone, methyl isobutyl ketone, xylene, ethylene glycol, n-butyl ether, alkylene glycol alkyl ethers such as ethylene glycol ethyl ether, and diethylene glycol isobutyl ether are specific examples of solvents. Suitable combinations of solvents include xylene/ethylene glycol ethyl ether, e.g., in a 1.5:1 to 0.3:1 weight ratio, and toluene/ethylene glycol ethyl ether.

A most preferred resin system comprises about 20 to 65 weight percent ethylene glycol ethyl ether, most preferably about 20 to 30 weight percent, and the balance an approximately stoichiometric blend of D.E.R. 331 brand epoxy resin, which is a liquid epoxy resin of the bisphenol A/epichlorohydrin type having an average epoxide equivalent weight of about 190, and Jeffamine AP22 brand polymethylene polyphenylamine, which has an average -NH equivalent weight of about 51.5.

The second essential step is the injection into the well bore of a liquid containing at least one chain-stopping compound to react with at least one component of the resin system to prevent any residual resin system material in the well bore from setting up in the well bore. Preferably, the second liquid includes a mutual solvent for the resin system components, for the chain-stopping compound, and for the reaction product of the resin system components with the chain-stopping compound.

A most preferred embodiment when employing the most preferred resin system hereinabove described is a 5 to 50 and more preferably 15 to 25 weight percent solution of monoethanolamine in an alkylene glycol alkyl ether such as ethylene glycol ethyl ether. The concentration of monoethanolamine in the glycol ether is not sharply critical.

Preferably, the set-inhibiting liquid contains from 0.2 to about 4.5% of a cellulose derivative, e.g., ethylcellulose, and 0.04 to about 0.75% of an inert particulate, by weight of the liquid. Quantities near the upper ends of the ranges are generally employed where the formation is relatively permeable, whereas lesser amounts are satisfactory where the formation is relatively tight. Most preferably, from 1.25 to about 3.3% of ethylcellulose and 0.18 to about 0.55% of silica flour are employed, e.g., about 100 to 300 lb of an approximately 6 parts ethylcellulose to one part silica flour mixture per 1,000 gal of fluid.

The following treatment has been proposed to a well operator wishing to plug off a zone in an offshore well producing water through perforations at a depth of 7,758 to 7,782 ft. The treatment can be carried out without use of a drilling rig. The well is a multiple completion zone well equipped with a 7", 26 lb/ft liner from 6,207 ft to total depth, and the zone of interest isolated by packers at 7,628 ft and 7,820 ft. After shutting off the perforations at 7,758 to 7,782 ft, the well operator wishes to perforate at another depth but still within the same area isolated by the packers at 7,628 and 7,820 ft. 2⅜", 4.7 lb/ft EUE tubing extends from the surface to the 7,628 ft packer, and another tubing string of the same size extends through the zone of interest to another zone below the 7,820 ft packer.

To divert treatment fluids from that portion of the liner extending from the bottom of perforations to the top of the lower packer, 1.25 bbl of heavy brine (e.g., 11 lb/gal calcium chloride brine) is first spotted in the well bore through coil tubing. To clean the formation, 500 gal of a 50:50 15% HCl:xylene acid dispersion as shown in U.S. Patent 3,794,523 is injected and displaced with 40 bbl filtered seawater and 10 bbl of diesel oil. Although not essential, diesel oil or similar hydrocarbon displacement fluid will be employed because the working time of the particular resin system is shortened somewhat by contact with freshwater, seawater, or acid. Alternatively, a lower alkyl alcohol may be used in lieu of part or all of the diesel oil. The foregoing optional steps are recommended to clean the formation and thereby achieve optimum sealing performance.

The resin system is to be prepared by bringing to the well site, a solution of 80 weight percent D.E.R. 331 brand epoxy resin and 20% ethylene glycol ethyl ether, and another solution of 60 weight percent Jeffamine AP22 brand polymethylene polyphenylamine in 40% ethylene glycol ethyl ether. The resin system is to be prepared by adding 0.4 part by volume of the polymethylene polyphenylamine solution to 1 part of the epoxy resin solution and admixing same for about 10 to 15 min.

Seventeen hundred gallons of the abovementioned internally catalyzed resin system, enough to penetrate radially at least about 3 ft around the well bore over the perforated interval, is to be injected down the tubing and displaced into the formation with 500 gal of a mixture of 1 part by weight monoethanolamine, 4 parts by weight ethylene glycol ethyl ether, and 150 lb of a 6:1 weight mixture of ethylcellulose and silica flour.

Finally, the set-inhibiting fluid may be displaced to depth with 22.8 bbl of filtered seawater, or a sufficient lesser volume to balance the formation pressure. The resin system may, if desired, be separated from the set-inhibiting fluid by a wiper plug to wipe the tubing substantially clean from the surface to the 7,628 ft packer and also to minimize mixing of the set-inhibiting fluid with all but the final portion of the resin system. Using the wiper plug, the theoretical minimum volume of set-inhibiting liquid to be employed would be the volume between the packer at 7,628 ft where the wiper plug would fall free and the surface of the heavy brine initially spotted below the perforations. However, an excess is preferably employed to assure an adequate volume to contact exposed well equipment.

In lieu of the filtered seawater, other suitable displacement fluids may be employed for the final displacement such as diesel oil, nitrogen, and the like. The well will then be shut in for about 12 hr to permit the internally catalyzed resin system to set up in the formation, after which time coil tubing will be inserted and the isolated portion of the well flushed clean. The new zone will then be perforated at the new location desired by the operator and conventional stimulation and sand control treatments carried out through the new perforations.

Solution of Neutralized Ionomeric Polymers

R.D. Lundberg, D.E. O'Brien, H.S. Makowski and R.R. Klein; U.S. Patent 4,183,406; January 15, 1980; assigned to Exxon Production Research Company describe the introduction of a fluid medium into a borehole in a subterranean formation to form a fluid flow blocking means.

In carrying out one embodiment of this process, a polymer solution is introduced into the well at the desired location for plugging the well bore. The polymer solution comprises a neutralized ionomeric polymer dissolved in a nonpolar organic liquid and a polar cosolvent. The ionomeric polymer can be characterized as having a backbone that is substantially soluble in the organic liquid and pendant ionic groups that are substantially insoluble in the organic liquid. The polar cosolvent solubilizes the pendant ionomeric groups such that the polymer solution introduced into the well has a viscosity less than about 20,000 cp. Upon mixing with water, the polar cosolvent in the polymer solution is taken up by the water and the polymer aggregates and increases the solution viscosity. The polymer solution upon mixing with only 5% water based on the volume of the polymer solution forms a gel having sufficient shear strength to form a plug in the well.

In a preferred method of plugging a well, a polymer solution comprising diesel oil, sulfonated EPDM (ethylene-propylene-diene terpolymers) and methanol is injected into the well bore down a drill pipe. The methanol concentration is less than about 15 weight percent of the polymer solution and the polymer concentration ranges from 0.1 to about 20 weight percent of the polymer solution. The water in the well bore mixes with the polymer solution, extracts the alcohol from the diesel phase, and causes the polymer to crosslink. The crosslinked polymer binds the diesel into a jellylike material, which can have a shear strength well above 30 lb/ft^2 (146 kg/m^2).

The aqueous fluid required to cause the polymer to crosslink may be injected before, simultaneously with, or after, the injection of the polymer solution. The aqueous fluid may also be aqueous formation fluid which flows from the formation into the well bore.

The polymer solutions are advantageously used as well treating fluids because the polymer solution is capable of changing from a low viscosity system into an extremely viscous gel with relatively small amounts of water. It is possible to achieve increases in the shear strength by factors of 10,000 or more by addition of only 5 to 15% water based on the polymer solution volume. After initial thickening is achieved, the amount of additional water mixed with the gel mass does not significantly affect the shear strength.

The ionomeric polymers may be incorporated into the organic solvent at a level of from 0.1 to 20 weight percent, preferably from 0.2 to 10 weight percent, most preferably from 0.5 to 5 weight percent, based on the organic solvent and the polar cosolvent.

Specific examples of preferred ionomeric polymers include sulfonated polystyrene, sulfonated poly-tert-butylstyrene, sulfonated polyethylene (substantially noncrystalline), and sulfonated polyethylene copolymers, sulfonated polypropylene (substantially noncrystalline), and sulfonated polypropylene copolymers, sulfonated styrene-methyl methacrylate copolymers, styrene-acrylic acid copolymers, sulfonated polyisobutylene, sulfonated ethylene-propylene terpolymers, sulfonated polyisoprene, sulfonated polyvinyltoluene, and sulfonated polyvinyltoluene copolymers.

The ionomeric polymers may be prepared prior to incorporation into the organic solvent. For example, preferably the acid derivative is neutralized immediately after preparation. Hence, if the sulfonation of polystyrene is conducted in solution, the neutralization of the acid derivative can be conducted immediately following the sulfonation procedure. The neutralized polymer may then be isolated by means well known to those skilled in the art, i.e., coagulation, steam stripping or solvent evaporation, because the neutralized polymer has sufficient thermal stability to be dried for employment at a later time in the process. It is well known that the unneutralized sulfonic acid derivatives do not possess good thermal stability and the above operations avoid that problem.

The most preferred organic solvents are those based on various oils or diesel fuel for economic reasons, convenience, availability and toxicity. In particular, diesel fuel is preferred because it is also an excellent solvent for the preferred polymers in the presence of a suitable cosolvent.

Specific examples of organic solvents to be employed with various types of polymers are: sulfonated polystyrene–benzene, toluene, ethylbenzene, methyl ethyl ketone, xylene, styrene, etc; sulfonated poly-tert-butylstyrene–benzene, toluene, xylene, ethylbenzene, styrene, aliphatic oils, aromatic oils, etc; sulfonated ethylene-propylene terpolymer–aliphatic and aromatic solvents, oils such as Solvent 100 Neutral and 150 Neutral and similar oils; styrene-acrylic acid copolymers–aromatic solvents, ketone solvents, tetrahydrofuran, dioxane, halogenated aliphatics, etc; and sulfonated polyisobutylene–saturated aliphatic hydrocarbons, e.g., diisobutylene, aromatic and alkyl substituted, aromatic hydrocarbons, chlorinated hydrocarbons, n-butyl ether, etc.

The polar cosolvent is incorporated with the organic solvent and ionomeric polymer to solubilize the pendant ionomeric groups. The polar cosolvent has a solubility parameter of at least 10.0, more preferably at least 11.0, is water miscible and may comprise from 0.1 to 40, preferably 0.5 to 20, weight percent of the total mixture of organic solvent, ionomeric polymer, and polar cosolvent. The solvent system of polar cosolvent and organic solvent in which the neutralized sulfonated polymer is dissolved contains less than about 15 weight percent of the polar cosolvent, more preferably about 5%. The viscosity of the solvent system is less than about 1,000 cp more preferably less than about 100 cp and most preferably less than about 10 cp.

The polar cosolvent is selected from the group consisting essentially of water-soluble alcohols, amines, di- or trifunctional alcohols, amides, acetamides, phosphates, or lactones and mixtures thereof. Especially preferred polar cosolvents are aliphatic alcohols such as methanol, ethanol, n-propanol, isopropanol, 1,2-propanediol, monomethyl ether of ethylene glycol, and n-ethylformamide.

The preferred levels of water are not believed to be critical provided that levels of greater than 5 volume percent are employed (based on volume of hydrocarbon-alcohol solution).

Various materials may be added to the polymer solution to increase gel strength, if desired. For example, the addition of clays, fillers (calcium carbonate, asbestos, zinc oxide, mica, talc), carbon blacks and the like can be employed to strengthen the gels.

Example: This example illustrates how the polymer solution may be used to selectively plug water-bearing zones in a formation. A production well penetrates a 60-ft-thick oil-bearing zone at a depth of 10,000 ft. A 10-ft-thick high-permeability water-bearing zone lies below the oil-bearing zone. In order to reduce the ratio of water-to-oil being produced, the well is given a treatment wherein about 500 gal of a polymer solution comprising diesel oil, 3 weight percent sulfo EPDM and 5 weight percent methanol is pumped into the tubing string and displaced down the tubing string with diesel oil or water. The polymer solution viscosity should have a viscosity less than 20,000 cp at 23°C.

The tubing string should be open-ended and the end should be adjacent the water-bearing zone. When the polymer solution starts to exit the tubing string, the annular space between the tubing string and casing should be closed to prevent return flow up the annular space. Continued pumping of fluids down the tubing string will force the polymer solution into the permeable water-bearing zone. About 1 hr after the polymer solution is displaced into the formation, the residual water in the formation pore space invaded by the polymer solution will cause gelation of the polymer. The gelation will block the invaded pore spaces and greatly reduce the permeability of the water zone. The well may then be returned to production without significant production of water from the water zone.

Polyacrylamide/Dialdehyde Gel

L. Vio; U.S. Patent 4,155,405; May 22, 1979; assigned to Societe Nationale Elf Aquitaine (Production), France describes a process concerned with preventing inflow of water into shafts or wells which have already been drilled or which are in the course of being drilled into the ground, particularly drilling operations which are carried out for exploiting oil or gas deposits.

The process consists in the impregnation of the ground to be made impermeable with a dilute aqueous solution of nonhydrolyzed polyacrylamide to which is added a dialdehyde, characterized in that the polyacrylamide has a molecular weight of at least 2,000,000 and preferably from 4,000,000 to 8,000,000, and the pH of the solution being at least 6.5.

In accordance with a preferred feature of the process, the impermeabilization of the soil or ground is achieved in a time which is longer than that in the prior art, namely, in 24 hr or more; it is to be noted that this operation takes less than 1 hr in the prior art. (See Belgian Patent 757,161 and U.S. Patent 3,975,276.)

In a general manner, the dialdehydes which can be used are water-soluble and of the type $OHC{-}R_n{-}CHO$ in which R is a divalent hydrocarbon radical, capable of comprising 1 to 4 carbon atoms and preferably CH_2; n may be from 0 to 6 or better still from 0 to 3. Glyoxal is particularly recommended. When the dialdehyde is present in the form of an oligomer, as is the case, for example, in the trimer of glyoxal, it is practical to use such an oligomer.

As regards the polyacrylamide, it may be used in the form of a homopolymer of the amide $CH_2{=}CR{-}CONHR'$ in which R may be H or CH_3, R' being H, $-CH_3$, $-CH_2OH$, ethyl, propyl, isopropyl, butyl, isobutyl or possibly another alkyl, preferably having at most 12 carbon atoms. It is likewise possible to use copolymers of such an amide with another unsaturated compound, as, for example, styrene, vinyl acetate, alkyl acrylate or hydroxylalkyl acrylate, alkyl methacrylate or hydroxyalkyl methacrylate, acrylonitrile, acrolein or others.

The preparation of the solutions which are intended for the gelling within the earth or material into which they are injected comprises the dissolution in water of the selected polymer, the addition of the appropriate proportion of dialdehyde, the homogenization of the medium and the adjustment of the pH to the desired value. The water may be soft or salty, particularly seawater or deposit water, because the presence of salt does not impair the gelling operation. The adjustment of the pH can be effected by means of a mineral base, an amine, an alkali phosphate or carbonate or any other basic compound which does not react with the substances which are present. In addition, it may be expedient to add an inert loading agent to the solution.

In one particular embodiment of the process, the gelable aqueous solution is formed in situ with the groundwater to be treated. In this case, the mixture of polyacrylamide with dialdehyde is prepared in the form of a suspension in a nonaqueous liquid, particularly a water-free hydrocarbon.

For this purpose, the adequate quantities of polyacrylamide, dialdehyde and a basic agent for the adjustment of the pH are dispersed in the liquid. This suspension, to which may possibly be added an inert filler, is injected into the ground which is to be made impermeable, and the particles of its components react with the water which is present in order to form a sealing gel. In this embodiment, it is highly practical to use the dialdehyde in the form of a solid oligomer which can be more easily dispersed in the nonaqueous liquid. Consequently, the trimer of glyoxal is particularly useful for this purpose.

As a general rule, for molar ratios of dialdehyde/amide higher than 0.5, the gelling procedure is more rapid and more complete, while it is slower when this ratio is lower than 0.5. With ratios lower than 0.2, the gelling procedure takes place very slowly and several days are necessary in order to obtain the gel.

Crosslinked Hydratable Polysaccharide

J.W. Ely; U.S. Patent 4,133,383; January 9, 1979; assigned to Halliburton Company describes a method of terminating the flow of formation fluids from uncontrolled wells, which comprises introducing a low viscosity fluid into the formation by way of at least one separate well penetrating the formation, the fluid having the property of becoming highly viscous after it has reached and flowed into the formation. The introduction of the fluid is continued until the fluid surrounds the portion of the formation adjacent the uncontrolled well and becomes highly viscous therein thereby terminating the flow of formation fluids to the uncontrolled well.

More specifically, the retarded gelling agent which is preferred for use is a hydratable polysaccharide crosslinked with a compound selected from the group consisting of dialdehydes having the general formula $OHC(CH_2)_nCHO$, wherein n is an integer within the range of 0 to about 3; 2-hydroxyadipaldehyde; dimethylol urea; water-soluble urea formaldehyde resins; water-soluble melamine formaldehyde resins; and mixtures of the foregoing compounds.

Preferred crosslinking compounds for forming the retarded gelling agent are dialdehydes wherein n is an integer within the range of 1 to about 3.

Examples of such dialdehydes are glyoxal, malonic dialdehyde, succinic dialdehyde and glutardialdehyde. Of these glyoxal is preferred.

Polysaccharides which are suitable for forming the retarded gelling agent are hydratable polysaccharides having a molecular weight of at least about 100,000 and preferably within the range of about 200,000 to about 3,000,000. Examples of such polysaccharides are hydratable galactomannan gums and derivatives thereof, hydratable glucomannan gums and derivatives thereof and hydratable cellulose derivatives. Of these guar gum, locust bean gum, karaya gum, hydroxypropyl guar gum and carboxymethylcellulose are preferred. The most preferred polysaccharide is hydroxyethylcellulose having an ethylene oxide substitution within the range of from 1.3 to about 3 mols of ethylene oxide per anhydroglucose unit.

When the hydratable polysaccharides are crosslinked with the abovementioned crosslinking compounds at concentrations within the range of 0.05 to about 100 parts by weight crosslinking compound per 100 parts by weight polysaccharide, the resulting retarded gelling agent is substantially insoluble in an aqueous fluid having a pH of less than about 7 at a temperature of less than about 100°F.

The pH of the aqueous fluid is adjusted to provide a rate of hydration or gelation time of desired duration whereby the aqueous fluid remains at a low viscosity for a period of time to allow penetration into the formation, etc. Preferably a water-soluble acid is added to the fluid to adjust the pH thereof to the desired level. Acids such as fumaric acid and sodium dihydrogen phosphate are preferred because of their buffering qualities.

The pH of the aqueous fluid is also a factor if the aqueous fluid is heated to temperatures above about 300°F. At such temperatures, the acid in combination with the high temperature rapidly degrades the hydrated polysaccharides and reduces the viscosity of the aqueous fluid. In order to overcome this problem, an encapsulated base can be included in the aqueous fluid which is released at about 300°F, thereby offsetting the degrading effect of the acid. A preferred encapsulated base is sodium bicarbonate coated with a paraffin having a melting point within the range of 150° to about 300°F.

The hydrated polysaccharides are most stable at high temperatures when a small amount of a stabilizing agent such as a water-soluble alcohol is present in the aqueous fluid. Suitable alcohols are represented by the general formula

$$C_nH_{2n+1}OH$$

wherein n is an integer within the range of 1 to about 5, and preferably within the range of 1 to about 4.

The most preferred alcohols are those falling within the above formula wherein n is an integer within the range of 1 to about 3.

The most preferred fluid having the properties mentioned above for use in accordance with the method is an aqueous fluid having a pH of less than 7 and having a retarded gelling agent mixed therewith comprised of hydroxyethylcellulose having an ethylene oxide substitution within the range of about 1.3 to 3.0 mols of ethylene oxide per anhydroglucose unit crosslinked with glyoxal at a concentration within the range of 0.05 to about 100 parts by weight glyoxal per 100 parts by weight of the hydroxyethylcellulose, the retarded gelling agent being present in the aqueous fluid in an amount in the range of from 10 to about 500 lb/1,000 gal of the aqueous fluid.

For use in formations at temperatures above about 200°F, the aqueous fluid preferably also contains a water-soluble alcohol selected from the group consisting of methanol, ethanol, n-propanol and isopropanol present in an amount in the range of 1 to about 10 and preferably 2 to about 7 parts by volume alcohol per 100 parts by volume of aqueous fluid.

For use in formations at temperatures above about 300°F, the aqueous fluid can additionally contain a base, e.g., sodium bicarbonate, coated with a paraffin having a melting point within the range of about 150°F to about 300°F, present in an amount in the range of from 5 to about 50 lb of base per 1,000 gal of aqueous fluid.

Titanium Dioxide Plugging Fluid

G.R. Avdzhiev and L.M. Ruzin; U.S. Patent 4,217,146; August 12, 1980 describe a plugging solution containing a hydrocarbon liquid and a weighting material, the weighting material being titanium dioxide, the listed-above components being taken in the following amounts in percent by weight: hydrocarbon liquid, 60 to 80; and titanium dioxide, 20 to 40.

The plugging fluid can be prepared by a method known to those skilled in the art, which comprises the following steps:

(1) A hydrocarbon liquid required for the fluid is prepared by distilling light fractions off highly viscous crude oil if such are present until an oil product is obtained having a temperature at the onset of boiling of 200° to 220°C.

(2) Titanium dioxide required for the fluid is prepared by finely crushing the starting titanium dioxide of any quality to a particle size of no more than 0.1 or 0.5 μ.

(3) After preparation, the initial components are weighed to provide the desired ratio and thoroughly mixed to form a homogeneous suspension.

This plugging fluid makes for reliable sealing of the annular space of upward intake wells when injecting coolant thereinto, as well as for precluding the penetration of coolant from the wells into excavations and reducing the heat release into those excavations due to heat conductivity, which results in the preservation of an adequate atmosphere in the mine and in the reduction of ventilation costs.

In addition, when used for sealing the annular space of upward development wells, the plugging fluid does not wash out during the recovery of oil from the wells thanks to its high density and adequate viscosity.

This provides for reliable sealing and heat insulation of the annular space of upward development wells, as well as precludes the possibility of the release of large amounts of heat into excavations and the violation of the temperature and gas conditions of the atmosphere in the mine.

Example: 240 g of crude oil having a viscosity of 52.0×10^6 m^2/sec at 20°C and a density of 905 kg/m^3, and 60 g of pigmentary titanium dioxide (at 20% of the solid phase) were used for preparing the fluid. Oil and pigmentary titanium dioxide were thoroughly stirred for 10 min with heating up to 70°C.

The resulting suspension had a viscosity of 56.4×10^6 m^2/sec at 20°C and a density of 1,075 kg/m^3.

Apparatus for Plugging Holes in Ground Strata

K.E. Baughman and E.N. Doyle; U.S. Patent 4,191,254; March 4, 1980; describe an apparatus for plugging voids in a ground stratum. Such an apparatus is self-contained, not requiring other equipment or elaborate procedures for its use.

The process provides a self-contained, sealed cartridge with separate chambers for containing the precursors of a polymeric system. The cartridge is designed so that the precursors are kept separated until it is in the vicinity of voids in a ground stratum which cause loss of circulation. When the cartridge reaches the vicinity of the voids to be plugged, the pressure about it causes a breakdown of the structure separating the precursors, allowing the precursors to mix, and as the precursors mix, the resultant foam is released from the cartridge into the voids to be plugged where it solidifies.

Referring to Figures 5.1a and 5.1b, a plugging material transporting, mixing and releasing apparatus **26** comprises a cartridge **28** made of a flexible, and preferably water-soluble, substance so that it will yield or collapse in response to the increased pressure from outside the cartridge.

Figure 5.1: Apparatus for Plugging Holes in Ground Strata

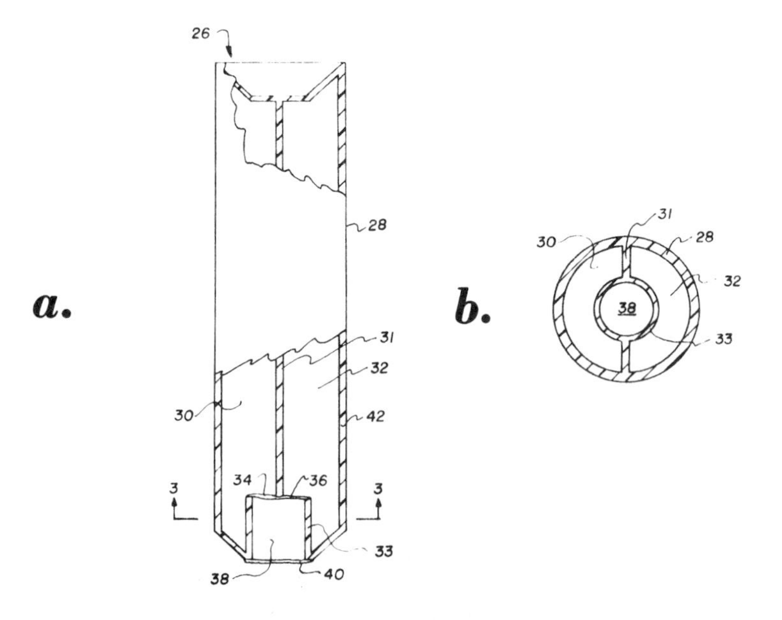

(a) Plugging material apparatus
(b) View of cartridge along line **3-3**

Source: U.S. Patent 4,191,254

The cartridge includes means for separating the precursors of a polymeric system such as storage chambers **30** and **32**, which are separated from each other by a wall or partition **31** extending through the center of cartridge **28**, a cylinder wall **33**, and membranes **34** and **36** located at the lower portion of cartridge **28**. Thus, when one precursor of a two-precursor polymeric system is placed in chamber **30**, the chamber in conjunction with membrane **34** completely encloses that precursor making it shelf-stable. Similarly the other precursor of a two-precursor polymeric system is enclosed in storage chamber **32** which is closed at the bottom by membrane **36**. The two precursors remain separate and stable until such time as the cartridge is to be used.

Membranes **34** and **36** are preferably made of a flexible material, such as a thin rubberlike material, which can be readily ruptured by the pressure of the material in chambers **30** and **32** when the well pressure causes collapse of cartridge **28**.

Cartridge **28** also includes a mixing chamber **38** formed by cylindrical wall **33**, membranes **34** and **36**, and a rupturable membrane **40** at the bottom of cartridge **28** for mixing the precursors of the polymeric system. Mixing chamber **38** is separated from the storage chambers by membranes **34** and **36** prior to their rupture. The outer wall **42** of cartridge **28** is made of a flexible material so that outside pressure on wall **42** will tend to collapse the cartridge and increase the pressure on membranes **34** and **36**. When sufficient pressure is applied to the outer wall of cartridge **28**, membranes **34** and **36** will burst and fill mixing chamber **38** with the two stored precursors which begin to mix. As they mix the pressure in the mixing chamber increases until membrane **40** is ruptured and the foaming substance flows out of the chamber.

Once the precursors of the polymeric system have begun mixing, they are forced out of the cartridge and into the well bore and the fissures to be plugged. Forcing the foam out of the cartridge can be accomplished in a number of ways such as including a blowing agent in one precursor, or mixing precursors which expand when mixing such as self-foaming precursors, or mixing components which expand when mixed with the water of the drilling mud or the formation. As used here, "precursor of a polymeric system" includes known blowing agents and catalysts. The foam can also be forced out of the cartridge with the force of the drilling mud.

The preferred method is to drop one or more cartridges which are cylindrically shaped and suitable for stacking down the drill string so that the cartridges will stack one on top of another at the bottom of the drill string.

The polymeric materials to be utilized in such cartridges may be any type of thermosetting polymer, such as polyurethane, epoxy, polyester, urea-formaldehyde, phenolic, etc. where two or more components must be intimately mixed together in order to form a polymer. The polymer may be used as a solid material in uses where the end product must be very strong, or the polymer may be a foamed polymer of any given density. Such formulations are well known in the art with all such thermosetting polymers.

Cartridges may also be made of a water-soluble material if desired, such as ethylene oxide polymer, or polyvinyl alcohol polymer, or they may be made of an oil-soluble polymer, such as styrene-maleic anhydride or other such polymer.

SAND CONSOLIDATION

Displacement of Rathole Fluids

Difficulties due to solids entrained in produced fluids have led to sand control methods which employ a variety of different synthetic resins for the consolidation of incompetent subterranean formations. These methods involve the injection of a liquid resin or a resin-forming material into the formation and permitting the resin to cure to an infusible state, thereby bonding the formation sand in place. Typically, a solvent is injected as a preflush into the formation to remove resident fluids and precondition the formation to be consolidated.

Often, the consolidation treatment is unsuccessful because of contamination of the resin during injection by fluids contained in the well bore. These contaminating fluids often exist in the casing rathole (i.e. that portion of the casing below a perforated zone adjacent the unconsolidated formation). Typical contaminating fluids include water-base completion fluids, seawater, and formation brine remaining from a well completion or workover operation.

During sand consolidation treatments, the resin being injected into the formation can displace a considerable amount of aqueous rathole contaminating fluid into the formation. This results in a continuous contamination of the resin during injection, resulting in poor consolidation. Moreover, a significant amount of resin accumulates in the rathole and is wasted.

C.M. Shaughnessy and W.M. Salathiel; U.S. Patent 4,137,971; February 6, 1979; assigned to Exxon Production Research Company describe an improved method of displacing unwanted rathole fluids. Briefly, the method involves pumping a dense nonaqueous liquid into a cased well bore to the level of the rathole fluid, followed by injection of a preflush solution and a sand consolidating resin into the formation. The dense nonaqueous fluid immiscibly displaces the aqueous rathole fluids which are removed from the formation in the vicinity of the well bore by the subsequent injection of the preflush solution. This effectively prevents contamination of the resin when consolidating the formation.

A dense nonaqueous liquid useful for displacing rathole fluids should be characterized by several properties. First, the displacing fluid should have a density greater than that of the rathole fluid or any of the subsequent consolidating fluids. This is so that once the aqueous rathole fluids are displaced, subsequent consolidating solutions will not in turn displace the nonaqueous displacing fluid. As a practical matter, the displacing fluid should have a minimum density of about 1.05 g/cc and preferably a density of about 1.10 g/cc or greater.

Second, the displacing fluid should be immiscible with the aqueous rathole fluid. This eliminates the problem of mixing between the incoming dense fluids and rathole fluids, and permits efficient gravity displacement from the top downwards. As a practical matter, this means that an expensive workover rig is not needed in order to introduce the displacing fluid into the well bore. The nonaqueous displacing fluid may be introduced simply by pumping the fluid into the casing at the surface, commonly referred to as a bullhead technique. Third, the displacing fluid should be miscible with the preflush solution, and preferably with the resin solution as well. This eliminates the problem of removing excess displacing fluid from the formation pore spaces. Any excess displacing fluid which enters the unconsolidated formation is removed from the formation in the vicinity of the well bore, along with the displaced aqueous rathole fluids, by the normal quantities of preflush and resin solution.

Certain other practical considerations enter into the choice of a suitable nonaqueous displacing fluid. First, to facilitate pumping of the displacing fluids and also to facilitate removal of excess displacing fluid from the formation, a suitable fluid should have a viscosity of less than 100 cp at formation temperatures, and preferably less than about 20 cp. This viscosity is similar to that of most other consolidation fluids. Second, the displacing fluid should be compatible with standard oil field and refining practices. Because halogenated hydrocarbons are known to be harmful to many common refinery catalysts, the use of such compounds is not preferred.

Suitable nonaqueous displacing fluids that exhibit the abovedescribed properties include o-nitrotoluene, carbon disulfide, dimethyl phthalate, nitrobenzene, and isoquinoline.

In a typical application of the process, the displacing fluid is pumped directly into the casing at the wellhead. As mentioned previously, the portion of the well bore from the perforations to the surface generally contains either an aqueous fluid, such as brine remaining from perforation operations, or produced fluids. Then when pumping operations are commenced, these fluids are forced into the formation.

The aqueous rathole fluids, however, will remain in place until the dense nonaqueous fluid reaches the level of the lowest perforations. The dense fluid flows down to the formation perforations where displacement of the aqueous rathole fluids takes place. Preferably, the displacing fluid is continuously pumped during the displacement process. This confines the displacement process to the rathole and also permits displacement to take place more rapidly, since the aqueous rathole fluids are dynamically forced into the formation through the perforations.

Typically, 10 to about 12 bbl of nonaqueous displacing fluid are required to fill a typical rathole of about 50 ft. Generally, adequate displacement rates are achieved by pumping at standard rates, i.e., about ½ bbl/min. Under these conditions, the rathole can be filled with the nonaqueous fluids and the aqueous rathole fluids displaced in about 20 min.

Once the dense nonaqueous fluid has been pumped and fills the rathole, a suitable preflush solution may be pumped into the casing at the wellhead. Since all water is now confined to the formation, injection of the preflush solution tends to displace all water (which would later contaminate the resin solution if left in place) radially outwardly from the formation volume to be consolidated. Moreover, by choosing a nonaqueous displacing fluid that is miscible with the preflush solution, any excess dense nonaqueous fluid is similarly displaced outwardly during injection of the preflush solution.

Preferably, the preflush solution whould have a density which is less than both the nonaqueous displacing and the aqueous rathole fluid. This facilitates displacement of both fluids by the preflush solution radially outwardly into the formation.

After injection of the preflush solution, a proper volume of resin solution is injected into the formation followed by the curing agent or catalyst dissolved in a suitable solvent. Normally, from 50 to about 75 gal of resin solution per foot of interval to be treated are used.

Since the nonaqueous fluid filling the rathole has a density greater than that of the resin solution, waste of the resin solution is limited to the small amount which may be located between the lower perforations and the top of the displacing fluids in the rathole. This amount will be significantly less than the amount lost when the resin solution is able to displace the rathole fluid. For example, the amount lost in the rathole when aqueous fluids are not displaced may be as much as 50 gal.

Injection of the solution containing the curing agent or catalyst displaces the resin from the pore spaces of the formation and imparts the desired permeability to the invaded pore spaces. The curing agent or catalyst is extracted by the resin solution that remains in contact with the sand grains. This causes the resin to harden, bonding the sand grains together to consolidate the formation.

Example: The following field test illustrates a specific procedure for performing the method. A well completed in an 8 ft interval at about 8,900 ft was not capable of sustaining production for long periods of time because of sand problems. A 1" work string was placed in the well and brine circulated therethrough to remove sand from the well bore. A slurry of 10 to 15 mesh sand and brine was injected through the perforations and pressure-packed in place. Diesel oil was then used to displace brine above the lowermost perforations. Following an injectivity test to insure that the perforations were open, pumping operations were commenced. The pumping schedule was as follows:

	Quantity	Type	Supplier
Acid solution, gal			
15% HCl	300	15% HCl	–
Mud acid	750	3% HF-12% HCl	–
Neutralizer	1,000	H_2O, NH_4Cl, NH_4HCO_3	–
Nonaqueous displacing fluid, gal			
Nitrotoluene	300	o-Nitrotoluene	Du Pont
Preflush solvent			
Ethylene glycol/isopropyl ether, gal	800	Solvent AC	Union Carbide
Thickening agent, lb	32	PVP(K-90)	GAF Corp.
Coupling agent, gal	4	Z-6020	Dow Corning
Resin solution, gal			
Epoxy resin	293	XD-7818	Dow Chemical
Acetone/cyclohexane	149	–	–
Spacer fluid, gal			
Refined oil	160	Mentor 28	Exxon Co.
Refined oil	80	Flexon 766	Exxon Co.
Catalyst solution, gal			
Refined oil	1,340	Mentor 28	Exxon Co.
Refined oil	660	Flexon 766	Exxon Co.
Catalyst (tertiary amine)	60	DMP-30	Rohm and Haas

The various liquids were injected through the work string and into the formation in the above sequence at a rate of about 20 gal/min. The final solution was displaced from the workover string with diesel oil and the well was shut in to permit the resin to cure. When the well was returned to production, produced fluids were sand free indicating initial success of the treatment.

Gelled Water/Epoxy Resin Composition

J.R. Murphey; U.S. Patents 4,216,829; August 12, 1980; and 4,199,484; April 22, 1980; both assigned to Halliburton Company describes a method for forming a consolidated pack or a gravel pack of sand coated with a preferred epoxy resin composition. For the gravel pack, the sand is suspended in a viscous gelled water and coated with the epoxy resin composition in the presence of the gelled water which is then used to transport and place the coated sand in the desired location and to hold the coated sand in place while the resin becomes tacky and consolidates into a porous permeable high-strength consolidated pack.

Setting of the epoxy resin composition can be controlled by accelerators or retarders to give the desired working and consolidation time. Preferably the aqueous gel is broken after, or as close to as possible, the placement of the coated sand at the desired location. In some cases this may be just before placement. This assures intimate contact of the coated sand particles and allows the aqueous fluid to separate from the sand to enter the formation or return to the surface through the well as desired.

The preferred neutral polymers or agents which are suitable for gelling the aqueous fluids are neutral polysaccharide polymers having a molecular weight in the range of about 100,000 to 2,000,000 and preferably about 500,000 to 1,200,000. These gelling agents should have a degree of substitution to produce the desired water solubility and gelling effect to produce a gelled aqueous fluid having a viscosity of at least 30 cp and preferably in the range of about 60 to 180 cp.

Particularly preferred gelling agents include the classes of ethylene oxide substituted hydroxyethylcellulose polymers having a degree of substitution in the range of about 1.5 to 3.5. A polymer must be soluble in an aqueous phase to some extent in order to produce the desired gelling effect. The amount of polymer, the degree of substitution and the particular polymer selected will vary with the aqueous fluid, the application parameters and the other additives selected.

Likewise, the optimum blend of surfactants for adsorption and coating of the epoxy resin onto silica surfaces will vary with the application parameter such as pH, temperature and other components present. However, the preferred surfactants are primarily linear having a molecular weight of about 160 to 600 and containing alkyl, aryl, hetero and unsaturated groups of radicals wherein each group or radical contains about 2 to 12 carbon atoms.

The amine quaternary-type cationic surfactants are preferred for the major portion of the surfactant blend used to control adsorption and coating of the epoxy resin. The noncationic surfactant can also be an amine-type surfactant but should be anionic or nonionic in overall character. It should be water miscible at ambient temperatures. This blend of cationic with a minor portion of noncationic surfactant can be varied depending upon the gelling polymer and the aqueous fluid selected. The aqueous fluid can be selected from distilled water, tap water, brine or seawater according to availability. Each of the components should be tested with the entire system and the parameter selected for compatibility.

For gravel packs, the sand particles used should be carefully graded as to size with all interfering contaminants removed. The preferred size would be in the range of about 10 to 60 U.S. Mesh series with the preferred sizes being 10 to 20, or 40 to 60, which is selected to closely match the formation particle size and distribution next to which the pack is to be placed.

The epoxy resin composition contains an epoxy resin with a hardening agent. The preferred class of epoxy resin is a polyepoxide type with an amine-type hardener. The amine hardener can also serve as a surfactant or even a cationic surfactant which aids in coating epoxy onto the silica surfaces. However, this is generally not the case. Optional components include diluents, retarders or accelerators. The preferred diluents are hydrocarbons or substituted hydrocarbons in which the epoxy resin is soluble and which are at least partially soluble in water.

A preferred class of diluents are the esters such as ethyl acetate, methyl formate, ethyl glycol acetate and other normally liquid low molecular weight esters and ethers including the ethers of ethylene glycol. Esters containing alkyl or aryl groups having about 2 to 18 carbon atoms per radical are preferred. For high-temperature applications, the substituted esters having higher boiling points and slower rates of hydrolysis are preferred. The retarders used in the epoxy composition are acid or acid-producing compositions, some of which can also serve as diluents.

The retarders should also be at least partially soluble in water and soluble in the epoxy resin composition. The retarder should be selected to produce the desired rate of hydrolysis or acid production according to the temperature and working times desired. The low molecular weight organic acid retarders are preferably produced by hydrolysis of an organic compound having a hydrolysis constant K of about 4×10^{-5}. The acid produced by hydrolysis should have at least two carbon atoms such as 2 to 5 carbon atoms and preferably 2 to 3 carbon atoms. The other half of the organic compound, (i.e., ester or ether) should not interfere with the coating or setting of the resin and preferably acts as a diluent for the resin. The other half of the hydrolysis product can have up to 18 carbon atoms.

The accelerators used with the composition are weak organic acids with additional water-soluble components. These water-soluble compounds can be low molecular weight inorganic or organic salts containing about 2 to 18 carbon atoms which are water-soluble and readily dispersed in the gelled aqueous fluid and are also at least partially soluble in the epoxy resin composition.

Examples of the diluents include ethyl acetate, ethylene glycol monoalkyl ether (C_{1-4}), acetone and C_{2-4} ketones. Examples of the hardening agents include most conventional amines, polyamines, amides and polyamides known to those skilled in the art. Examples of the retarders include methyl and ethyl esters of low molecular weight alkyl acids (C_{2-3}) and the esters of the above diluents. Examples of the accelerators include salicylic, hydroxybenzoic, citric, fumaric, oxalic and maleic acids.

An essential component of the overall composition and process is the gel breaker. The gel breaker can be selected from three classes depending upon the application temperature, working time, pH concentration limits and contaminants which might be encountered. These three classes include (1) an enzyme-type breaker such as cellulase for substituted cellulose gelling agent, (2) a low molecular weight organic hydroperoxide such as tert-butyl hydroperoxide or an alkyl hydroperoxide containing about 2 to 18 carbon atoms and a surfactant including both tertiary and quaternary amine, or (3) a combination of an organic hydroperoxide with a cupric ion supplying salt for low-temperature applications and/or surfactant containing both tertiary and quaternary amines.

A particularly preferred composition uses an aqueous HEC (hydroxyethylcellulose) gel and a polyepoxide resin composition which can be used over a temperature range of about 100° to 170°F for a practical application temperature or slightly higher than about 220°F. The preferred aqueous gels as described herein are formulated to produce a working time or to break the aqueous gel within a period of about 1½ to 2 hr. The preferred catalyst or epoxide hardener for use with the preferred epoxy and other additives is methylenedianiline. Other hardeners can also be used as described herein. An accelerator such as salicylic acid is preferred with the methylenedianiline at temperatures of about 140°F and lower.

Amino functional silanes and surfactants are also preferred to promote resin adsorption and coating onto the sand in the presence of the HEC gel. For retarding the curing of the epoxy resin, acetic acid can be used at higher temperatures. Ethyl acetate and higher boiling esters such as ethyl glycol diacetate can also be used. Additional esters which can be used include the methyl esters of acrylic and fumaric acid and similar strength organic acids which have some water solubility and some solubility in the epoxy resin composition. Ethyl glycol diacetate can be used to prolong working time of the methylenedianiline-catalyzed resin at temperatures of about 170°F and above since the diacetate has a boiling point of over about 300°F, while the ethyl acetate ester has a boiling point of about 160°F.

The preferred aqueous fluid for dispersing the sand and epoxy can contain from about 2 to 10% and preferably 5 to 10% of an alkali or alkaline earth salt such as the halides and an ammonium halide such as ammonium chloride. In addition, seawater can be used if care is taken to avoid calcium precipitation with some of the components of the system.

Salts which can be present in the aqueous gel include sodium chloride, calcium chloride, potassium chloride, calcium bromide, ammonium chloride and buffering agents such as fumaric acid and ammonium carbonate. The aqueous gel is preferably buffered so that the pH is in the range of about 6 to 7 for greater predictability of the gel breaking time and epoxy resin hardening time.

The cationic surfactants and the surfactant blend should be present in the aqueous fluid in the amount of up to about 1½% or at least a quantity sufficient to coat the silica surfaces of the sand particles and saturate the aqueous gel. Minimum amount of cationic surfactant in the aqueous fluid should be about 1.8 to 2.5 g of cationic surfactant per 1,800 g of 40 to 60 mesh sand.

The surfactant blend should contain at least about 20 to 25% cationic surfactant and up to about 60 to 75% cationic surfactant or have a preferred ratio of cationic to noncationic surfactant of about 2 to 3.

After the sand particles and epoxy resin composition are thoroughly mixed with the gelled aqueous fluid the breaker composition is added. The total mixture is pumped into the desired location either behind perforations or in a cavity or void space adjacent the well bore and formation to be consolidated. The gel breaking mechanisms should be timed according to the working time and temperature requirements so that the aqueous gel breaks immediately after reaching the desired location where the resin coated sand is to be deposited. This breaking of the aqueous gel immediately after placement aids in separating the aqueous fluid from the resin-coated sand thereby permitting easy removal of the aqueous fluid either through or around tubing in the well or by forcing the aqueous fluid into the formation.

Breaking of the aqueous gel immediately after placement also permits the resin-coated sand particles to pack in close proximity and tightly against each other in the formation to form a stronger permeable porous pack. Setting of the epoxy resin becomes tacky after the gravel or sand particles are placed in the desired location and forced together in a tight pack so that the coated sand grains will have maximum contact when the resin becomes tacky and begins to set thereby producing a stronger permeable porous pack.

Enediol Breaker Component for Epoxy Resin Gravel Pack

A description of the art in the field of consolidated gravel packing is found in U.S. Patents 4,074,760 and 4,101,474 by C.T. Copeland and V.G. Constien.

The process of Copeland and Constien represents a significant breakthrough in the art of consolidated gravel packing whereby consolidated gravel packs of a quality meeting high commercial standards could for the first time be formed from slurries having an aqueous carrier fluid.

In an improvement on this process of Copeland and Constien, D.R. Carpenter and Constien discovered that polyvalent metal cation contamination of a slurry such as that disclosed by Copeland and Constien could be controlled by use of a chelating agent. The process of Carpenter and Constien is the subject of U.S. Patent 4,081,030.

In each of the foregoing processes, it was shown that the aqueous carrier fluid could include a viscosity builder, i.e., a gelling agent, to improve the solids carrying capacity of the slurry. It was also shown that the fluid could contain a breaker for the gelling agent. Enzyme breakers and ammonium persulfate were specifically mentioned.

The breakers contemplated at that time, however, have not proved particularly satisfactory. Although such slurries are operable, results with such breakers are somewhat erratic. Frequently, the slurry sustains relatively high resin loss and the resulting compressive strength of the consolidated pack is considerably reduced. Consequently, breakers had not been regularly employed in the commercial practice of the processes of Copeland et al and Carpenter et al.

V.G. Constien and P.E. Clark; U.S. Patent 4,220,566; September 2, 1980; assigned to The Dow Chemical Company describe a slurry suitable for use in emplacing a permeably consolidated particulate mass in communication with a permeable subterranean formation, of the type containing an epoxy resin, a particulate material, a coupling agent, a quaternary ammonium halide surfactant, an aqueous carrier fluid containing a viscosity builder, and a breaker to reduce the viscosity of the carrier fluid after a period of time. The improvement in the slurry lies in the use of a compound having an aromatic or α-carbonyl enediol functional group as the breaker component of the slurry.

Suitable quaternary ammonium halides correspond to the formula:

$$\left[\begin{array}{c} R_1 \\ | \\ R_4-N-R_2 \\ | \\ R_3 \end{array}\right]^+ X^-$$

R_1 and R_3 are each independently a lower alkyl, a hydroxy substituted lower alkyl, or a polyoxyethylene alkyl moiety of the formula $-(CH_2CH_2O)_nH$ where n is 2 or 3. By lower alkyl is meant a straight chain or branched 1- through 4-carbon moiety. Preferably, R_1 and R_3 each contain at least one hydroxyl substitution. R_2 is (a) an 8- through 18-carbon moiety which may be saturated or unsaturated, branched or straight, but not cyclic, (b) hydroxy substituted lower alkyl, or (c) a polyoxyethylene alkyl moiety of the formula $-(CH_2CH_2)_nH$ where n is 2 or 3. R_4 is an aliphatic hydrocarbon moiety which may be branched or straight, saturated or unsaturated, or, R_4 may be an alkylaryl moiety. R_4 in either case has from 7 to 19 carbon atoms. Preferably, R_4 is an alkylaryl moiety, most preferably benzyl or alkyl substituted benzyl, e.g., dodecylbenzyl. A particularly preferred subgenus of surfactant are alkyl (C_{8-18}) bis(2-hydroxyethyl)benzylammonium chlorides and mixtures thereof. Another preferred subgenus includes compounds corresponding to the formula:

$$\left[C_{12}H_{25}-C_6H_4-CH_2-N\begin{array}{l} (CH_3)_{3-n} \\ (CH_2CH_2OH)_n \end{array}\right]^+ Cl^-$$

where n is 1, 2, or 3.

Except for the breaker component, one skilled in the art will be able to carry out the process by referring to the patent of Copeland et al and Carpenter et al. However, further developments within the scope of the process there disclosed have been made. In particular, it was shown that an accelerator such as salicylic acid, benzoic acid, phenol, etc., could be employed. The particular accelerator most preferred is dimethylaminomethyl phenol. Although not essential, it is normally used where the well has a bottom-hole static temperature of about 130°F or less.

Also, based on a discovery by Constien and J.R. Colemen, it is contemplated that the coupling agent be prereacted with a stoichiometric excess of the curing agent to form a mixture of the curing agent and an adduct having one moiety which promotes adhesion to the particulate and another moiety which reacts with the epoxy resin. Thus, for example, 54 g of 3-glycidoxypropyltrimethoxysilane may be reacted with 150 g of polymethylene polyphenylamine in a nitrogen atmosphere at 70°C until substantially all of the epoxysilane is reacted as can be determined by chromatographic monitoring of the reaction, e.g., about 5 hr. The reaction product may be diluted with a suitable solvent such as ethylene glycol ethyl ether if desired.

The enediol employed as a breaker may be a compound containing an

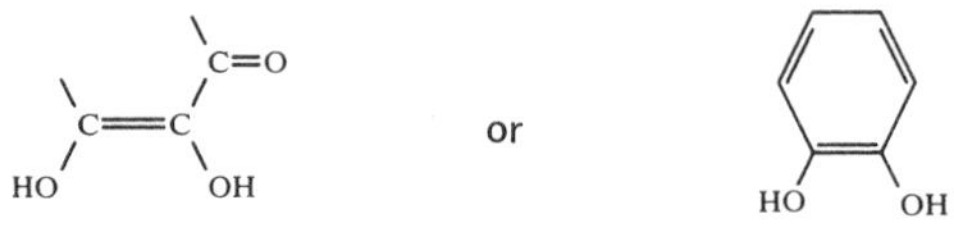

C=O
C=C
HO OH

or

HO OH

group (cis- or trans-), which is sufficiently soluble in the aqueous medium so that an amount effective to break the gel can be dissolved in the fluid at the temperature of the gel. Suitable compounds include, for example, D- and L-ascorbic acid, dihydroxymaleic acid, catechol, catechol derivatives wherein the enediol functional group remains underivatized such as 1,2-dihydroxytoluene, rhodizonic acid, 6-desoxy-L-ascorbic acid, L-ascorbic acid 6-palmitate, 2,3-dihydroxyacrylaldehyde (i.e., reductone), reductic acid, croconic acid, and the like, all of which are known compounds prepared by known methods. Soluble salts of the aforementioned compounds may also be employed.

Typically, the concentration of the breaker by weight of water is from about 0.005 to 0.05%. A preferred range, particularly where ascorbic acid is employed, is about 0.01 to 0.025%, and most preferably about 0.012% (1 lb/1,000 gal). When the breaker is incorporated in the slurry itself and no gelled pad fluid is employed, the above quantities are based on the carrier fluid portion of the slurry. Preferably, however, the breaker is incorporated in a gelled pad fluid which is injected either ahead of, behind, or preferably both ahead of and behind, the slurry. In the latter embodiment, the foregoing quantities of breaker are based on the total amount of water in the gelled pad as well as in the carrier fluid portion of the slurry.

In addition, there may optionally be included a slightly basic pH buffer, such as sodium bicarbonate, potassium dihydrogen phosphate/disodium hydrogen phosphate mixtures, ammonia/ammonium chloride, and the like. By controlling the pH immediately after addition of the breaker to from about pH 7 to a pH approaching the pK of the second ionizable hydrogen of the enediol group, the breaking action can be delayed as desired. Once the gel begins to break acid is generated and the generated acid lowers the pH thereby promoting an increasingly rapid rate of break thereafter.

For breakers of this type to function properly, there must be present a small amount of dissolved oxygen, e.g., to the extent of at least about 1 ppm. However, this is not a significant limitation as most water sources inherently have sufficient dissolved oxygen.

Example: An experimental field test was carried out under the supervision of laboratory personnel in a well having a bottom-hole static temperature of 105°F. The well had been producing a very viscous oil which carried sand out of the formation. The procedure tested was as follows:

(a) 1 gal of γ-aminopropyltriethoxysilane and 11 gal of a

solution of, by weight, 80% D.E.R. 331 brand epoxy resin of the bisphenol A/epichlorohydrin type having an average epoxide weight of about 190, 12% xylene, and 8% ethylene glycol ethyl ether were admixed about 6 hr before final preparation of the slurry.

(b) 12 bbl of a kerosene/aromatic solvent preflush were injected to clean the well bore.

(c) 1.5 gal of the alkylbis(2-hydroxyethyl)benzylammonium chloride surfactant blend described in Example 6 of Copeland et al (Surfactant Blend) were added to 10 bbl of 1% KCl brine and injected as a spacer.

(d) Fluid for a pad and for use as the carrier fluid in the slurry was prepared by admixing 33.3 lb of hydroxyethylcellulose in 14.5 bbl of 1% KCl brine and then adding 17.6 lb of Na_4EDTA. The fluid was then split into two portions, one of 12 bbl for use as a pad fluid and one of 2.5 bbl for use as the carrying fluid portion of the slurry. 2.5 gal of Surfactant Blend were added to the 12 gal pad fluid, and 1 gal of Surfactant Blend was added to the 2.5 bbl carrying fluid aliquot.

(e) The resin-coupling agent mixture prepared in step (a) was admixed with 6 gal of a 40 weight percent solution of methylenedianiline curing agent in ethylene glycol ethyl ether and 0.5 gal of dimethylaminomethyl phenol accelerator.

(f) 1,900 lb of 20 to 40 mesh sand was added to the 2.5 bbl of carrying fluid prepared in step (d).

(g) The resin system of step (e) and the carrier-sand mixture of step (f) were admixed for 15 min to form the slurry (about 5 bbl total volume).

(h) 0.5 lb ascorbic acid was dissolved in about a ½ gal of water, and this was added to the 12 bbl of pad fluid.

(i) 8 bbl of the pad fluid were injected, followed by the slurry of step (g), followed by the remaining 4 bbl of pad fluid. The foregoing sequence of fluids was displaced with 1% KCl brine. The well was shut in for 24 hr to permit the slurry to consolidate. The excess hardened slurry was drilled out and production resumed with virtually no sand production.

Urethane Prepolymers

E.W. Janssen and J.H. Simpson; U.S. Patent 4,139,676; February 13, 1979; assigned to Minnesota Mining and Manufacturing Company describe a method for consolidating superficial aggregate material comprising the steps of:

(a) Contacting the top surface of the aggregate with a fluid agent consisting essentially of a water-insoluble, moisture-curable NCO-terminated prepolymer which, after curing in the presence of moisture, has unfilled film properties as follows: a tear strength of at least 50 pounds per lineal inch, an elongation of at least 60%, a tensile strength of at least 50 psi at 25% elongation, a tensile strength of at least 75 psi at 50% elongation, and an ultimate tensile strength of at least 500 psi.

(b) curing the prepolymer in the presence of sufficient moisture to form a substantially nonfoamed polymer binding the aggregate together.

The fluid agent (i.e., prepolymer and liquid vehicle, if desired) used has the advantage of being a one-part system, requiring no expensive or elaborate mixing equipment for preparation thereof or for its application to the aggregate. Although the presence of a catalyst is not required, one or more catalysts may be used, if desired, to obtain faster curing of the prepolymer.

Water-insoluble NCO-terminated prepolymers, sometimes referred to as urethane prepolymers, can be selected from those known in the art. In general, the NCO prepolymers are the reaction product of an equivalent excess of at least one organic polyisocyanate with one or more organic compounds having a plurality of hydroxy, thiol, or amine groups, the molar excess (molar ratio greater than one) being needed to obtain the isocyanate termination. The prepolymers generally have an average molecular weight ranging from 400 to about 10,000, and preferably from 600 to 3,000. The equivalent ratio of isocyanate moiety, –NCO, to active hydrogen will be at least 2/1, and preferably at least 2.1/1 to 2.5/1, and can be as high as 4/1 or higher.

Polyisocyanates which can be used to prepare the isocyanate-terminated prepolymers described above include conventional aliphatic and aromatic polyisocyanates. The preferred polyisocyanates to be used will be aromatic because the prepolymers made therefrom will generally react faster with moisture.

The preferred class of water-insoluble prepolymers are those of the formula:

$$Y_1\left[\left(\underset{\displaystyle CH_3}{\overset{|}{C}}H{-}CH_2{-}O\right)_n\overset{\displaystyle O}{\overset{\|}{C}}NH{-}R'{-}(NCO)_p\right]_z$$

where Y_1 is an active hydrogen-free residue of a compound having a plurality of active hydrogen atoms, e.g., propylene glycol or propylenediamine; R' is tolylene or p,p'-diphenylmethane, n is the number of oxypropylene units shown, p is 1 to 2 and z is equal to the functionality, e.g., 2 or 3, of the compound from which Y_1 is derived.

The prepolymer is preferably first dissolved in a solvent at a concentration of about 25 to 90 weight percent so as to adjust the viscosity of the material into the range of about 5 to 200 cp.

Solutions more viscous than this do not percolate into superficial aggregate as readily as desired.

Conventional catalysts, ultraviolet light stabilizers and antioxidants, and colorants may also be added to the solution in minor amounts if desired. Upon application of the fluid agent (i.e., prepolymer with or without solvent, catalysts, and other additives) to the superficial aggregate material to be consolidated, the prepolymer reacts with moisture present in such aggregate (or absorbed from the air) to form a substantially nonfoamed polymeric matrix adhesively binding the aggregate into a unitary integral mass.

The polymeric matrix is aesthetically pleasing, water-permeable, water-insoluble and tough. Since the matrix will deform and yield, (i.e., is not brittle) it will not fracture easily. Consequently, the matrix is resilient and extremely durable. The matrix also exhibits good compressive strength even when soaked with water.

Generally, the amount of prepolymer used is in the range of 0.5 to 20 weight percent or more, preferably in the range of 2 to 10 weight percent, based on the weight of the aggregate.

Calcium Hydroxide and Calcium Salt

J.H. Park; U.S. Patent 4,232,740; November 11, 1980; assigned to Texaco Development Corp. describes a process by means of which a solid mass with appreciable mechanical strength and sufficient permeability to permit flow of fluids therethrough may be formed in a portion of the formation adjacent to a well bore penetrating the subterranean earth formation. The permeable mass restrains the undesirable flow of particulate formation material such as sand into a production well during oil recovery from the well.

The process involves contacting unconsolidated aggregates such as sand or gravel with an aqueous solution which is saturated with calcium hydroxide, and also contains from 0.1 to 10.0 and preferably from 1.0 to 5.0% by weight of calcium chloride or other calcium salt having solubility in water greater than the solubility of calcium hydroxide, and further contains from 0.1 to 10.0 and preferably from 1.0 to 5.0% by weight of an alkalinity agent, preferably sodium hydroxide, the alkalinity agent having solubility in water considerably greater than the solubility of calcium hydroxide. Other soluble calcium salts which may be employed are calcium acetate, calcium ferrocyanide, calcium formate, calcium isobutyrate, calcium ethyl methyl acetate, calcium nitrate, calcium propionate, calcium thiocyanate, and mixtures thereof.

Other alkalinity agents are potassium hydroxide, lithium hydroxide and ammonium hydroxide. As calcium hydroxide is removed from the solution as the result of the cementing reaction, the more soluble salt, e.g., calcium chloride and the alkalinity agent, e.g., sodium hydroxide react to form additional calcium hydroxide, thereby maintaining the concentration of calcium hydroxide dissolved in the treating liquid at a level sufficient to maintain the cementing reaction.

The degree of improvement in the cementing reaction is surprisingly greater than one would expect from maintaining the calcium and hydroxyl levels constant. Apparently the presence of the more soluble calcium salt and or the alkalinity agent accelerates the cementing reaction considerably.

The sand grains which are cemented together may be sand which is naturally occurring in the formation being treated. In another embodiment, the well is enlarged in the zone where the sand-restraining barrier is to be formed, and sand of a preferred particle size and size range is introduced into the formation, and then the treating fluid is injected thereto to cement the sand grains together.

Example: A tar sand deposit approximately 75 ft thick is located under an overburden layer of 350 ft. An exploitation is planned using steam and caustic drive process. This process requires the establishment of a communication zone by fracturing and subsequent treatment to enlarge the fracture into a stable communication path, followed by injecting a mixture of 85% quality saturated steam and approximately 1% caustic soda into the communication path. In order to stabilize the well bore of the injection well, and especially to prevent sanding up of the production well, it is desired to apply sand control treatments to both the injection well and the production well.

It is determined that the sand in the portion of the formation where the sand control barrier is desired is unsuitable for use in forming the barrier for several reasons. The particle size of the sand is much too fine and the particle size range too great. Furthermore, the sand is contaminated with clay and viscous petroleum coating which is tenaciously adhering to the sand grains. Accordingly, it is desired to form a cavity adjacent to the injection well and production well in order to provide space for forming the sand barrier.

An injection string is inserted into the well and positioned so that the bottom of the string is approximately adjacent to the bottom of the interval to be treated. Steam is injected into the well for approximately 12 hr to raise the temperature thereof sufficiently to cause flow of petroleum, after which hot water containing 1% caustic is injected into the string to flow past the surface of the zone to be treated and back up to the surface of the earth. The hot water emulsifies the viscous petroleum present in the tar sand and carries both the emulsified bitumen from the tar sand and sand suspended therein to the surface of the earth. After approximately 2 hr of this treatment, a cavity extending an average of 4 ft into the tar sand from the center of the well is created, which is sufficient for this process.

A slurry is formulated by suspended 20 to 40 mesh frac sand in water and this slurry is pumped down the tubing. The injection rate is maintained at a level which ensures the slurry is being pumped into the cavity, but not so excessive that large amounts of the sand-water slurry are returned to the surface of the earth. Sand filters out of the slurry against the formation face, and as a consequence of this forms a pack of essentially clean frac sand against the formation face. As injection pressure rises, indicating the occurrence of sand in the bottom of the well, the tubing is raised during the course of forming the sand pack, in order to obtain uniform packing of the sand against the formation face, completely filling the cavity.

Air is injected slowly into the tubing to displace fluid from the well bore initially, and then to force water remaining from the slurry out of the pack and into the formation. The injection pressure is maintained at the minimum level which will cause approximately 200 scfh to be pumped into the injection string. This is continued for 6 hr, to thoroughly remove water from the pore spaces of the sand pack, and to accomplish some drying of the sand pack. The air temperature is increased by approximately 200° during the last 3 hr, in order to raise the temperature of the sand mass, to increase the rate of the cementing reaction.

Air injection is terminated and 30 bbl of a liquid solution comprising field brine having added thereto approximately 1% by weight calcium hydroxide or lime, plus approximately 5% by weight calcium chloride and 5% by weight sodium hydroxide is formulated and injected into the sand mass. This fluid is allowed to remain in the sand mass for 24 hr, during which time the first stage of cementing of the sand grains together is accomplished. Hot air is again injected into the tubing to displace the spent liquid out of the pore spaces of the sand mass, and also to reheat the sand prior to injection of the next treatment of the treating liquid which is injected to resaturate the sand pack. After the third quantity of treating liquid has remained in the pack for 24 hr, the well is put on production to determine whether sufficient cementing has occurred to prevent flow of sand into the well bore. Only a small amount of sand flows back into the well during a 24 hr test, which indicates that the cementing reaction is nearly complete. Hot air injection is again applied to the sand pack for 4 hr, and a fourth 30 bbl slug of treating liquid, identical to the previous slugs, is injected and allowed to remain in the sand mass for 24 hr.

This fluid is then displaced by injecting air, and it is determined that a stable sand control barrier has been formed adjacent to the well. Both the injection well and production well are treated in identical fashion, prior to initiating injection of steam and caustic for the steam emulsification drive viscous oil recovery process.

WELL CEMENTING

WELL CEMENTING COMPOSITIONS

Lignosulfonate Derivatives

The subterranean geological formations penetrated by well bores for production of petroleum and gas have been at increasingly greater depths, encountering in the process an increasingly rigorous environment including significantly higher temperatures. In addition, an increasing number of wells requiring cementing are also disposed in offshore salt water environments so that the cementing compositions must manifest a compatibility with, or tolerance to salt.

The cementing compositions are used particularly for sealing or cementing the annular space in a well bore between the casing of the well and the formation surrounding the casing. In practice, the cementing composition is incorporated in a slurry, using, desirably, and by way of illustration, where an offshore well is being cemented, seawater to form the slurry. The slurry is pumped down through the well casing, into the formation and up the outside of the casing to effect the requisite seal.

W.J. Detroit; U.S. Patent 4,219,471; August 26, 1980; assigned to American Can Company describes lignosulfonate derivatives and alkali metal salts thereof derived from sulfite waste liquor that has been subjected to alkaline oxidation, hydrolysis and partial desulfonation with subsequent resulfonation; and followed by one or more addition and double decomposition and reactions thereof; as well as the method of cementing subterranean geological formations penetrated by well bores utilizing these additives in well cementing compositions.

The lignosulfonates of the process are derived from alkaline oxidized, hydrolyzed, desulfonated and subsequently resulfonated lignosulfonates, wherein the resulfonated lignosulfonates have substituted therein, as the resulfonation units, moieties of the formula: $-(C_xH_{2x})SO_3H$, wherein x has a numerical value from 0 to 3, thus including sulfoalkyl groups of the formula $-(C_yH_{2y})SO_3H$, wherein y has a value of 1 to 3 as well as sulfonyl radicals $-(SO_3H)$; and the alkali metal salt derivatives thereof; the resulfonated lignosulfonate thus formed containing between about 1.50 and 15 wt % of total sulfur in combined organic sulfonic sulfonate form; the lignosulfonate prior to resulfonation having a relative molecular size, substantially, of from 1,000 to 20,000.

The addition and condensation products are secured by reaction of the foregoing lignosulfonate starting material with:

(a) a halocarboxylic acid, and more particularly, a carboxylic acid of the formula, $M_p(C_nH_{2n})COOH$, wherein M is a bromine or, preferably, a chlorine atom; p has a value of 1 to 3, and preferably 1; and n has a value of 1 to 7 inclusive, for example, 5-chlorocaproic acid, 3-bromobutyric acid, 2-chloropropionic acid, 4-bromocaprylic acid, trichloroacetic acid or preferably monochloroacetic acid; or indeed, the corresponding iodo- and fluoro;

(b) a halocarboxylic acid as aforesaid and sequentially a hydroxy-substituted γ or α-lactone containing from 4 to 6 carbon atoms and preferably a polyhydroxy-substituted lactone of the molecular formulas $C_4H_6O_4$, $C_5H_8O_5$ or $C_6H_{10}O_6$ or mixtures thereof;

(c) a lactone as aforesaid;

(d) a halocarboxylic acid as characterized hereinabove; and sequentially, the derivative of a sugar acid-containing spent sulfite liquor derived from subjecting spent sulfite liquor to reaction with an alkali metal, and preferably sodium cyanide to form the cyanohydrin of the available reducing sugars present therein, followed by hydrolysis thereof; or

(e) the foregoing derivative of the spent sulfite liquor.

The reaction of the resulfonated, oxidized lignosulfonates with a halocarboxylic acid is carried out in an aqueous alkaline medium having a pH preferably of at least 8. The alkaline medium is provided by an ammonium, alkali metal or alkaline earth metal base or mixtures thereof, including, for example, carbonates, bicarbonates, or where appropriate in view of the cation employed, amines. Preferred are the hydroxides of the cations sodium, lithium, potassium and calcium, as well as, but less desirably, strontium and barium. The reaction takes place at a temperature desirably of 20° to 110°C and at ambient pressure for a period of 2 to 6 hours, and most desirably at a temperature of about 80° to 100°C for about 5 hours. The parameters of time, temperature and pressure are not, however, narrowly critical.

In a significantly preferred embodiment however, the foregoing carboxyalkylated product (and most desirably the carboxymethylated derivative) secured is further reacted with one or more of the foregoing hydroxy-substituted lactones, and preferably glucono-δ-lactone, in an amount by weight of 5 to 20%, and preferably about 7.5% of the lactone to total lignosulfonate halocarboxylic reaction product to provide a further product, believed to be an addition reaction product of the lactone and the acid condensation reaction product of the previous step; a product, characterized by an ability to impart an even more significantly enhanced stability to temperature and pressure, predictability, salt tolerance and resistance to gelation to hydraulic cements combined with extension in the period in which settling of the cement occurs even at the extremely elevated temperatures found in well bores at depths of 10,000 or 16,000 feet, or more. The lactone is conveniently introduced into the reaction product mixture of lignosulfonate and halocarboxylic acid.

The hydroxylated lactones are similarly reacted with the lignosulfonate starting materials without the intermediate reaction with halocarboxylic acid in a further embodiment in which, however, the resulting product, while superior to the lignosulfonate starting material as a cement retarding agent is nevertheless less efficacious in securing predictable cement retardation with high salt tolerance than the chloro- or bromocarboxylic acid-lignosulfonate condensation product and significantly less than the acid condensation-lactone reaction product.

Whether added to the resulfonated lignosulfonate unreacted with halocarboxylic acid or introduced into the acid condensation reaction product mixture, the reaction goes to completion over a relatively abbreviated period of 0.5 to 1.5 hr and preferably about 0.5 hr, and is undertaken desirably at ambient pressure in a temperature range of from 75° to 100°C, and preferably 80° to 85°C.

Example: A commercially available product Marasperse CBO sometimes referred to as Opcolig A, (American Can Company) provides alkaline oxidized, hydrolyzed, partially desulfonated lignosulfonate material useful in the practice of this process. Marasperse CBO generally has the following typical analysis and physical characteristic features:

Typical Analysis (Moisture Free Basis)

pH, 3% solution	8.5-9.5
Total sulfur, %	1.0-1.5
Sulfate sulfur as S, %	0.1-0.25
Sulfite sulfur as S, %	0.05
Sulfonic sulfur as S, %	0.85-1.2
CaO, %	0.02-0.05
MgO, %	Trace-0.03
Na, %	6.3-7.5
Reducing sugars, %	0
OCH_3, %	12.4-13.0
Sodium lignosulfonate, %	99-99.5

Physical Characteristics

Usual form	Powder
Moisture content (max.), %	8.0
Color	Dark brown-black
Bulk density, lb/ft^3	43-47
Solubility in water, %	100
Solubility in oil and most organic solvents, %	0
Surface tension, 1% soln., dynes/cm	51.4

Resulfonated derivatives containing about 5.2 wt % of organic sulfonic sulfur (based on composition weight) are made in large scale preparations by the sulfomethylation of the foregoing oxidized, hydrolyzed, desulfonated lignosulfonates Marasperse CBO obtained by the process described in U.S. Patent 2,491,832.

The foregoing lignosulfonate starting material, Marasperse CBO liquor, was sulfomethylated according to the following procedure wherein the following components are employed in the concentrations recited:

	Solids (lb/100 lb finished product)	Liquid (gal, U.S. measure)	Liquid (~lb)	6,000 Gel Solids (lb/gal)	Batch (lb solids)
Marasperse CBO liquor	71.29	5,760	53,450	2.50	14,400
Sodium hydroxide	2.97	105	1,300	5.7	600
Formaldehyde, MW 30	6.68	105	3,650	3.33	1,350
Sodium bisulfite, MW 104	22.28	–	4,500	–	4,500
Total	103.22	6,430	62,900	3.24	20,850
Finished product	100	4,250	44,880	4.75	20,200

To the lignosulfonate starting material there were added NaOH to a pH of 10.5±0.2; 9% formaldehyde based on liquor and NaOH total solids; and slowly with agitation 30% of sodium bisulfite based on liquor and NaOH total solids. Agitation was continued for 30 minutes, and the reaction mixture cooked for 3 hr at 170°C (338°F), 110 psi.

To the sulfomethylated product so obtained was slowly added, with agitation, 10% by wt of chloroacetic acid based on sulfomethylated product solids at about 170°F. Agitation and heat were applied for a period of 20 minutes whereupon 11% NaOH based on sulfomethylated starting material solids was added. The reaction mixture was heated to 194°F (90°C) and reacted for a further period of 5 hr at this temperature.

Glucono-δ-lactone in a concentration of 14% based on sulfomethylated starting materials was added to the reaction product mixture and the reaction continued for a period of 30 minutes at 175° to 185°F (80° to 85°C).

The product was secured, to which NaOH was added sufficient to achieve a pH of 9.0 to 9.5. The product is spray-dried to a solids content by weight of 48 to 50% (and is very fluid when hot) to provide a powder or is prepared as a liquid by addition of water to provide a product having a 40% solids content.

Additive for Control of Setting Time

E.E. Bodor and J.T. Payton; U.S. Patent 4,223,733; September 23, 1980; assigned to Texaco Inc. describe a method of and composition for the retardation of the setting time of a cement-water slurry.

The method comprises cementing a zone in an oil well, penetrating a subterranean formation by injecting down the well and positioning therein opposite the zone to be cemented a hydraulic cement aqueous slurry composition comprising dry hydraulic cement, sufficient water to form a pumpable slurry and a cement retarding additive consisting of from about 0.2 to 1.0% by wt, based on the dry hydraulic cement component, of a zinc salt and from about 0.2 to about 1.0% by wt of a water-soluble ammonium, alkali or alkaline earth metal salt of an alkaryl sulfonic acid, as hereinafter defined, and allowing the composition thus positioned to set to a monolithic mass.

The oil well cement additive used in the method and cement composition is an admixture of a water-soluble or dispersible zinc salt or salts and a water-soluble ammonium or alkali metal or alkaline earth metal salt of an alkaryl sulfonic acid such as a C_{6-13} alkylbenzenesulfonic acid salt, preferably a C_{6-9} alkylbenzenesulfonic acid salt, or the corresponding alkyl naphthalene sulfonic acid salt, preferably a C_{6-13} alkyl naphthalene sulfonic acid salt.

Suitable zinc salt components include zinc chloride, zinc oxide and zinc sulfate and particularly zinc chloride. The zinc salt component is admixed with the cement component in an amount preferably from about 0.5 to 1% by wt.

Suitable salts of the alkaryl sulfonic acid component of the admixture include the ammonium, sodium and calcium salts of an alkaryl sulfonic acid, particularly a salt of an alkaryl sulfonic acid commercially available as CFT-2 (Halliburton).

The salt of an alkaryl sulfonic acid component is admixed with the cement component of the composition, preferably from about 0.5 to 1% by wt.

A preferred cement component is a class H portland cement having a density of about 94 lb/ft^3 and an approximate chemical analysis as follows: silicon dioxide, 22.4%; aluminum oxide, 4.8%; ferric oxide, 4.1%; calcium oxide, 64.8%; magnesium oxide, 1.1%; sulfur trioxide, 1.7%; loss in weight on ignition about 0.5%; and having a minimum fineness of at least 1,500 cm^2/g, measured by the Wagner turbidimeter.

The amount of water employed to make up the hydraulic cement slurry is not critical and generally the amount of water necessary to give a settable cement composition having the required characteristics will be in an amount of from about 25 to 60% by wt, based on the weight of the composition. The amount of water employed should be only such as is sufficient to produce a pumpable slurry.

A preferred quantity of water for slurry formation is from about 40 to 55%, preferably about 42 to 48%.

Carbon Black Additive

J.P. Gallus; U.S. Patent 4,200,153; April 29, 1980; assigned to Union Oil Company of California describes a method for cementing a high temperature well in which a hardenable slurry formed from a water-containing liquid vehicle, cement and a minor amount of a carbon black cement additive is introduced through the well into a confined space communicating therewith. The hardenable slurry is allowed to set and harden under the high temperature and pressure conditions in the well to form a hardened cement mass having an improved resistance to degradation of its ultimate compressive strength, density and permeability. Preferably, the carbon black additive is employed in an amount effective to impart to the hardened cement mass an ultimate compressive strength of at least about 1,000 psi and an ultimate permeability less than about 1 millidarcy under the conditions encountered in the well.

The hardenable slurry may be formed by mixing the cement, carbon black additive and water-containing liquid vehicle in any order. Preferably the carbon black additive is dry mixed with the cement prior to bagging and the cement system thus formed can be handled in the usual manner for oil well cements. Conventional additives normally mixed with or used with oil well cements may be incorporated in the cement system.

The carbon blacks preferred for use in the method are those carbon blacks having relatively high carbon contents, such as from about 95 to 99.5 wt % carbon, and a correspondingly low volatile material content, such as from about 5 to about 0.5 wt %. The particle size distribution and surface area of the carbon black employed is not believed critical, and carbon blacks having a mean particle size between about 15 and 250 millimicrons and surface areas between about 5 and 350 m^2/g are believed suitable. Furnace blacks and thermal blacks having the aforementioned properties are particularly preferred.

Preferably the cement additive comprises between about 0.01 and about 1 wt % of the cement system, i.e., of the total dry solids, and good results are obtained when the carbon black cement additive comprises between about 0.1 and 0.5 wt % of the cement system.

In preparing cement slurries utilizing the cement system of the process, the percent of water by weight of cement will typically range from about 36 to about 46% to form a slurry of pumpable consistency. The slurry thus formed is pumped in conventional fashion into the portion of the well to be cemented.

Examples 1 through 4: Four cement systems are prepared by admixing API Class G cement with various amounts of a furnace black known as Sterling R (Cabot Corporation). Typical properties of this furnace black include a fixed carbon content of about 99 wt %, a volatile material content of about 1 wt %, a surface area of about 25 m^2/g and a mean particle size of about 75 millimicrons.

A hardenable slurry is formed with each of the cement systems by admixing it with water in an amount equal to about 37 wt % of the API Class G cement. The slurries are poured into separate molds, the cavities of which define 2" cubes. The molds are filled to overflowing; the excess slurry is leveled off with a straight edge; and the openings of the molds are sealed with metal plates. The molds are then placed in an autoclave and are therein maintained at a temperature between about 200° and 300°F and a pressure of about 3,000 psig for about 24 hours in order to partially cure the slurries into hardened cubes.

The hardened cubes are removed from the molds and are then exposed to flowing geothermal steam having a temperature of about 464°F for a period of 3 months. Thereafter the cubes are cooled to room temperature. The cubes of Examples 2, 3 and 4 are recovered intact, however the cube prepared in Example 1 using neat API Class G cement is severely cracked indicating a substantial increase in permeability and a substantial reduction in compressive strength. Thus, Example 1 demonstrates that neat API Class G cement is unsuitable under the conditions of this test.

A 1" diameter by 1" long core is cut from each of the cubes of Examples 2 through 4 and is used to measure the permeability and compressive strengths of the cement masses. The compressive strength tests are conducted in accordance with API Specification RP 10B, section 6. The permeability tests are conducted utilizing a gas permeameter which consists of a pressure plate and O-ring which is connected to a source of gas under pressure. The gas is introduced into the pressure chamber formed between the pressure plate and the face of the core and the rate of pressure drop is measured. The pressure drop rate is then converted to permeability in millidarcies. The results of these tests are summarized below:

	. . . .Cement System (wt %). . . .		Compressive	
Ex. No.	API Class G Cement	Sterling R Furnace Black	Strength (psi)	Permeability (millidarcies)
1	100.0	0.0	. . . . . .not determined.	
2	99.9	0.1	4,000	<1
3	99.5	0.5	3,500	<1
4	95.2	4.8	1,113	15.8

It is generally recognized throughout the field of cementing wells that the maximum acceptable permeability of an oil well system is about 1 millidarcy and the minimum acceptable compressive strength is about 1,000 to 2,000 psi. All of the cubes, regardless of the cement system from which they are formed, have a permeability of less than about 1.0 millidarcy prior to the initiation of the steam tests. From the results summarized above it can be seen that the cement systems of Example 2 and 3 form cement masses which exhibit substantially no measurable permeability. The cement system of Example 4 shows a substantial increase in permeability. With the higher permeability there is a substantially greater chance that a cement mass formed from the cement system of Example 4 will fail before the cement masses of Examples 2 and 3. Likewise, while the compressive strengths of the cement masses of Example 2 through 4 exceed 1,000 psi, the cement masses of Examples 2 and 3 have compressive strengths over three times the compressive strength of the cement mass of Example 4.

Prevention of Flow-After-Cementing

Problems have occurred in gas cutting or channeling through the cement slurries placed in wells. In cementing casing in a well, gas may flow behind the casing and through the cement slurry placed there to the surface of the earth, or may flow into lower pressured formations which communicate with the well where such gas is usually lost. The term "flow-after-cementing" has been used to characterize this phenomenon.

J.U. Messenger; U.S. Patent 4,235,291; November 25, 1980; assigned to Mobil Oil Corporation describes a process whereby a low-density calcined shale cement is blended with attapulgite in an amount no greater than about 2 wt % based on cement and water in an amount to form a thixotropic cement slurry having a density within the range of 11.5 to 13.5 lb/gal and having essentially zero water separation. The slurry is injected down the well and maintained there and allowed to set. Sodium chloride and calcium chloride may be included in the slurry as an accelerator. The amount of sodium chloride and calcium chloride is selected to obtain a desired thickening time for the slurry.

Normally, sodium chloride is included in an amount to provide a concentration no greater than about 10 wt % based on fresh mixing water and calcium chloride in an amount no more than about 3 wt % based on cement. The water used for forming the cement slurry may be fresh water or brine such as seawater.

This process is particularly applicable for cementing pipe in a well that penetrates an active gas zone to prevent flow-after-cementing. In cementing pipe in a well in accordance with this process, a low-density calcined shale cement is blended with attapulgite and water as previously described and is injected down the well and into the annulus formed about the pipe and maintained there and allowed to set. This process is particularly applicable for use in cementing offshore wells inasmuch as seawater may be used and, in fact, is preferred for forming the cement slurry. Seawater is readily available at such locations and when used as mixing water for forming a slurry with calcined shale cement and attapulgite produces a slurry having essentially zero water separation.

A particularly effective cement slurry for mitigating the problem of flow-after-cementing is formed by blending a calcined shale cement such as that calcined shale cement known as Trinity Lite-Wate cement with about 0.5 wt % attapulgite based on cement and 8.9 gal of seawater per 75 lb sack of cement. This blend gives a 12.5 lb/gal slurry having a yield of 1.62 ft^3/75 lb sack of cement. This slurry may be accelerated as previously described by including therein calcium chloride in an amount no greater than about 3 wt % based on cement. Sodium chloride may also be used as an accelerator, in which case the sodium chloride should be used in an amount no greater than about 7 wt % based on seawater in order for the sodium chloride content of the slurry to be no greater than about 10 wt % based on fresh water.

Well Cementing in Permafrost

W.N. Wilson; U.S. Patent 4,176,720; December 4, 1979; assigned to Atlantic Richfield Company has found that drilling fluid actually used in drilling a well bore in permafrost can be employed as a major constituent in cementing a permafrost well bore.

By this process the drilling fluid previously used is conserved and, therefore, replaces a substantial amount of cementing material which would otherwise be transported to the drill site at a substantial expense.

More specifically, this process comprises cementing a string of pipe or any other desired item in the permafrost region of a well bore by recovering at least part of the aqueous drilling fluid used in drilling a well bore in the permafrost region, not necessarily the same well bore, and analyzing the thus recovered drilling fluid for its solids content. If the solids content of the drilling fluid is already in the range of from about 2 to about 16 vol % based on the total volume of the drilling fluid, it is suitable for use as a base for the cementitious material. If the solids content is not within the above-noted range, it should be adjusted to within the range by the addition of any suitable particulate solids. The particular type of solids, e.g., sand, gravel, clay, etc., present and the size gradation of the solids is not as important as the total volume percent of solids present.

Once the drilling fluid with a proper solids content is obtained, there is mixed therewith an additive selected from the group consisting of lignosulfonate, lignite, tannin, and mixtures thereof, in an amount of from about 0.05 to about 0.4 lb/bbl per equivalent lb/bbl of bentonite as determined by the methylene blue test as set forth by the American Petroleum Institute (API) RP13B Section 9. The additive is preferably chrome lignosulfonate. At the same time that the additive is mixed with the drilling fluid and/or after the additive is mixed with the drilling fluid, cementitious material which will harden in from about 30 to

40 hours (depending on the amount of lignosulfonate present) while sitting essentially quiescent at about 40°F is mixed with the drilling fluid in an amount of from about 50 to 250 lb of cementitious material per barrel (42 gal/bbl) of drilling fluid plus additives. A wide variety of cements meeting the 30 to 40 hour at 40°F test are available and well known to those skilled in the art, a particularly desirable cement being one which is a high alumina cement. Preferably the cementitious material is a high (at least about 40 wt %) alumina calcium aluminate cement. A particularly suitable commercially available cementitious material is fondu.

Also essentially contemperaneous with and/or after mixing of the additive with the base drilling fluid and before, or at the same time as, the addition of the cementitious material, sufficient base is added so that the pH of the final mixture is in the range of from about 9 to about 12.

This final mixture is then used as the cementing material for the permafrost well bore and is simply pumped into the permafrost region of the well bore to be cemented and allowed to set until it hardens in the well bore. A known friction reducer can be used to improve pumpability.

Example: Fresh water containing 8.7 vol % (based on the volume of the whole drilling fluid) of 200 mesh and smaller bentonitic clay solids and 0.265 vol % based on the volume of whole drilling fluid of sand was taken from a well drilled in permafrost in Alaska. To this drilling fluid was first added 6 lb/bbl of chrome lignosulfonate with mixing. Thereafter, 1.5 lb/bbl of sodium hydroxide and then 130 lb/bbl of Lafarge fondu was added. All mixing was carried out at ambient temperature and pressure until an intimate mixture of all materials was achieved by observation. The API fluid loss value for the resulting cementitious mixture was about 10 cubic centimeters for 30 minutes.

The resulting mixture hardened while setting under quiescent conditions at 40°F in about 40 hours. The resulting mixture was hardened into test cylinders 3" in diameter and about 8" long and was tested for its elastic modulus under pressure, the pressure simulating the permafrost overburden pressure that the cement would encounter in place of a well bore. Results were as follows:

Confining Pressure, psi	Average Elastic Modulus, psi
400	92,223
200	77,810
100	64,075

Additional tests under the same confining pressures were conducted with similar cylinders and the results were:

Confining Pressure, psi	Average Elastic Modulus, psi
400	82,134
200	74,304
100	65,673

Preshearing of Water-Swellable Clays

J.U. Messenger; U.S. Patent 4,202,413; May 13, 1980; assigned to Mobil Oil Corporation has found that the effectiveness of water-swellable clays as extenders to increase the yield of cement is greatly enhanced by preshearing an aqueous mixture of the water-swellable clays prior to blending the water-swellable clays with the cement. This preshearing of the aqueous solution of water-swellable clays enhances their effectiveness as cement extenders greatly over that of either using the water-swellable clays in dry form with the cement or in prehydrating or aging the water-swellable clays prior to mixing the clays with the cement. It was also found that 1% of either attapulgite or bentonite vigorously presheared for a predetermined time before being mixed with the cement is equal to about 8% dry blended bentonite.

This enhanced effect can be had when using brine as mixing water. When using brine as mixing water, attapulgite may be mixed with the brine and the attapulgite brine solution presheared for a predetermined time and thereafter mixed with the cement to form a cement slurry. When using bentonite, best results are obtained by forming a mixture of the bentonite in fresh water and preshearing the bentonite freshwater suspension for a predetermined time and thereafter mixing this presheared bentonite freshwater solution with brine prior to mixing with cement therewith to form a cement slurry.

In an embodiment of this process, a suspension of attapulgite, or other clays that are swellable in brine, and brine such as seawater is formed, presheared and used as mixing water for forming a cement slurry. Additives may be included in the brine, such as calcium chloride and sodium chloride. The attapulgite is added to the brine in an amount to provide a concentration of attapulgite when mixed into a cement slurry within the range of about 0.1 to 3.0 wt % based on cement, and the aqueous suspension of attapulgite is subjected to vigorous shearing for at least one-half hour. Thereafter the aqueous suspension of attapulgite is blended with dry cement to form a pumpable cement slurry having attapulgite therein in a concentration within the range of 0.1 to 3.0 wt % based on cement. This cement slurry is then injected into a well for use therein.

In accordance with another embodiment, a lightweight cement slurry weighing about 13.0 lb/gal is formed by blending Class G portland cement with about 2% attapulgite which has been presheared in seawater having about 2 wt % calcium chloride based on cement added thereto. This lightweight cement slurry is particularly applicable for cementing shallow casing strings such as conductor and surface casing in wells drilled in cold regions. This lightweight cement slurry will set and develop at least 100 psi of compressive strength in 24 hours at 40°F. The calcium chloride is added to the slurry to accelerate the setting thereof. The calcium chloride should be added to the seawater before adding the attapulgite thereto.

In accordance with still another embodiment, clays that are swellable in fresh water as exemplified by bentonite may be used as a cement extender for slurries which are formed with seawater. In this embodiment there is formed a suspension of freshwater swellable clay exemplified by bentonite in fresh water and this suspension is presheared as previously described. The bentonite is added to the freshwater in an amount to provide a concentration of bentonite when mixed into a cement slurry within the range of about 0.1 to 4.0 wt % based on cement. Preferably the amount of fresh water used in carrying out this embodiment is an amount of at least about one-quarter to one-third of the total amount of water required for forming the cement slurry. Desirably a minimum amount of fresh water is used in locations such as offshore locations where fresh water is scarce and brine is readily available. The minimum amount of fresh water which can be used is controlled by the thickness of the bentonite suspension. Normally this minimum amount of fresh water is about one-quarter of the total water required in forming the slurry. The use of less fresh water would produce a bentonite suspension that is too thick to readily pump. The freshwater-bentonite suspension should not be further sheared after it is added to the seawater because such further shearing will reduce the cement-extending characteristics of the bentonite. The freshwater-bentonite suspension should be added to the seawater rather than the seawater being added to the freshwater-bentonite suspension. Addition of seawater to the freshwater-bentonite suspension may result in a suspension which is too viscous to pump. The preshearing of the bentonite in fresh water in accordance with this embodiment enhances the effectiveness of bentonite as an extender for cement to the effect that only about half as much bentonite is required as contrasted to that which would be required where the bentonite is only prehydrated in fresh water.

SPACER AND CHEMICAL WASH FLUIDS

Chemical Wash with Fluid Loss Control

Among the problems associated with drilling muds, is that the liquid phase of a drilling mud tends to flow from the well into exposed permeable formations with the result that mud solids are filtered out on the wall of the well and a filter cake is formed thereon. Mud filter cakes are detrimental in the completion of wells in that they interfere with obtaining a good cement bond between the wall of the well and the conduit, or casing, positioned in the well. Also, drilling muds frequently contain components which are incompatible with fluids which one may desire to inject into a well containing a mud. For example, it has long been recognized that if certain cement slurries containing free polyvalent metal cations, especially calcium, are brought into contact with muds containing clay or certain polymers, a very viscous and detrimental plug can form in the vicinity of the mud-cement interface.

High pressures required to move such a plug can rupture tubing, or make it necessary to stop pumping to avoid rupturing the tubing with the result that appreciable volumes of cement are left inside the tubing. Also, the high pressures can cause fracturing of the formation, thus causing loss of cement to the formation and formation damage. Another example of mud-cement incompatibility is that lignins, frequently used as dispersants in high density muds, can cause excessive retardation in cements if the cement becomes comingled with the mud.

For these reasons, various techniques have been developed for the removal of drilling muds from a borehole, particularly in the context of injecting a fluid into the borehole where the fluid is not compatible with the mud, and even more particularly, in the context of cementing. A common technique is to employ a "spacer" or "chemical wash". Although it is not always clear in the literature whether a particular fluid is a spacer or a chemical wash, a spacer is generally characterized as a thickened composition which functions primarily as a fluid piston in displacing the mud. Frequently, spacers contain appreciable quantities of weighting materials and also include fluid loss control agents. Chemical washes, on the other hand, are generally thin fluids which are effective principally as a result of turbulence, dilution, and surfactant action on the mud and mud filter cake. Chemical washes may contain some solids to act as an abrasive, but the solids content is generally significantly lower than in spacers because chemical washes are generally too thin to have good solids carrying capacity.

J.R. Sharpe and D.L. Free; U.S. Patent 4,207,194; June 10, 1980; assigned to The Dow Chemical Company have found that fluid loss control can be imparted to aqueous based, water-thin chemical washes of the type containing at least one of (1) at least one surfactant to remove water based drilling muds from a borehole or (2) at least one surfactant to enhance the bonding of cement to the walls of a borehole or to the walls of a conduit placed in the borehole, by including in such chemical washes, an effective amount of a fluid loss additive of the type disclosed in U.S. Patent 3,827,498 (and its divisionals, U.S. Patents 3,891,566 and 3,898,167), namely, an additive comprising a mixture of at least two oil-soluble particulate resins, one of which remains hard and friable, and the other of which is soft and pliable (at the temperature to be encountered in the well).

The resins provide effective fluid loss control at concentrations of at least about 2 pounds (total weight of both resins) per 1,000 gallons of fluid, i.e., at a total concentration in the wash of at least about 0.025 wt %. Preferably, the wash contains from about 0.02 to about 0.4 wt % pliable resin and from about 0.02 to about 0.5 wt % pliable resin. Little additional fluid loss control is realized at higher concentrations.

In the best mode contemplated for carrying out this process, from about 0.5 to about 2 parts by volume of a resin formulation, hereinafter described, is mixed with from about 2 to about 8 parts by volume of a surfactant formulation hereinafter described, and about 164 parts by volume water (i.e., about 0.5 to 2 quarts resin formulation and 0.5 to 2 gallons surfactant formulation per barrel of chemical wash), to provide a wash containing by weight, about 0.15 to 0.6% active surfactant; about 0.04 to 0.2% pliable resin; and about 0.05 to 0.27% friable resin.

The resin formulation employed in the best mode is comprised, by weight of (1) about 25 to 40% pliable resin dispersion such as Elvax D112 brand ethylene/vinyl acetate copolymer dispersion in water (50% solids) or Picconol A102 brand 50% thermoplastic aliphatic hydrocarbon resin dispersion in water; (2) 15 to 25% friable resin such as Piccomer 150 brand alkyl aromatic hydrocarbon resin; (3) 25 to 40% water or a water-soluble solvent as a carrier, e.g., glycerin, aqueous ammonium chloride, and the like; (4) about 3% silicon-type antifoaming agent; and 5 to 20% anticoagulant for the pliable resin, such as an adduct of di-sec-butylphenol with ethylene oxide, cocobetaine, and the like.

The surfactant formulation employed in the best mode comprises, by volume, about 4 parts water and about 1 part n-propyl or isopropyl alcohol containing from about 50 to 75 wt % reaction product of triethanolamine with a mixture of fatty acids.

If desired, the chemical wash may contain small amounts of other functional additives which do not affect its performance; for example, a dye aids in recognition of wash returned from the borehole.

A sufficient amount of chemical wash is injected in a borehole to be cemented in advance of the cement slurry, to adequately thin the mud so that substantially no viscous plug is formed along the leading edge of the cement. As those skilled in the art will realize, use of greater amounts of chemical wash, within reason, results in a cleaner borehole, and consequently better cement jobs. Generally, from about 5 to 30 barrels of wash are satisfactory, though the actual amount employed will vary depending on the volume of the annulus, the effectiveness of a particular surfactant with a particular mud, and the like. Smaller volumes of the chemical wash are required, however, than an otherwise identical wash containing none of the resin fluid loss material, since the volume of wash lost to the formation will be minimal.

Oleyl Amide Dispersant System

L.L. Carney; U.S. Patent 4,233,162; November 11, 1980; assigned to Halliburton Company describes substantially universal spacer fluids having excellent temperature and pressure stability which are capable of spacing cements from substantially all mud systems employed in the drilling of oil and gas wells.

The spacer fluids of this process are basically freshwater-in-oil emulsions containing approximately equal volume parts of a normally liquid hydrocarbon oil and water with about 15 to 40 lb/bbl of an emulsifier, about 0.5 to 10 lb/bbl of a particular type surfactant dispersant; and optionally, a weighting material in an amount effective to impart a density to the spacer fluid of from about 8 to 20 lb/gal.

The emulsifiers of the process include a particular type of fatty acid amide, which can be in liquid form or adsorbed on a powdered solid carrier such as lime and diatomaceous earth. The preferred fatty acid amide is an oleyl amide thought to be a nonionic emulsifier and is preferably present in an amount in the range of from about 1 to 20% by wt.

A specific preferred emulsifier includes a first type of oleyl amide, oleic acid and dimerized oleic acid adsorbed on a powdered carrier selected from the group consisting of lime and diatomaceous earth. The oleyl amide is present in an amount in the range of about 1 to 20% (preferably 2 to 10%) by wt of emulsifier. The dimerized oleic acid is present in an amount in the range of about 2 to 20% by wt of the emulsifier. The oleic acid is present in an amount in the range of about 5 to 15% by wt.

The first type of fatty acid amides useful as the basic emulsifier can be prepared as the reaction product of fatty acid and a low molecular weight alkyl secondary amine such as diethanolamine. The fatty acid or oleyl amide herein referred to is considered the primary emulsifying component in the compositions, and preferably is derived from reacting a fatty acid containing about 12 to 18 carbon atoms with an amine. The amide reaction product preferably contains about 16 to 22 carbon atoms, and from 1 to 2 amide groups. The most preferred amide is prepared by condensing oleic acid with diethanolamine.

Dispersing agents of the process for dispersing hydrocarbon materials in aqueous solutions contain a surfactant-dispersant comprised of mixtures of a second type of fatty acid amide and a waste lignin liquor, wherein the waste lignin liquor is a product of the sulfite process or the Kraft process used in the wood pulping industry. The lignin is sulfonated with a sulfur content preferably about 1 to 3% by wt.

While the preferred surfactant-dispersant agent is produced by combining a particular type of oleyl amide with waste lignin liquors, fatty acid amides other than oleyl amides are useful herein. Such amides are prepared from saturated and unsaturated fatty acids having in the range of about 14 to 18 carbon atoms per molecule. Such acids include but are not limited to oleic acid, linoleic acid, linolenic acid, stearic acid, palmitic acid, myristic acid and myristoleic acid.

The most preferred single emulsifier composition for use in the spacer fluids consists essentially of powdered slaked lime in an amount of about 68.1 wt %, about 4.9 wt % oleyl amide, undistilled oleic acid present in an amount of about 5 wt %, red oil present in an amount of about 5 wt %, and undistilled dimerized oleic acid present in an amount of about 10 wt %.

For fluid loss control with the spacer fluids, it may be desirable to include a conventional particulated solid asphaltic resin which is added to the spacer fluid composition in an amount in the range of about 1 to 30 (preferably 1 to 20) wt %, based on the weight of the emulsifier. Usually about 10 to 14% of an asphaltic resin will provide adequate fluid loss control.

Suitable asphaltic resins include those which have a melting point between 250° and 400°F. They may be kettle bottoms, air blown resins or naturally-occurring resins. Where such asphaltic resins are used, as little as about 55 wt % of the lime carrier can be utilized.

A surfactant dispersant which has been found to be particularly suitable and highly effective in the spacer fluid compositions, and which constitutes the preferred dispersing agent for use in accordance with the process, is produced by reacting oleyl chloride with N-methyltaurine. This reaction product, referred to as an oleyl amide, is mixed with waste sulfite liquor in an amount in the range of about 25 to 75 wt % of the sulfite liquor-oleyl amide mixture. The oleyl amide reaction product and the sulfite liquor are preferably mixed in about equal amounts.

In preparing the emulsifier compositions, the liquid oleyl amide can be sprayed onto the powdered lime or diatomaceous earth while mixing so that the oleyl amide is adsorbed. The oleic acid and dimerized oleic acid (preferably undistilled dimerized oleic acid) are mixed together and heated to a temperature of from about 120° to about 150°F to facilitate their easy adsorption. The heated liquid mixture is then sprayed on the lime or diatomaceous earth while mixing so that it is adsorbed thereon, and the lime or diatomaceous earth remains in a free-flowing powdered state. The dispersing agent used and particulated asphaltic resin are next combined with the mixture to form the final composition. Conventional mixing and blending apparatus can be utilized for carrying out the abovedescribed procedures, and the final emulsifier composition produced is a dry free-flowing powder which can be shipped and stored in paper sacks or other conventional dry material containers.

Example: A 21,000 ft well was drilled in a California location, and has a bottom hole circulating temperature of 310°F. In completing the well, the mud used in drilling the well is displaced through the annulus between the casing and the well bore by the use of a spacer fluid interposed between the mud and the cement following the spacer fluid for purposes of cementing the casing. The mud employed has a density of 18 lb/gal, and is a lignosulfonate-containing water base mud. The cement used is an API Class G cement having a weight of 14.4 lb/gal, and is modified by the inclusion of small amounts of sodium chloride, silica flour, iron oxide and an appropriate retarder.

The spacer fluid formulated in accordance with the process and used between the cement and mud has a weight of 18.2 lb/gal achieved using barium sulfate as a weighting material. The spacer fluid emulsion contains water and oil in a 50 to 50 volume ratio, and has incorporated therein 25 lb/bbl of an emulsifier which contains 68.1 wt % of powdered slaked lime having adsorbed on its surface, 4.9 wt % oleyl amide derived from the condensation of oleic acid with diethanolamine, 5 wt % undistilled oleic acid, 5 wt % red oil, and 10 wt % undistilled dimerized oleic acid. The spacer fluid also contains 4 lb/bbl of the preferred surfactant dispersant. Finally, the spacer fluid used contains 12 wt % of an asphaltic resin incorporated to reduce fluid loss.

The mud and cement utilized in the California cementing run are not compatible. 50 barrels of the spacer is used between the mud and cement.

The critical velocities at which the transition from laminar to turbulent flow occurs, are measured and evaluated for the drilling mud alone, the spacer fluid alone, a mixture of the spacer fluid and the cement used, and a mixture of spacer fluid and the drilling mud. The temperature at which the critical velocity tests are carried out is 200°F. The critical velocities are measured for various annulus sizes as determined by the difference in the hole diameter and the casing outside diameter. The results of these runs are set forth in the table below.

	Critical Velocity (fpm)					
	Hole Diameter – Casing o.d. (inches)					
	1	2	3	4	5	6
Drilling mud	470	370	320	290	270	250
Spacer fluid	355	230	175	145	125	115
Spacer fluid + 50% cement	325	240	205	180	165	150
Spacer fluid + 50% mud	340	250	210	185	170	155

The results of the critical velocity measurement indicate that substantially lower critical velocities are obtained for mixtures of the spacer fluid with either cement or mud, than in the case of the mud alone. There is, therefore, no detrimental increase in viscosity, and a greater ease in achieving turbulent flow is realized where any significant mixing of the spacer fluid with either the mud or the cement occurs.

Sulfonate/Oleyl Amide Dispersant System

J.L. Watson; U.S. Patents 4,217,229; August 12, 1980; and 4,141,843; Feb. 27, 1979; both assigned to Halliburton Company describes a high stability, nondamaging spacer fluid containing readily available weighting agents such as calcium carbonate and iron carbonate dispersed in water using a polymer viscosifier, a salt inhibitor, a primary sulfonated dispersant and a secondary fatty acid amide dispersion. Conventional high density weighting agents can also be used. The spacer fluid is stable over a temperature range of about 32° to 300°F for extended periods with densities ranging from 11 to 17.6 lb/gal.

A preferred class of viscosifiers for the high stability fluid is the water-soluble polysaccharides and especially the substituted nonionic cellulose polymers, such as hydroxyalkylcellulose or cellulose ethers in which the hydroxy alkyl groups have 2 to 3 carbon atoms. Other substituents can be present or used which produce a water-soluble cellulose which does not adversely react in the high density fluid system. The substituted cellulose should be hydratable in the high density fluid. The preferred cellulose class can be represented as a series of anhydroglucose units shown as follows:

(1)

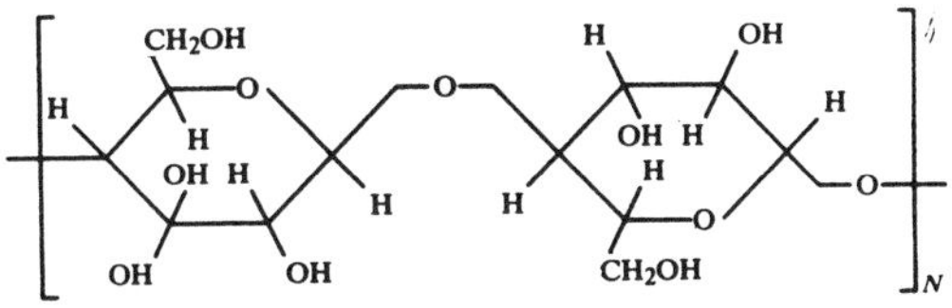

The portion in brackets is two anhydroglucose units, each having three reactive hydroxyl groups. N is an integer which would give the desired polymer molecule length and preferably an aqueous viscosity of about 105 to 130 viscosity units of consistency at 72°F (approximately equal to centipoise, cp) on a VG meter at 300 rpm with a 2.5% aqueous solution in fresh water.

When the cellulose polymer is treated with sodium hydroxide and reacted with ethylene oxide, an ether substituted cellulose such as hydroxyethyl ether or hydroxyethylcellulose is produced shown as follows:

(2)

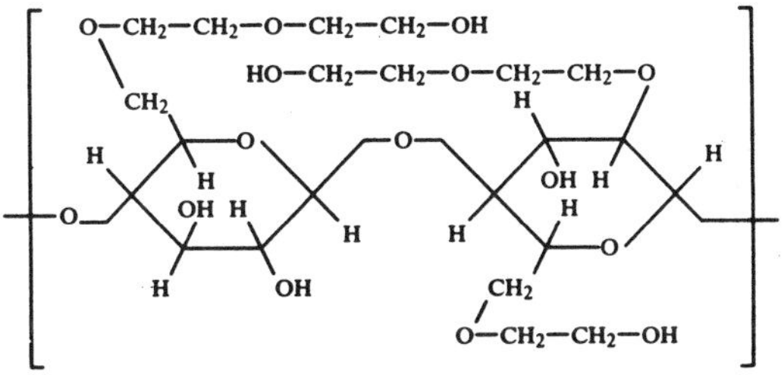

The hydroxyethylcellulose or HEC shown has three of the six hydroxyl groups substituted by ethylene oxide; therefore the degree of substitution (or DS) is three of six or 1.5 per anhydroglucose unit. The preferred DS for cellulose polymer viscosifiers is about 1.0 to 3.0.

The above formula also shows that two of the substituted hydroxyl groups have two mols of ethylene oxide and one has 1 mol of ethylene oxide; therefore, the ratio of mols of ethylene oxide to anhydroglucose unit (or MS) ratio is 5 mols for two units or 2.5. The preferred MS ratio for HEC polymer viscosifiers is 1.5 to 3.0.

Dispersants used for compositions of this process are of two principal types. Either or both types of dispersants can be used over the full density range, but the primary dispersant which is referred to as sulfonate dispersant is preferably used for density up to about 16 lb/gal. For higher density fluids where more than one type of weighting agent and/or high loading of weighting agent is used, the secondary dispersant which is referred to as oleyl amide dispersant (see previous process) is used. Generally, less than about 0.4% by wt (i.e., 2.5 lb/42 gal bbl) of primary dispersant and less than 0.45% by wt or 3 lb/bbl of secondary dispersant is used in the high stability fluid. For low densities (e.g., 12 pounds per 42 gal bbl) and under certain conditions the dispersant can be considered optional, but generally it is used for easier mixing and to improve suspension characteristics of the aqueous base fluid which may be water or brine. The maximum concentration of dispersant is usually determined by economics and the density desired, but is preferably less than about 1.5% by wt of resulting fluid.

The preferred class of primary dispersants is the product of a low molecular weight aldehyde and a naphthalene sulfonate salt. A preferred sulfonate dispersant combined with polyvinylpyrrolidone (i.e., PVP) is described in U.S. Patent 3,359,225. Up to 10% PVP can be used with the naphthalene sulfonate and can be in the form of an alkali or alkaline metal salt, but preferably is a sodium or potassium salt. Other conventional dispersants such as lignosulfonates, sulfonated lignites, gluconic acid, delta-lactone and lignin liquor can be used alone in some cases and in combination with a primary sulfonate dispersant.

A preferred class of secondary dispersants is fatty acid amides produced by the reaction of saturated or unsaturated fatty acid halides having about 14 to 18 carbon atoms per molecule with a low molecular weight amino sulfonic acid having about 1 to 6 carbon atoms. The sulfonic acid can have alkyl and/or aryl radicals having 1 to 6 carbon atoms and 1 or more sulfonic acid groups or salts thereof. A preferred amide is the reaction product of oleyl chloride and a C_3 sulfonic acid or sodium salt, N-methyl taurate. This preferred oleyl amide is also mixed with approximately 25 to 75% by wt, but preferably equal amounts of lignin liquor. This lignin liquor is a waste product of the sulfite process or the Kraft process used in the wood pulping industry. This lignin is sulfonated with a sulfur content of preferably about 1 to 3% by wt. Other fatty acids which can be used are linoleic, linolenic, stearic, palmitic, myristic, myristoleic and mixtures of fatty acids. The amide dispersant can be used as a liquid or adsorbed on a relatively inert particulate carrier such as diatomaceous earth.

The water or aqueous base used to prepare the high stability fluid can be fresh water or brine containing one or more salts up to saturation. Fresh water is preferred as the base material because it is easier to mix the ingredients if a particular order is used. A salt or inhibitor is usually added as the last component where possible. The salt serves to inhibit clays which may be encountered. The polymer viscosifier also acts as an inhibitor. Alkali metals, alkaline earth metals and ammonium salts are preferred cations of the salt, especially sodium, magnesium, potassium, and/or calcium halides such as chlorides, bromides or combinations thereof. The salt concentration should be about 0.5 to 15% by wt, and preferably about 1 to 6% by wt.

The weighting agents used are relatively inert finely divided particulate materials having a particle size with at least 80% by wt between about 2 to 50 microns. Preferably, all or at least 90% of the material will pass through a 200-mesh U.S. standard sieve screen. The particulate weighting agent should also have a specific gravity of at least 2.4 and preferably about 2.5 to 3.8. A preferred class of weighting agents are considered acid-soluble in aqueous acids such as acetic, hydrochloric, nitric, sulfurous, sulfuric and phosphoric. This class includes calcium carbonate, iron carbonate and the iron oxides. Higher specific gravity weighting agents having a specific gravity of 4.0 to 7.0 such as barium sulfate and lead sulfite can be used in combination with the acid-soluble weighting agents. The acid-soluble weighting agents preferably have a particle size distribution so that at least 80% is between 2 to 20 microns in size with a mean size of about 4 to 10 microns.

The fluid is preferably mixed in a turbine blender, but can be mixed successfully using a jet mixer or ribbon blender.

When using a ribbon blender, a sufficient agitation to pull the viscosifier into the fluid is necessary or it will water wet, forming balls of polymer that will not easily disperse.

The order of addition for weights through 16.0 ppg (pounds per 42 gallon barrel) is as follows: water, sulfonate dispersant, weighting agent, viscosifier and inhibitor, such as potassium chloride. The order of addition for weights from 16.0 through 17.6 ppg are as follows: water, sulfonate dispersant, weighting agent, oleyl amide dispersant, high density weighting agent such as iron carbonate, viscosifier and inhibitor. The oleyl amide dispersant and iron carbonate can be added together to minimize foaming or a defoaming additive can be used.

High Temperature Spacer Fluid

R.M. Beirute; U.S. Patent 4,190,110; February 26, 1980; assigned to The Western Company of North America describes a single phase water-based spacer composition which is stable even at elevated temperatures and is useful with both water-base and oil-base muds and which can be prepared from attapulgite clay. Stabilization of the attapulgite clay in aqueous suspension is effected through the addition of water-soluble zinc salts such as zinc oxide, for example, and the adjustment of the pH of the composition to a value of about 7 or above, and preferably between from about 8 to 10. In addition to the pH adjusted, zinc salt-containing, attapulgite clay-water suspension, well cement retarders, water-wetting surfactants, cement defoamers, and weighting agents can be employed to tailor the spacer composition for use with specific oil muds, cement compositions, and borehole conditions.

The spacer compositions make possible a method for implacing the cement composition in the well bore containing either oil-base or water-base drilling fluids, by interposing a sufficient amount of the spacer composition to insure that the well cement and drilling fluids will not come in contact with one another. The spacer compositions of the process, when employed in such a method, have the advantages that they are compatible with oil-base mud, water-base mud, and well cements. Further, these compositions are stable at temperatures of at least about 500°F BHCT (bottom hole circulating temperature). The compositions are capable of being weighted in a range of from about 8 to 22 lb/gal, thus making them useful with a wide range of drilling muds. The compositions have good fluid loss control and, when the spacer compositions comprise a surfactant, provide a water-wetting effect which insures maximum bonding of the well cement in the annulus between the casing and borehole wall.

The amounts of attapulgite clay and water-soluble zinc salts present in the spacer compositions will vary depending upon the desired weight of the spacer composition. Generally, however, it can be stated that attapulgite clay is added in quantities of up to about 56 lb/bbl of spacer when the desired weight of the spacer composition is 10 lb/gal, and can be present in as little as about 4½ lb/bbl when the weight of the composition is desired to be about 20 lb/bbl. A study of Figure 6.1a will indicate to one skilled in the art the preferred amounts of attapulgite clay per barrel of spacer composition, as a function of the desired final weight (lb/gal) of the spacer composition.

Figure 6.1: Preferred Spacing Fluid Composition

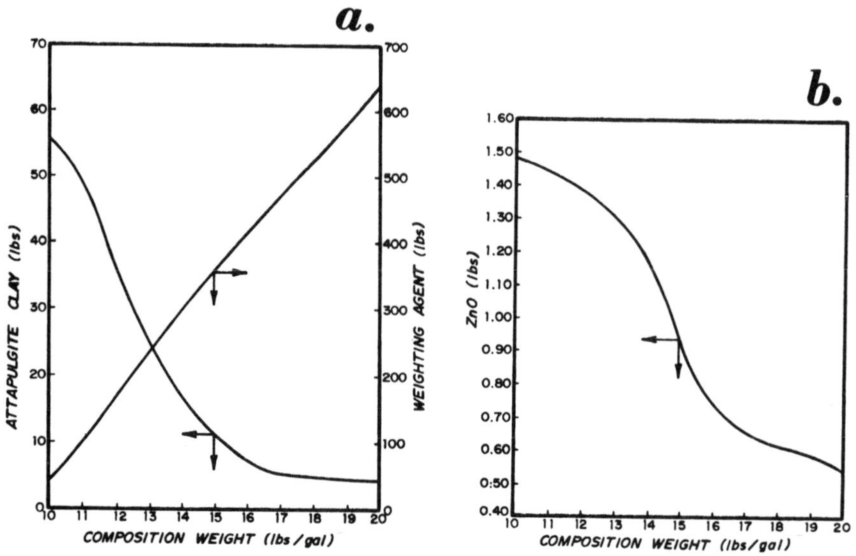

(a) Amount of attapulgite clay and barite (weighting agent) as a function of composition weight

(b) Amount of ZnO as a function of composition weight

Source: U.S. Patent 4,190,110

The zinc-containing stabilizing compound is preferably zinc oxide but can be suitable water-soluble zinc salts such as zinc chloride and zinc carbonate which are effective to stabilize the attapulgite clay-aqueous suspension at the pH composition indicated above.

The amount of zinc oxide necessary to stabilize the attapulgite clay-aqueous suspension of the spacer composition varies as a function of the desired weight of the spacer composition. Thus, a study of Figure 6.1b demonstrates that within a spacer composition weight range of 10 to 20 lb/gal, zinc oxide in the amount of from about 1.5 to 0.54 lb/bbl of spacer composition will provide desirable results.

COMPANY INDEX

INVENTOR INDEX

U.S. PATENT NUMBER INDEX

NOTICE